The Fortifications of Pompeii and Ancient Italy

The fortifications of Pompeii stand as the ancient city's largest, oldest, and best preserved public monument. Over its 700-year history, Pompeii invested significant amounts of money, resources, and labor into (re)building, maintaining, and upgrading the walls. Each intervention on the fortifications marked a pivotal event of social and political change, signaling dramatic shifts in Pompeii's urban, social, and architectural framework. Although the defenses had a clear military role, their design, construction materials, and aesthetics reflect the political, social, and urban development of the city. Their fate was intertwined with that of Pompeii.

This study redefines Pompeii's fortifications as a central monument that physically and symbolically shaped the city. It considers the internal and external forces that morphed their appearance and traces how the fortifications served to foster a sense of community. The city wall emerges as a dynamic, ideologically freighted monument that was fundamental to the image and identity of Pompeii. The book is a unique narrative of the social and urban development of the city from foundation to the eruption of Vesuvius, through the lens of the public building most critical to its independence and survival.

Ivo van der Graaff is Assistant Professor of Art History at the University of New Hampshire, Durham. He earned his MA in Mediterranean Archaeology from the University of Amsterdam and his PhD in Art History from the University of Texas at Austin, with a focus on Greek and Roman art and architecture. Dr. van der Graaff has participated in archaeological research projects in the Netherlands, Belgium, Greece, and Italy. He co-directs and collaborates on projects examining the ancient Bay of Naples and Etruria.

The Fortifications of Pompeii and Ancient Italy

Ivo van der Graaff

Routledge
Taylor & Francis Group

LONDON AND NEW YORK

First published 2019 by Routledge

2 Park Square, Milton Park, Abingdon, Oxfordshire OX14 4RN

52 Vanderbilt Avenue, New York, NY 10017

Routledge is an imprint of the Taylor & Francis Group, an informa business

First issued in paperback 2020

British Library Cataloguing-in-Publication Data
A catalogue record for this book is available from the British Library

Library of Congress Cataloging-in-Publication Data
Names: Graaff, Ivo van der, author.
Title: The fortifications of Pompeii and ancient Italy / Ivo van der Graaff.
Description: 1st edition. | Abingdon, Oxon ; New York, NY : Routledge,
    [2019] | Includes bibliographical references and index.
Identifiers: LCCN 2018012142 | ISBN 9781472477163 (hardback : alk.
    paper) | ISBN 9780429462009 (e-book)
Subjects: LCSH: Fortification—Italy—Pompeii. | Pompeii (Extinct city)—
    Buildings, structures, etc.
Classification: LCC UG430.P558 G73 2016 | DDC 623/.1937—dc23
LC   record available at https://lccn.loc.gov/2018012142

ISBN: 978-1-4724-7716-3 (hbk)
ISBN: 978-0-367-66554-8 (pbk)

Typeset in Times New Roman
by Apex CoVantage, LLC

For Caitly, Clio, and Nicholas

Contents

Figures

Unless otherwise specified all the images are by the author or under license and are reproduced here by concession of the Ministero dei Beni e delle Attività Culturali del Turismo—Parco Archeologico di Pompeii

Color plates

Acknowledgments

As always with such projects, this book began as an idea thrown in the air with a colleague and friend, in this case Michael Thomas, while walking back to our hotel from dinner in Pompei several years ago. The topic of fortifications had been long on my mind, but it was only then that I realized how a diachronic study of city walls with regard to a single site was lacking. The foundation of the book then took shape under the guidance of John Clarke and Penelope Davies at the University of Texas (UT) at Austin. The research work continued at the Center of Advanced Study for the Visual Arts in Washington, D.C., to arrive at the point it is today. To these institutions and in particular this trio of mentors I owe a profound debt of gratitude for supporting and seeing this project through to its final publication. I have to thank Nassos Papalexandrou, Rabun Taylor, and Andrew Riggsby for their guidance in the initial stages of this project. A special thank you goes to Paul Jaskot who generously read a draft of the manuscript, engaged, and acted as a mentor through the later stages of the project.

Many of the ideas in this book have taken shape through the conversations, drafts, and support from colleagues along the way. Without this back and forth and even simple observations during walks around the Pompeian walls, I could never have completed this book. They include Rebecca Ammerman, Mattia Biffis, Lea Cline, Caitlin Earley, Ingrid Edlund-Berry, Steven Ellis, Jess Galloway, Elaine Gazda, Regina Gee, Michael Greenwood, Alvaro Ibarra, Sandra Joshel, Gretchen Meyers, Nayla Muntasser, Jennifer Muslin, Lorenzo Pericolo, Fabrizio Pesando, Lauren Hackworth Petersen, Eric Poehler, Austin Shumate, Ann Steiner, Robert Vander Poppen, Andrew Wallace-Hadrill, Greg Warden, Iain Boyd Whyte, Paul Wilkinson, Kathy Windrow, Richard Woolley, and the anonymous peer reviewers of the first draft. To this list I must add the librarians who have helped procure the primary materials and work with the images, in particular Jacqueline Protka and Otto Luna.

I have benefited immensely from the generosity of the people and the projects on the ground at Pompeii. I owe much to the Director of the Ufficio Scavi of Pompeii, Dottoressa Grete Stefani, and Michele Borgongino, who helped me navigate through the archives, library, and the magazzini of the site. To Grete I also owe the rediscovery of the Porta Marina Minerva, a statue that had languished in a corner of the storage rooms since the late 1800s. A particular note goes to Giovanni Di Maio with whom I travelled to Velia and Paestum. His expertise in the geology of Campania supplied precious insights into the building blocks of Pompeii. Of course, none of this work would have been possible without the generous permits and support granted by the succession of directors at Pompeii while this book took shape, including Pier Giovanni Guzzo, Antonio Varone, Teresa Elena Cinquanta-quattro, and Massimo Osanna. To them I owe particular thanks for opening the doors to the archives, library, and the archaeological site of Pompeii, as well as the permission to reproduce many images in this volume. Particular gratitude goes to Steven Ellis, Eric Poehler, and more generally the team of the Pompeii Archaeo-logical Research Project: Porta Stabia, including Allison Cartmell Emmerson. Their generous support and suggestions have shaped new ideas on the develop-ment of Porta Stabia and the adjacent fortifications. The project also supported the plans of the gates presented in this volume by supplying some critical measure-ment data needed to correctly scale and orient the drawings.

Funding for the book and research has come from a variety of venues. I carried out the groundwork with the help of the Department of Art and Art History at UT Austin. I was able to spend time in Pompeii largely because of my participation in the Oplontis Project, which supported part of my stay during my research endeav-ors. A fellowship from the Center for Advanced Study of Ancient Italy supported a year of work needed to finalize the first draft. Further research funds came from generous grants supplied by the Center for Advanced Study in the Visual Arts in Washington, D.C.; I am eternally indebted to Therese O'Malley, Elizabeth Crop-per, and Peter Lukehart for their confidence in my work.

The images for this volume have come from a variety of sources. I have to thank Lucas Kukler for the reconstruction drawings of the fortifications. Timothy Liddell produced the scale plans of the gates at great pains since a proper survey of the fortifications was outside of the premise and means available for this volume. The gate plans come from a number of sources, including Maiuri's original drawings for the Porta Ercolano, Vesuvio, Nola, and Stabia. A base drawing generously supplied by Florian Seiler and Heinz Beste of the German Archaeological Institute enhanced the plan of the Porta Vesuvio reproduced here. I thank Eric Poehler and the Pompeii Mapping and Bibliography Project for the base drawings of the Porta Marina, Nocera, Sarno, and the other gates reproduced here. As a result of this collection of plans used to produce the final versions, I must add a disclaimer that there might be some error in the drawings, which I hope can be corrected in future architectural surveys. John Clarke generously supplied other images as taken by his late partner Michael Larvey. Most images are otherwise personal photographs. Funding for the reproductions, including the color plates, drawings, plans, and permits, has come from the Center for Advanced Study in the Visual Arts, the College of Liberal Arts at the University of New Hampshire, and private sources.

Immense gratitude goes to my family as an incessant source of support through my endeavors. My wife Caitlin is the foundation upon which I have managed to pursue a life in academia and build a family. She and our children are the source of inspiration that keeps me working every day. My parents have supported my life decisions at every turn. They have also accompanied me in some of my research trips and have taken it upon themselves to travel to a few locations for photographs of the images in this volume. My father, in particular, has acted as a field photographer for a number of the reproductions. Finally, I thank my sisters, who, as siblings often do, always knew what I wanted to do before I did and encouraged me to pursue my endeavors.

Defenses and walls are as old as warfare itself. They have served to unite and ostracize, protect and divide as elements of cultural and social identity. Yet for all their presence at ancient sites such as Pompeii, walled circuits are neglected as tourists shuffle past ignoring one of the largest and oldest public monuments in the city. The aim of this book is to examine the overlooked ring enclosing the city. Yet considering the size of ancient Pompeii, the volume of associated publications, and the many unanswered questions that surround the Pompeian fortifications, this volume can be only the beginning on the topic, as research is gathering pace on fortifications in general and on early Pompeii in particular. The pages that follow are current to the early fall of 2017, and already I know of several publications that would deserve more scrutiny. Further efforts will need systematic campaigns of documentation and excavation that are not the intention of this volume. Some questions will have to remain open for the reader to ponder and I hope to pursue answers.

Introduction

If our conclusions are just, not only should cities have walls, but care should be taken to make them ornamental, as well as useful for warlike purposes, and adapted to resist modern inventions. For as the assailants of a city do all they can to gain an advantage, so the defenders should make use of any means of defense which have been already discovered, and should devise and invent others, for when men are well prepared no enemy even thinks of attacking them.[1]

With this single statement, Aristotle implies that city walls functioned as a civic monument as well as a military structure. Centuries later, Vitruvius would second this observation, stating that all public buildings, including fortifications, should possess strength, utility, and beauty.[2] Yet because of their martial character, we tend to think of fortifications primarily in terms of their military function. This despite the fact that city walls are arguably some of the largest investments in terms of labor, materials, and design that any community would put forward. This book takes Aristotle and Vitruvius as premises to equate the monumental aspect of fortifications with their military function in Pompeii and on the Italian peninsula in antiquity. It poses a new series of questions regarding their social and architectural role within an active urban matrix. Far from static defensive structures, city walls emerge as carrying multivalent meanings critical to the definition of a community.

The exceptional preservation of the Pompeian fortifications, their associated urban matrix, and the wealth of information concerning the social structure of the city are a natural focus for this book. During the roughly 600 years of their existence, the Pompeian defensive circuits would see relatively little in terms of military action. Sullan forces certainly attacked the circuit during the siege of 89 BCE, and it probably functioned as a deterrent in the Second Punic War and the revolt of Spartacus. For all their military character, the day-to-day functioning of the defenses occurred largely in peacetime. Factors such as patronage, political ideologies, and ideas of communal identity would continually influence the appearance of the fortifications. In turn, the city walls, through sheer scale and presence, would be a critical structure in the spatial and social development of Pompeii. Preserved graffiti, inscriptions, frescoes, mosaics, and reliefs provide a unique glimpse on how individuals perceived the defenses in the final years of the city.

Pompeii built three main circuits in the course of its 700-year history (see plate 1). The first two, known as the Pappamonte and the Orthostat fortifications – so named after their construction material (pappamonte) and technique (orthostats) – have survived in a fragmentary state. Dating to the sixth and fifth centuries BCE, they mark a first genesis of the city before a period of urban contraction set in. In the late fourth/early third centuries BCE, Pompeii built its third main circuit, also known as the Samnite enceinte, because the city was now in the Samnite sphere of influence. The new circuit marked the beginning of a period of sustained development, which would only end with the eruption of Vesuvius in 79 CE. The Samnite circuit would be among the largest and oldest public structures in the city. At about 3.2 kilometers long and 8 meters high, the enceinte was a grand affair built in a classic *agger* construction where an outer wall acted as a revetment to a tall earthen embankment. It featured at least one postern and eight gates, seven of which (the Porta Stabia, Nocera, Sarno, Nola, Vesuvio, Ercolano, and Marina) are still open today (see plate 2).

The Samnite circuit would be the object of patronage and continuous renewal. In the late third and the late second centuries BCE, the walls would receive two further upgrades: The first involved widening and adding height to the *agger* with the construction of a new internal wall, whereas the second upgrade consisted of building the towers. These events would be pivotal moments in the architectural, social, and historical development of the city. Each upgrade occurred using locally quarried construction materials such as travertine, tuff, and *opus incertum* – a concrete faced with lava stones – that implicitly tied the fortifications with the territory of the city.

In antiquity fortifications would include a measure of ornamentation and aesthetics that would link them with the identity and social structure of a community. The application of specific construction techniques and materials developed emotional and historical ties that would change significantly in a similar way that John Onians has identified for the canonical classical architectural orders as applied through the ages.[3] Such a development is part of a wider trend on the Italian peninsula where materials such as the tuff present in Southern Etruria – easily sculpted into regular ashlar masonry – and the limestone commonly used in polygonal walling techniques in communities along the Apennine Mountains naturally tended to stimulate the development of regional identities.[4] Towns and cities would enhance these basic formal aspects with further embellishments, including the degree of stone dressing (embossing/rustication), the juxtaposition of construction materials, and other sculptural embellishments as a reflection of the community.

Through its continuous presence and upgrade, the Pompeian circuit acquired architectural, urban, social, and political roles. It would operate in terms of what Aldo Rossi defines as an "urban artifact": buildings that endure within an urban matrix to shape its development and change meaning with each successive generation through preservation and alteration. From this point of view, the enceinte, although an inherently defensive structure, would acquire an artistic value because it conditioned the developing urban matrix, while the city would condition the development of the defenses in a reciprocal process.[5] In this framework, the fortifications would develop along Marvin Trachtenberg's concept of "building in time" where a structure receives continuous modifications and in each event acquires a new meaning.[6] Although the upgrades to the Pompeian defenses occurred in bursts of feverish activity, their continuous modifications made them dynamic urban artifacts that reshaped their function, meaning, and memory for each successive generation.

Fortifications would be the markers of urban space that conditioned a viewer and a city to define its urban layout (armature) and image. Their static line would shape the urbanization process, making the city walls a defining element of civic space and image. Kevin Lynch describes five units characteristic of modern urban frameworks: paths, edges, districts, nodes, and landmarks. Paths (i.e., streets and walkways), nodes, and landmarks create channels dictating individual movement. Along their edges, buildings and districts, commonly defined by landmarks or other characteristics such as construction techniques, define the image and concept of a city.[7] William MacDonald has shown that such paths, nodes, and landmarks formed the basic skeleton, or armature, of ancient cities.[8] City walls and gates were the formative elements of an armature, anchoring streets and roads and greeting regional arteries heading toward the heart of the city. Such a concept was already apparent to Vitruvius, who urged architects to plan fortifications and public buildings first in the basic layout of a new town.[9] Once built, defenses would seldom move. Instead they could change their meaning as urban artifacts or change their form through patronage.

The role of the Pompeian defenses as a civic monument offers a unique window into how fortifications acquired an inherent symbolism for the city. Surprisingly, given the importance of urbanization processes to the formation of communal identities, this topic has received little attention.[10] Studies have tended to focus on the early Roman empire, with the Augustan and Julio-Claudian emphasis on the (re)construction of cities in Italy and the western empire after the civil wars. These emperors operated in a well established tradition that was a fundamental premise for urbanization, communal genesis, citizenship, and notions of the ideal city. Some of this symbolism associated with fortifications stretches back to Aristotle and finds an origin with Archaic and Classical ideas of the Greek polis. Cities could not exist as independent communities without communal defenses as the lowest common denominator of identity for the *polis*.[11] Such a basic definition must have accompanied the process of urbanization and taken root in Greek and Etruscan cities contemporary to Pompeii.

On the Italian peninsula, the symbolism associated with city walls stems from a tradition where fortifications tied into architectural, ideological, social, and ethnic developments. Yet its development remains little understood. Part of the reason is that regional studies of defensive systems and fortifications of (pre-)Republican Italy remain underrepresented. This is particularly the case for the area of ancient Samnium, whose population would be crucial to the development of Pompeii.[12] The better published Roman, Etruscan, and Greek fortifications are therefore a natural departure point for this study.[13] Another issue is that ancient accounts on city walls tend to focus on their role in battle or their layout rather than on their physical appearance.[14] Ancient authors such as Philon of Byzantium, Vitruvius, and Apollodorus Mechanicus write about the fundamentals of offensive and defensive siege tactics, construction techniques, and architectural design.[15] Philon and Apollodorus discuss weaponry and siege tactics, whereas Vitruvius focuses on construction. Battle strategies are the focus of Aineias the Tactician (Aineias Tacticus) who wrote a treatise on the practicalities and tactics useful to endure a long siege.[16] Their collective writings span from the Hellenistic Age to the height of the Roman Empire – Apollodorus Mechanicus may be Apollodorus of Damascus working under Trajan.

Anthropological approaches help define the interaction between fortifications and communities. Warfare, both offensive and defensive, has always been a main factor in the drive toward urbanization and the organization of society. Dominant patterns of warfare – whether seasonal or year-round affairs, whether conducted by raiding parties or complex armies – influenced settlement location. The social significance of warfare, military architecture, and defense is evident for Republican Rome: About 10 percent of freeborn men annually joined the army, and military expenditure reached as high as 80 percent of state budgets.[17] In Italy, early warfare was mostly seasonal, consisting of simple raids into enemy territory. Such offensive tactics resulted in the establishment of small defensive enclosures. Rather than facing war parties, those living in the countryside could withdraw into the safety of a refuge and bring with them movable possessions such as domesticated animals and food. In the beginning, these safe havens may not have featured fortifications, relying instead on the natural strength of a position to discourage direct assaults. People grouped within their confines could neutralize the threat of surprise attacks and enhance overall security.[18] Raiding parties would usually avoid them, preferring instead to pillage the immediate territory.

Strongholds would become essential psychological reference points in the landscape, offering safety in times of danger. A society placed their most prized assets, including gods, wealth, and magistrates, within their defenses to keep them safe.[19] Often located in inaccessible places within broad territories, strongholds tended to dominate a landscape militarily and physically. Militarily, the dominant position of strongholds made them an active staging point to protect the territory.[20] Defenders were often at an advantage, with wide lines of sight enabling them to predict enemy movement, set devastating ambushes, and launch effective counterattacks to defend critical unmovable elements such as fields, pastures, water sources, hunting areas, sources of raw materials, and trade routes.[21] If it would come under direct attack or siege, a particularly well defended strongpoint allowed a relatively small band of soldiers to hold out against a numerically superior enemy almost indefinitely, allowing communities to retain their cultural autonomy. The construction of defenses around a strongpoint signaled territorial assertions that intertwined with the emotional and historical ties binding a community to their region. They physically formed a symbolic boundary between the collective dwelling, its territory, and the outside world.

Fortifications were the single architectural element negotiating between order and chaos, acting as the symbol and signifier of social balance and hierarchy.

Their presence in a settlement were critical factors in the genesis of a communal identity and a driving force in the process of urbanization and the development of civic institutions. The concept of a city-state, or *polis*, is that of an institutionalized and centralized microstate consisting of a defended town, its immediate hinterland, and a stratified society inhabiting it.[22] Its origins lie in the need for small communities to unite in a mutual necessity for defense. Fortifications increased the sense of security against external dangers. They protected a community from human and natural threats such as outsiders, enemy armies, war parties, brigandage, banditry, natural disasters, disease, and wild animals. Fortifications thus had a broad protective role to keep out the unwanted "other," creating a symbolic psychological boundary protecting the civilized community against uncivilized dangers.[23] City walls created a sense of hegemony within their confines and associated territory, acting to harmonize the discrepancy between the two opposites of danger and security.

Internal social, economic, and political factors influenced the manner of defense and the appearance of fortifications. For instance, (semi)nomadic and pastoral societies often relied on a series of predetermined natural strongpoints within a territory or along transhumance routes to find refuge in times of danger. Depending on their position and relationship to the landscape, these refuges had the potential to become nuclei for habitation or spaces of religious importance. Settled agricultural societies relied on a strongpoint or a refuge that often evolved into cities and towns. Communities then decided to build fortifications either in response to a direct threat or, if they gained enough wealth to construct them, in anticipation of future trouble.[24] Several different types of government may be evident in the settlement and defense pattern in a specific region. Aristotle mentions that a single citadel was characteristic of tyrannical, monarchical, and oligarchic models of government where the fortified area enabled control of the population. A city located in a plain was characteristic of a democracy, offering much more freedom for commerce and mobility for its citizens. A ruling aristocracy would have a number of strongholds dotting the landscape where each aristocrat or group of families controlled their dependents through private fortified enclosures.

These Aristotelian ideas have found resonance in the archaeological record.[25] Fortifications built in late Bronze Age Greece were mainly concerned with the private interests of select elites.[26] The presence of an *acropolis* with an associated settlement suggests a hierarchical society led by a king, tyrant, or oligarchy. Client villages surrounded the fortified centers, which protected the population from outside threats and the elite from social upheavals. This setup differs from fortifications built in early Greek colonial foundations where enceintes protected the entire community as a reflection of the evening out of social classes. The *acropolis* therefore quickly lost its defensive importance; in some cases, cities dispensed with it altogether.[27] Changing politics and the appearance of new social ladders led to similar developments on the Greek mainland. The rise of the Athenian democracy shifted defensive concerns from a territorial network of strongholds reflecting the land possessions of the hoplite class to the construction of the long walls as a reflection of the growing influence of tradespeople, artisans, and the new defensive emphasis on its navy. In Etruscan Italy, similar patterns developed that centered initially on the *arx* located on hilltops and later settlements that spread out through the landscape as a reflection of an oligarchic social structure. As opposed to Greece, the *arx* would remain a separate space from the city because of its continued religious function. From here priests, that is, the elite, would take divinations and, through religious association, maintain their legitimacy and political power.[28]

Fortifications also entailed a degree of social control. Authorities could attempt to control any kind of social upheaval from the position of the walls. Ruling elites or a tyrant could use city walls as a base to contain a rebellion. Diodorus Siculus recounts how in 405 BCE the tyrant Dionysius I of Syracuse, when faced with a rebellion of the populace, retreated to the island of Ortygia, leaving the city to the rebels. Dionysius then sealed off the island with an extra wall and further fortified its *acropolis*. From this position of strength, he eventually managed to put down the rebellion.[29] On occasion, walls also functioned to separate two different populations. In the fifth century BCE, Greek settlers had founded a colony called Neapolis – later redubbed Emporiae/Empúries – in modern Spain. The tensions with the local inhabitants led the Greek population to build a wall and keep it guarded at all times. Spaniards could not go into the city, whereas Greeks would exit only under escort until the two communities eventually fused in the early first century BCE.[30] The fortified wall here was a clear divider, separating the two populations, psychologically, linguistically, and culturally for hundreds of years. When applied to foreign politics, social control translated to the deliberate implantation of colonies inside enemy territory. Troop concentrations within their confines could contain revolts from fortified positions or move quickly into the surrounding territory. The walls of a colony were symbolic emblems of the mother city to both the colonists and the local population. As such, they could also act as instruments of mutual assimilation.[31]

Fortifications also afforded a measure of social and economic control through policing and taxation as part of their role as formalized filtering points between the city and its countryside. Because they regulated access to the settlement, city gates were the formal elements reflecting the state. Gates functioned as toll points where authorities could tax those entering and exiting the city and control the movement of the population. As the Roman Empire grew and relegated its defense to distant provinces, many gates within the empire would acquire secondary entrances as a measure to prevent traffic buildup and congestion. In examples such as Carsulae and even later at Timgad, gates would stand isolated without fortifications as symbolic markers of the city.[32] Authorities added embellishments reflecting the political and social structure that the fortifications represented – in the Augustan period, city gates would start to take on the form almost of triumphal arches.

Authority over the walls essentially meant control of the settlement. This aspect was symbolic enough that the Roman army awarded a special *corona muralis*, or mural crown, as a distinction to the first individual to scale an enemy wall.[33] The act of reaching the battlements meant that troops had breached the city; control of its defenses signaled defeat. Aineias the Tactician, writing on how to survive a siege, repeatedly warned the reader – that is, future military commanders – that a loyal body of troops manning the walls was imperative to the control of the city. Internal plotters could sabotage control of the walls and thereby the city by giving an enemy knowledge of the defenses. Gatekeepers, sentinels, and guards at the most vulnerable parts of the wall must be drawn from the wealthiest citizenry: They had the most to lose in defeat, and the enemy could not corrupt them.[34] A single episode among many in Livy's history of the Second Punic War further elucidates the point. He describes the citizens of Henna in Sicily demanding the keys of the town gates from the Roman soldiers occupying the walls since they considered themselves freemen in alliance with Rome.[35] The implication here is that the keys to the gates and therefore to the fortifications symbolize the liberty or subjugation of the town.

As a critical element to communal defense, city walls would also carry religious symbolism often in the form of apotropaic devices meant to keep out the evil eye and malevolent spirits from the settlement. The Bronze Age gates ornamented with lions of Hattusa and Mycenae, as well as the Lamassu protecting Assyrian palaces, are prominent examples where such apotropaic devices would help shape the identity of the settlements they protected. Inevitably, they would also connect with the political hierarchy of the city. In Italy, such devices would find a place in shrines and busts at city gates meant to protect liminal spaces. Fortifications would also rank among the sacred things (*res sacra*) that fell under the protection of the gods. These associations stretched back to the Etruscans and the foundation of cities where a ritually marked furrow symbolically traced the extent of the future settlement. The religious significance of the gates, the symbols they carried, and the tutelary divinities of cities would intertwine with elite patronage to protect a community from visible and invisible threats.

Fortifications built by single city-states were symbols of social and military independence. Any construction event from the refurbishment of single gates to the building of single towers or entire enceintes would equate to such ideals. Much of this symbolism came from the inherent difficulty in capturing fortified cities, which would require costly and time-consuming sieges. Each construction episode was an act that carried an element of deterrence.[36] Master tacticians such as the Chinese general Sun Tzu expressly warned commanders to avoid sieges as much as possible. Attacking armies would have to plan for supplies as well as build the necessary siege equipment on site, in a process that could take months and lead to prolonged exposure in enemy territory.[37] For these reasons, most cities would fall to internal treachery or ruses meant to avoid direct attack – the Trojan horse being perhaps the most celebrated example.[38] From this position of strength, city walls could become bargaining chips in diplomatic negotiations. Thucydides recounts how Athens razed the walls of Tanagra and compelled Aegina and Samos to raze their own during the Peloponnesian War as a measure of their submission and to avoid future negotiations with the cities.[39] City walls had acquired such a symbolic function that, by razing them, a conquering army essentially disarmed a city: With the fortifications went its independence. Urban enceintes were the emblems of civic identity, symbolizing the community almost as its crown.

Notes

1 Arist. *Pol.* 7.1330b.
2 Vitr. *De arch.* I.3.2.
3 Onians 1988, 8–18, 27–38.
4 Lugli 1968, chps. 1 and 2.
5 Rossi 1982, 32–35.
6 Trachtenberg 2010, 11–13, 16–22.
7 Lynch 1960, 46–90.
8 Macdonald 1988, 5–110.
9 Vitr. *De arch.* I.3.1.
10 For a recent discussion on the symbolism of defenses, see S. Müth et al. 2016, 126–158. Otherwise, the list for the Augustan period grows rapidly: Richmond 1933, 149–174; Kähler 1942, 1–108; Gabba 1972, 73–112; Jouffroy 1986; dedicated proceedings, Colin (ed.) 1987; Rosada 1990, 364–409; Rosada 1992, 124–139; Pinder 2011, 67–79.
11 See McK. Camp 2000, 47–51.
12 Only S.P. Oakley 1995 has systematically tackled the fortifications of ancient Samnium.
13 Winter 1971 and Lawrence 1979 are still fundamental for Greek fortifications. Blake 1947 and Lugli 1968, although pioneers for Italy, included fortifications in surveys of construction techniques. For Republican Italy, Gabba 1972, 73–112; Jouffroy 1986; Brands 1988; Miller 1995 are seminal works. Recent volumes of conference proceedings dedicated to the fortified city have propelled the field: Camporeale (ed.) 2008; Attenni and Baldassare (ed.) 2012; Bartoloni and Michetti (ed.) 2013; Attenni (ed.) 2015; Fontaine and Helas (ed.) 2016; Frederiksen et al. (ed.) 2016; Müth et al. (ed.) 2016.
14 Lawrence 1979, 53–66; Camporeale 2008, 15–36.
15 Philon of Byzantium translated by Lawrence 1979, 67–107; Vitr. *De Archit.* I.5; Whitehead 2010 translates Apollodorus Mechanicus.
16 Whitehead 2001 translates Aineias the Tactician.
17 Cornell 1995, 122, 130.
18 Gat 2002, 133.
19 Garlan 1968, 245; Lawrence 1979, 113.
20 Ain. *Tact.* xvi, 16; Hansen 2000b, 162.
21 Rowlands 1972, 448.
22 Rowlands 1972, 447; Hansen 2000a, 19; Gat 2002, 12.
23 Van Gennep 1961, 18–25; Rowlands 1972, 48; Turner 1977, 94–130; Müth 2016, 162.
24 Lugli 1966, 27–32.
25 Arist. *Pol.* 7,1330A.35ff; Cherici 2008, 37–66.
26 Winter 1971, 289–290.
27 Winter 1971, 33–34, and 57.
28 Torelli 2000, 197–205; Cherici 2008.
29 Diod. Sic. *Bib. hist.* 14.7.
30 Livy *Ab ur. con.* XXXIV. 9; Aquilué et al. 2006, 24.
31 Pinder 2011, 72.
32 Stevens 2016, 295–297.
33 Aul. *Gel.* 5.6.16. The crown featured representations of the battlements of a wall.
34 See Aen. Tact. books vi, xi, xiv, and xxii.
35 Liv. *Ab ur. con.* XXIV. 37. 5–10.
36 Ducrey 2016, 332–336.
37 Sun Tzu 3.2.
38 See Ain. Tact. x.25–14.
39 Thucydides *Pel. War* 1.108; 1.117; 3.50; Lawrence 1979, 113; Kern 2000, 135–162 and 323–351.

1 Prolegomena to the city walls of Pompeii

The looming presence of the Pompeian defenses has shaped the understanding of the genesis of Pompeii in a manner that no other single monument has been able to do. Most modern hypotheses concerning the formation of the city have rested on their sequencing and dating. Gates such as the Porta Ercolano have remained exposed ever since their first discovery in the late 1700s, thereby influencing the views of countless visitors and scholars. Although the excavations of Pompeii have focused on recovering the city as it froze in 79 CE, the city walls were one of the few monuments where investigations could continue to shed light on the genesis of the city. The reasons for this approach were multiple. The fortifications of Pompeii represent the first tangible signs of an organized community preoccupied with defense. Understanding the line of the defenses meant that authorities could map the limits of the city – or so they thought at the time – and detail the succession of construction techniques used in fortifications and by extension the city. Given their location on the fringe of Pompeii, excavators could reach lower strata to date the site without destroying ancient floors or streets. Due to their size, large parts of the defenses would remain intact despite extensive investigations. Because the city gates anchored the urban grid, logic dictated that the walls were the oldest structure in Pompeii. These combined factors mean that scholars have long contemplated the fortifications, in the process producing many divergent theories concerning the development of Pompeii. Although a broad common consensus now exists for their construction sequence, many of the debates are still open to question.

Despite the importance of Pompeii's fortifications, no summary of their recovery exists. We know little about their state of preservation at the time of excavation and subsequent conservation. The sheer scale and the eclectic nature of the construction techniques present in the remains mean that the monument has resisted systematic evaluations. The Porta Stabia, one of the earliest gates, is encased in much later Roman concrete. Farther east, the circuit emerges near the Porta Nocera to display travertine masonry up to the amphitheater, where it turns into concrete again. After the amphitheater, the stretch of wall up to the Porta Ercolano displays a socle of travertine with surmounting courses of tuff masonry of varying height. Rough sections of later Roman concrete often interrupt the curtain. A tract remains buried at the height of Tower IX, whereas the walls emerge sporadically from beneath later housing built on the western side and from structures between the

Temple of Venus and the Doric Temple. The remains of eleven towers built at the turn of the first century BCE are unevenly spaced throughout the circuit, with concentrations on the northern and southern sides. The location of a twelfth remains contested. The piecemeal excavations, the divergent recording standards of past investigations, and the patchwork of masonry complicate the debate concerning the development of the walls. The fundamental lack of knowledge for the early city adds to the problem.

The city walls and the theories on the urbanization of Pompeii

With only 2 percent of Pompeii excavated below the level of 79 CE, the various theories associating the development of the city with the fortifications are approximations.[1] The debate divides into two main camps: those favoring a grand central design of the city and those who see it growing from an old core, also known as the *Altstadt* – or old city. On occasion, these two camps have combined into a third where the grand design and the old core represent different periods of development, each following independent stages of design. The original nineteenth-century hypothesis, known as the grand Pompeii theory, envisioned the city as founded in a single coherent urban design in the sixth century BCE. Its defenses consisted of a travertine-faced *agger*-type fortification – the equivalent of the first Samnite enceinte built in the late fourth century BCE that visitors see today.[2] Understanding the dynamics and reasons for such a massive beginning to Pompeii were daunting, leading to some more nuanced approaches to the single foundation theory. Giuseppe Fiorelli and Antonio Sogliano hypothesized the presence of earlier defenses where a *fossa-agger* system, composed of an earthen embankment and reinforced with a surmounting palisade and an exterior ditch, first enclosed the city. Although not in the form or date they had envisioned, the eventual discovery of the Pappamonte and the Orthostat enceintes partially validated Fiorelli and Sogliano's hypothesis. Based on the orthogonal plan of the city, Sogliano also suggested the presence of an early gate on the western end of the via di Nola as a logical counterpart of the Porta Nola.[3] Although Maiuri later discredited this idea, the recent identification of a postern beneath the House of Fabius Rufus (VII.16.22) confirmed Sogliano's hypothesis almost a century after his first assumptions.[4] Such long-term debates are typical of Pompeian studies.

In 1913, Francis Haverfield proposed a new model where Pompeii first developed around a smaller core centered on the Forum that was preserved in the irregular layout of the streets in this corner of the city.[5] A few decades later, Armin Von Gerkan elaborated the theory, identifying the vicolo dei Soprastanti, the via degli Augustali, the vicolo del Lupanare, and the via dei Teatri as the limits of an old core that he called the *Altstadt* (see plate 2).[6] This perimeter of streets was a relic of a path that followed a vanished fortified *agger*. Gates opening onto regional roads would become the major arteries of the later city. After the construction of the outer fortification in the late sixth century BCE, the new city, or *Neustadt*, would expand into its current layout. Hans Eschebach refined this model by identifying two phases of development for the *Altstadt*. The first settlement developed as an *urbs quadrata*, or squared city, with four neighborhoods on each side of the central Forum. This core gradually expanded in a second phase into the limits of the *Altstadt*. Eschebach also projected an earlier fortification defending the *urbs quadrata* after excavations identified the remains of a robbed out wall in the vicolo del Lupanare.[7] Much of this theory relied on Amedeo Maiuri's discovery of a hypogeum beneath the Terme Stabiane, which he believed was an Etruscan tomb. Ancient custom strictly dictated that funerary structures must be located outside the city walls. John Ward-Perkins was a vocal critic of the notion of the *urbs quadrata*; it has since been dismissed because Maiuri's hypogeum is likely not a tomb.[8]

The *Altstadt* theory, in its various nuances, dominated the hypotheses concerning the development of early Pompeii until Stefano De Caro excavated and dated the Pappamonte fortification to the sixth century BCE. This discovery essentially reversed the trend and returned to the idea of a grand foundation.[9] However, De Caro did not dispense with the *Altstadt* altogether, envisioning it instead as a main habitation core with a surrounding swath of open terrain protected by the Pappamonte circuit. Rather than an elaborate stone fortification, a simple palisade, road, or drainage ditch would explain Von Gerkan's *Altstadt* perimeter being preserved in the later streets.[10] Since then, archaeological excavations in Regio VI have further refined the *Altstadt* theory, indicating that it actually represents a retreat of the urban expanse into smaller, more defensible, confines in the fifth century BCE.[11] After the construction of the first Samnite enceinte in the late fourth/early third century BCE, Pompeii gradually expanded back out into the *Neustadt* after Campania entered the Roman sphere of influence.

The construction materials used in the circuits have played a critical part in ideas concerning the development of Pompeii. Giuseppe Fiorelli first defined the architectural history of Pompeii associating it broadly with the main historical developments occurring on the Italian peninsula. He described three distinct periods connected to the use of construction materials: limestone (travertine) as the oldest, tuff associated with the Samnite city, and concrete and brick dating to the Roman period. August Mau later refined these periods to five by dividing the Roman period into the early colony, the early empire, and finally the post–62 CE earthquake reconstructions.[12] The framework still forms the basic premise for the architectural history of Pompeii, although recent approaches rightly blur the rigid periodization because materials and techniques, once invented, can cross over into later periods. The identification of a separate limestone phase is particularly debatable due to the paucity of remains.[13] Nevertheless, Fiorelli's strict periodization almost certainly swayed Overbeck and Mau when they first formulated the construction sequence of the fortifications in 1875 and later influenced Maiuri when he reframed their development in the late 1920s (see Figure 1.1).

The use of *opus incertum* in the towers, the gate vaults, and the curtain wall in the late second century BCE is similarly part of a debate concerning the development of Pompeii and a broader discussion on the genesis of concrete in Roman architecture. A principle problem with these hypotheses is that the practice of dating masonry is a notoriously imprecise endeavor.[14] The earliest Pompeianists,

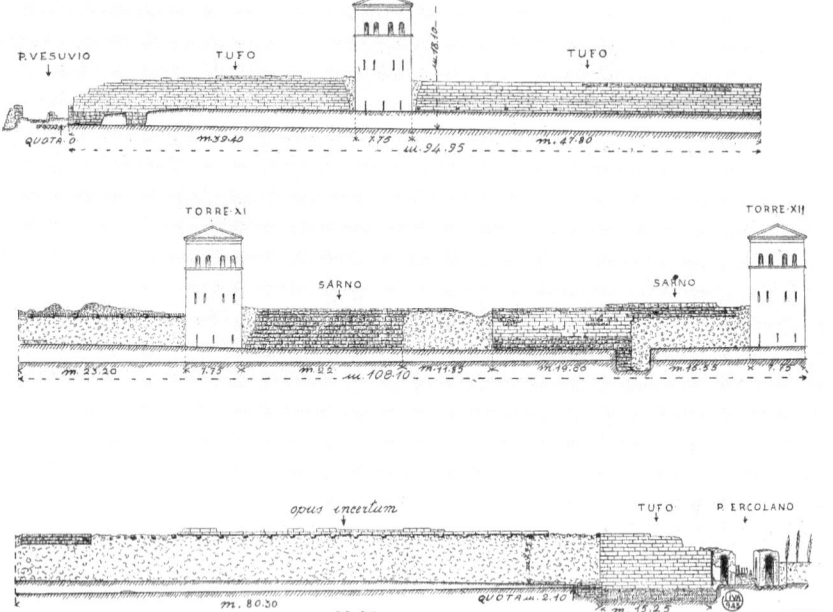

Figure 1.1 Maiuri's drawing of the north side of the fortifications between the Vesuvio and Ercolano gates detailing the various types of masonry. (After Maiuri 1943, fig. 1)

such as Romanelli, Mazois, and more recently Richardson, viewed the application of *opus incertum* in the towers and the curtain wall as repairs needed after the Sullan siege as a response to the unrest caused by Caesar's assassination.[15] Niccolini and Overbeck-Mau dismissed this notion, believing that the long tracts of concrete are too extensive to represent the damage of a single siege.[16] Soon afterward, Sogliano discovered a graffito with the letters L.SUL(A) in Tower XI. He imagined a victorious Roman soldier who scratched it into the plaster in the aftermath of the siege.[17] Mau subsequently suggested that the peace following the Second Punic War led to the abandonment of the fortifications and their spoliation for building material. A panicked restoration at the outbreak of the Social War led to the *opus incertum* repairs and the construction of the towers. A key argument for Mau was the urban development that had occurred in the southwest portion of the city in the late second century BCE.[18] Private houses had encroached upon the walls to exploit the views toward the Monti Lattari and the Sorrento peninsula, thereby weakening the defenses. Mau did not account for the fact that the houses initially used the fortifications as part of grand open terraces. Troops could still use these spaces in a sector where natural defenses needed little further fortification.[19] Perhaps Maiuri offered the most measured approach by proposing that the repairs using *opus incertum* can represent both pre- and postsiege construction activity.[20]

Recent research confirms this viewpoint, although the evidence tends to indicate that most sections of the curtain wall built in concrete represent postsiege repairs.

Excavating the fortifications

The long debate surrounding the construction of the fortifications is the result of their equally complex history of recovery. Their first discovery is mired in the murky history of the early digs when recording techniques and requirements were far less exacting. Much of the earliest archaeological evidence is lost, but some of it survives in fragments throughout the published excavation journals known collectively as the *Pompeianarum Antiquitatum Historia* (PAH), Giuseppe Fiorelli's compilation of excavation diaries written between 1748 and 1861.[21] The light it sheds on the current state of the fortifications is somewhat dim because it lacks basic descriptions, drawings, and plans. Nevertheless, it supplies a glimpse of the state of the fortifications during their recovery and of the decisions that led to the remains exposed today. From this account, the fortifications emerge as a continuous presence shaping the history of Pompeian excavations and modern perceptions of the city and its limits.

The excavations first touched the walls at the Porta Ercolano on the northwestern tip of the city. The PAH entry of September 17, 1763, mentions work occurring near the "puerta de la cuidad," suggesting that workers had already noted its presence.[22] Giuseppe Fiorelli gives the following dates for the full exposure of the gate: September 14–October 12, 1763, April 1–July 12, 1764, and June 29–November 11, 1769.[23] The following entries in the PAH are less precise in locating the excavation, often describing the works occurring around the gate.[24] The PAH entry of September 15, 1764, details how excavators had tunneled their way to the exterior of the gate. On October 14, they proceeded to uncover its full structure using a systematic top-down approach in one of the first recorded instances of the method widely used today.[25]

There are only hints at any sort of statuary that may have adorned the Porta Ercolano. Unfortunately, as is the case for much of the journals, the entries concentrate on the finds recovered during the excavations rather than on describing any architectural remains or embellishments on the walls. A series of entries dating to 1769 mention the recovery of statue parts near the gate, including pairs of heads, arms, and hands with missing fingers that have since vanished. They may have belonged to any number of nearby tombs or buildings.[26] In two early descriptions of Pompeii, authors Adams and Clark discuss fragments of bronze drapery recovered near the gate that they believed belonged to a tutelary deity of the city set on a pedestal next to the gate. The bronze fragments have disappeared, but the pedestal referred to in the text probably belongs to the unfinished tomb on the north side of the gate.

The records remain patchy for the next decades of excavation. Entries in 1766 and 1768 briefly report the discovery of the city walls near the soldier quarters, now known as the *Quadriporticus* (VIII.7.16), in an effort to chart their course.[27] Subsequent years saw the gradual exposure of the fortifications on the north side of city as the excavations progressed in Regio VI. The extent of the works remains elusive because the notebooks mention the fortifications only in passing as part of the citywide excavations.[28] In 1782, explorations started on the southern end of the city. The location remains rather vague with the reports signaling efforts to find the fortifications near the Temple of Isis.[29] The city walls are actually quite far from the temple, suggesting that the work involved the nearest section of fortifications by the *Quadriporticus*. Andrè De Jorio, author of one of the earliest guidebooks on Pompeii, mentions excavations occurring in the area that year.[30] He describes the recovered remains of the city walls as razed by earthquakes and reused for their material, indicating that he was referring to the section of demolished wall curtain west of the Porta Stabia. Entries between March 15 and May 3, 1787, describe the continued excavation of a trench around the city.[31] Once again, the extent and location of this trench is unclear. Presumably, the entries refer to the area just east of the Porta Ercolano, where diggers began their efforts. After this episode, the sources fall silent, but excavations seem to continue under General Championnet during the short-lived Parthenopean Republic of 1799. For reasons that remain unclear, the general was particularly interested in finding the gates of the city. By the end of the year, excavators had found the top of the Porta Nola and had identified the remains of the Porta Sarno and Nocera.[32]

Up to this point, the exposure of the walls was largely piecemeal. The policy changed radically when Queen Caroline Bonaparte (Murat) set out to find the entire circuit during the French occupation of the Kingdom of the Two Sicilies between 1808 and 1815. The Queen decided to chart the walls and gates so that authorities could expropriate terrains within the perimeter and diggers could follow the ancient roads to understand the urban layout.[33] Under the plan, the fortifications would function again as a security perimeter to keep out thieves and looters from the site.[34] Given the infancy of archaeology at the time, the policy of defining the limits of the site was quite innovative but also exploitative since it aimed to bring the ruins under full control of the king. The policy would inadvertently set the stage for the modern perception that the enceinte marked a clear edge to the city. This notion has changed only in recent decades when the excavations outside the Porta Marina made it clear that the urban area could extend well beyond the city walls.

Starting in 1810, authorities assigned groups of soldiers and some private contractors, including a certain Pirozzi, to chart the fortifications.[35] The queen herself took an active interest in the excavations, often visiting the site to supervise progress and personally pay for the works.[36] Understanding the development of these excavations is difficult; the entries in the journals remain vague, and the topography of Pompeii was still very unclear. Between August 1811 and May 1812, work concentrated on the stretch between the Porta Nola and Ercolano, where excavators uncovered the unspecified remains of four towers and one destroyed gate. The four towers mentioned could be any of five, numbers VII–XII, now known in this area, and the gate is most likely the Porta Vesuvio.[37] Between September 6, 1811, and September 12, 1812, continued excavation work on the houses in Regio VI included digging a narrow trench to expose the exterior side of the fortifications.[38]

On October 5, 1812, groups of soldiers started uncovering the walls on the east side of the city in three unspecified locations.[39] The PAH mentions the discovery of the Porta Nola as a new gate on May 8, 1813, despite the earlier identification of the gate under General Championnet.[40] Work progressed rapidly. June 5 marks the recovery of the inscription and keystone of the arch.[41] By September 9, 1813, with much of the gate uncovered, the digging changed focus to follow the via di Nola into the city.[42] De Jorio reports on a concentrated effort in the eastern sector near the amphitheater between May 1812 and June 1813 that exposed three towers and a gate, presumably the Porta Sarno and Towers V, VI, and VII.[43] The PAH offers a slightly different version of the events, describing how workers reached the Porta Sarno on March 31, 1814, and exposed it by May 5.[44] Curiously, these entries omit to mention the towers. Instead, they detail the discovery of unspecified posterns that today are absent in the ruins. Perhaps excavators failed to recognize the towers and instead interpreted their rear and side doors as posterns.

Work near the amphitheater turned west after reaching Tower V to recover the enceinte on the south side of the city. Progress is again murky. De Jorio reports the discovery of four towers and a gate in the area spanning the amphitheater and the *Quadriporticus* between April 1813 and September 1814.[45] The number of towers is slightly odd considering what we now know of Towers IV, III, and II in this section. The author may have included Tower V in his count or the yet unlocated Tower I. De Jorio describes the Porta Nocera, a structure that Championnet had already identified in 1799, as largely excavated by the end of 1814.[46] The PAH, however, mentions that the gate only began to emerge on January 12, 1815, after excavators uncovered the street leading up to it. Reports dating to roughly a month later indicate the full exposure of the Porta Nocera, noting that much of it lay ruined.[47] Efforts subsequently moved back to the northern sector of the fortifications to find the Porta Vesuvio. The degree of this campaign's success remains unclear, but excavations stopped on March 12, 1815, when worker gangs moved to uncover the amphitheater.[48] Only the entries of January 16 and 23, 1819 mention a renewed effort to unite two wall sections uncovered in the northern sector, but the precise location is unclear.[49]

By the time the digging officially ceased on April 30, 1815, the operation had uncovered two-thirds of the circuit,[50] including at least five city gates and an unspecified number of towers. With the exception of the individual exposure of larger structures, the work was largely superficial, touching the top of the walls to chart their course. Only occasionally did diggers actually reach the level of 79 CE on either side of the fortifications.[51] With much of the circuit's course known, further operations languished. A slow inexorable process began. Many sections were abandoned, slowly buried beneath the massive dirt piles, known as the *cumuli borbonici*, coming from digs elsewhere in the city.[52] A significant gap exists between what the nineteenth-century excavations uncovered and the ruins exposed today. None of the reports mentions any sort of embellishments on the walls, although the current remains still display patchy plaster revetment on the towers and gates. Sadly, much of it is gone after succumbing to its initial exposure to the elements in the 1800s.[53]

Despite the extensive recovery of the walls, two of the seven gates remained unknown. Their recovery would be crucial to understanding the layout of the city. Excavations began at the Porta Stabia on April 22, 1851, with the entry of 10 June marking its discovery. Roughly nine months later, the work ended with a large part of the structure exposed.[54] It subsequently gained notoriety because of the debate on the translation of an Oscan inscription found on a slab in the gate court that centered on reading the lettering and its accurate translation.[55] The exterior edge of the gate remained unexcavated until 1853, when Giulio Minervini, seeking to publish a full plan, dug a tunnel to find the outer corner of the eastern bastion.[56] Starting in 1874 through the end of the century, excavations resumed to expose the area in front of the gate. In addition to the various tombs, excavators recovered a spout fashioned from a tuff block carved in the shape of a lion's head. At first, they identified it as the keystone to the gate vault, until August Mau recognized it as a spout that he believed was one of many that functioned to drain the wall-walk.[57] Simpler versions of such spouts are indeed still present in the fortifications, but the lion head was a unique example that has since vanished.[58] It may have functioned as a distinctive marker for the gate.

Across the city, the Porta Marina was the last unknown gate. Giuseppe Fiorelli began targeting the area on March 5, 1861, because a previous campaign had detected it in 1844 as part of a project to provide a new entrance to the ruins. Nevertheless, the structure remained buried and virtually unknown.[59] A month after the operation began, workers had fully defined the gate. The reports omit any further details and mostly describe the recovery of an outer niche holding the fragments of a statue of Minerva.[60]

In February 1898, Antonio Sogliano, in an effort to date the towers more accurately, began to expose Tower X on the north side of the city. Toward late August 1901, the operation ceased after extending west to clear the curtain wall between Towers X and XI and to (re)expose Tower XII.[61] Work resumed 1906 to investigate the *agger* south and west of Tower XI, but Sogliano abandoned the excavations and never published the results. The graffito with Sulla's name located on the first window on the right of Tower X mentioned previously was among his most important discoveries.[62] Although the graffito provides no real association with the events of the siege, it prompted Sogliano to date the construction of this tower and the others to before Sulla's conquest of Pompeii.[63]

During his work on the towers, Sogliano also led a campaign to recover the nearby Porta Vesuvio in 1902. Its full extent and plan remained unknown despite its position on a major artery in the city and the excavations conducted here in the early 1800s. The recovery of the water *castellum* next to the Porta Vesuvio generated the most interest, and the study of the gate lost importance.[64] Sogliano published only a brief report on the gate that included the description of a small *sacellum* in the eastern bastion.[65] Work continued intermittently in front of the gate until 1910, when Giuseppe Spano published and described the tombs currently visible in the area but omitted the gate.[66]

Spano would return to the Porta Stabia with a minor intervention in October of 1911. His aim was to restore running water to the ancient fountain next to the gate. The intervention included the removal of a few of the basalt blocks that compose the road pavement through the gate in order to lay the pipe needed to supply the water. Among the artifacts recovered on October 7 were the fragment of an inscription and a sestertius dating to the time of Claudius.[67] At just 27 centimeters long, the broken fragment of the inscription contained the following text:

SVb
TRIBV
GRAD

According to the report, the inscription may have once read *Sub(structiones), tribu(nalia), grad(us)* and possibly referred to the construction of the seating in the nearby theater, but the full context of this find remains ambiguous.

Following these isolated campaigns, Amedeo Maiuri conducted the most systematic excavations to date between May 1926 and October 1927 (see Figure 1.2). He understood the importance of the fortifications and their capacity to shed light on the early history of Pompeii. The extensive publication of 1929 is still a fundamental starting point for any investigation into the walls of Pompeii.[68] The campaign included trenches at the Porta Stabia, Nola, Vesuvio, Ercolano, and the systematic exploration of Tower XI. The campaigns revealed the Orthostat wall and furnished the evidence that allowed him to formulate the construction sequence largely accepted today.[69]

Maiuri subsequently made the recovery of the abandoned and reburied fortifications his personal quest. He extended the excavations of the Palaestra in the 1930s to include the adjacent fortifications between Tower III and the amphitheater. Among the most important finds, although he never elaborated on the discovery, was the identification of the Pappamonte fortification and the recognition that it predated the Orthostat wall.[70] Maiuri next focused on reexcavating the walls between Porta Vesuvio and Ercolano, publishing the results in 1943. The debris from the excavations of the city, along with the collapse of the original trench dug to find the walls in the nineteenth century, had reburied the fortifications in this area.[71] These circumstances were true for much of the circuit. They explain why late-nineteenth-century Pompeianists, such as Johannes Overbeck and August Mau, had to rely on the drawings and reconstructions of the French architect François Mazois, published several decades earlier when writing their descriptions of the defenses.[72] This forced blind reliance led to a series of common misconceptions regarding the layout of towers and their decorations. Maiuri's investigations led to a new more elaborate reconstruction of the towers. They also led to the discovery of two fill layers that had raised the terrain in front of the walls first in the early Imperial period and later after the earthquake.[73]

By the 1950s, Maiuri devised a systematic plan to free the fortifications of Pompeii from the masses of excavation backfill by selling it for the construction of the new Napoli–Salerno highway and the leveling out of nearby agricultural fields. The operation included uncovering most of the southern fortifications spanning the *Insula Occidentalis* and the Porta Sarno. This was no small feat. It resulted in exposing the area near the Porta Marina and the southwestern quarter of the city. It

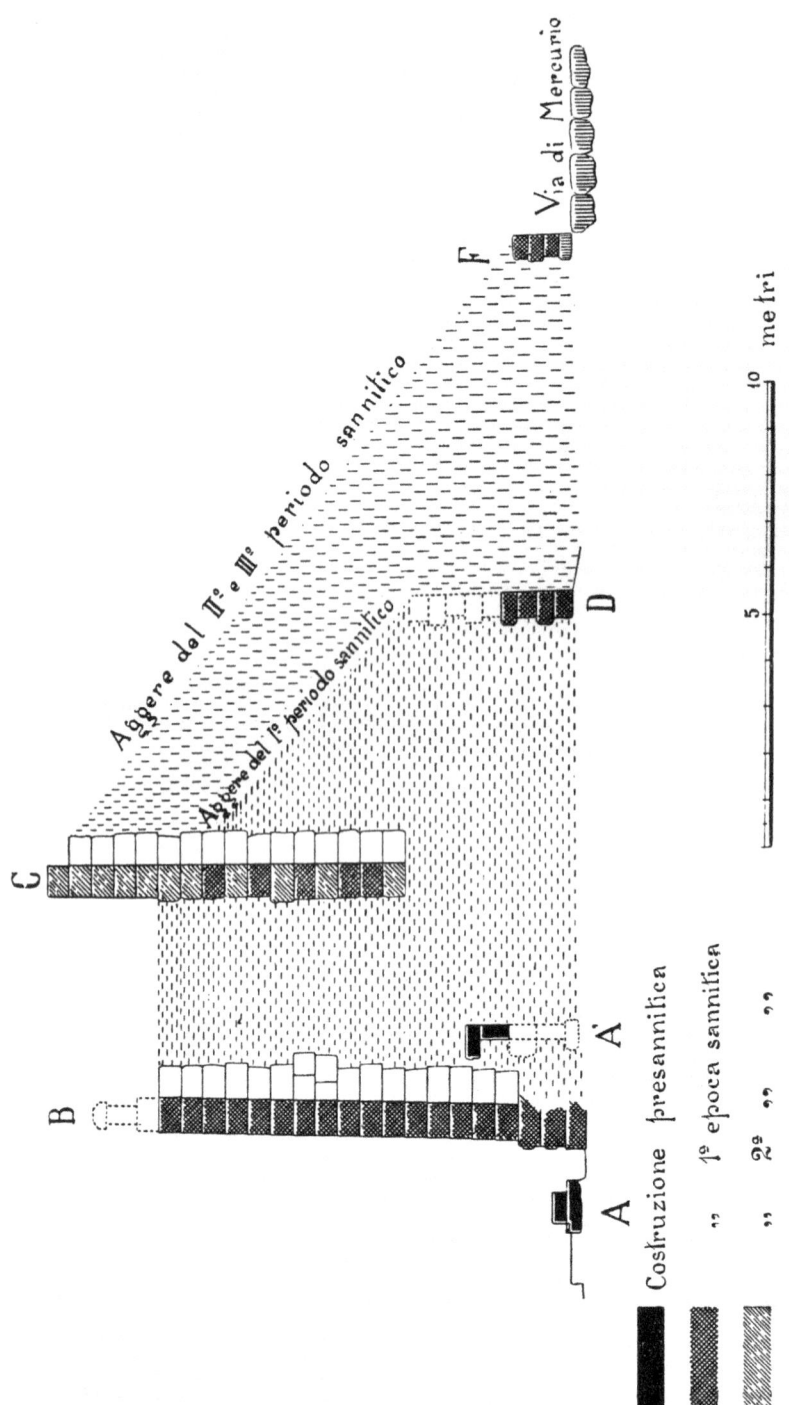

Figure 1.2 Phases of the walls as described by Maiuri. (After Maiuri 1929, fig. 12.)

also proved the existence of the Porta Nocera, which by now many doubted.[74] The scale of this operation hints at just how much the fortifications had disappeared since their nineteenth-century recovery. Most of the circuit exposed today is the result of this effort. Some areas, including the stretch between the Porta Stabia and Tower II, as well as the environs of Tower IX, remain buried under the debris.[75] Unfortunately, the monumental task produced only a small publication, and for the most part the state of the walls as recovered during this operation remains little understood.[76]

In the 1970s and 1980s, excavations targeted separate sections of the walls. Work at the Porta Sarno (re)exposed the gate between 1972 and 1976.[77] Early in the 1980s, the Porta Nocera area received renewed attention under the guidance of Stefano De Caro. The excavations sought to understand Maiuri's discovery of the Pappamonte wall found near Tower III and answer questions concerning the dating and phases of the fortifications. Recovered ceramics dated the Pappamonte wall to the first half of the sixth century BCE, and its destruction to the early decades of the fifth. The results assigned a relative date to the Orthostat wall to a period spanning the construction of the Pappamonte and the Samnite enceintes between the sixth and early fourth centuries BCE. Remains recovered near the Porta Ercolano and a revision of Maiuri's data enabled De Caro to prove that the course of Pappamonte wall foreshadowed that of later enceintes.[78] Until this campaign, scholars had dated the *Neustadt* to the fourth century BCE, when builders also erected a new fortification around the city. The discovery of the Pappamonte wall meant that Pompeii existed within its current boundary at a much earlier date.

Later in the decade, Cristina Chiaramonte Treré would excavate the curtain wall between Porta Nola and Tower VIII. Her results mirrored Maiuri's findings further west, which had identified post–62 CE earthquake debris dumps and evidence of post-Sullan repairs in the circuit. The recovery of Tower VIII was another major contribution that produced rare decorative elements and fragments of an inscription. Materials from the trenches significantly revised the construction of the first Samnite enceinte to the late fourth/early third centuries BCE. These results also allowed her to push the construction of the Orthostat wall to before the fifth century BCE because, she argued, this was the most plausible length of time for its existence between the Pappamonte and the Samnite fortifications.[79] These dates are those currently accepted for the genesis of the Pompeian defenses and, by extension, the main phases of development for the early city.

Between 1993 and 2002, the Japanese Institute of Paleological Studies conducted a study to establish the existence of the so-called Porta Capua on the northeast side of Pompeii.[80] The excavations of the 1800s had detected a structure that was possibly a gate, but its nature and even existence were unclear. The importance of assessing the presence of the gate correlated to the orthogonal layout of the city and its urban development. Giuseppe Fiorelli, a firm believer in the existence of a Porta Capua, defined it as a counterpart to the Porta Nocera and projected a street on the axis between the two gates. Although we now know that this street did not exist, the theory had led Fiorelli to divide the city into the administrative regions, or *regiones*, according to gates and streets still used today.[81] Rather than a gate, the team recovered the remains of Tower IX and traces of a defensive ditch that once extended in front of the building. Deep excavations revealed traces of the Pappamonte and Orthostat walls, whereas the recovered materials confirmed an early third century BCE date for the Samnite fortification.

Further results significantly complicate Maiuri's strict separation of phases according to construction techniques, the notion of a uniform defensive line around the city, as well as the assumption that the walls had remained untouched since the eruption. Surprisingly, the *agger* was absent, making this the only tower with a back door opening directly onto a street. The absence of the *agger* suggests that engineers reused the inner face of the Orthostat wall as a terracing structure rather than replacing it with the elaborate internal wall built as part of the much later tuff reinforcement of the fortifications.[82] The results also demonstrated that the tower and the adjacent curtain represent sections rebuilt in the early colony, thereby disputing the generally accepted notion that all the *opus incertum* segments in the fortifications dated to the precolonial period.[83] The investigations also recovered a medieval occupation layer inside the tower and evidence for a contemporary spoliation of the adjacent masonry for building material. Parts of the tall towers may have stuck out through the volcanic debris, making them targets for occupation well after the eruption. Crowbar marks identified at the Porta Sarno signal that similar processes occurred elsewhere along the circuit, but their extent and date remain unquantifiable.[84] It is a guess whether similar acts of spoliation may have occurred after the nineteenth century.

A series of targeted interventions have explored the fortifications in recent years. The Pompeii Archaeological Research Project: Porta Stabia (PARP:PS), run by the University of Cincinnati, has excavated at the Porta Stabia as part of a wider investigation into the adjacent urban district (see Figure 1.3). Important conclusions include the identification of a road passing through the area of the gate associated with the Pappamonte and Orthostat enceintes. Four further phases are part of the gate's current version that began in the late fourth century BCE, together with the Samnite fortification. The excavations also discovered an altar buried beneath two niches carved in the eastern passageway. The lowest niche and the altar are associated with the first two phases of the gate. In a later phase, around 100 BCE, workers built the *opus incertum* vault in conjunction with the towers. In a final phase, engineers built the causeway, repaved the road, and added a drain in the Augustan period or soon thereafter.[85]

The study campaign revised one important element in the construction sequence of the gate. A comparison between period photographs dating to the turn of the twentieth century against a detailed drawing that Giuseppe Fiorelli published in 1873 attest that a restoration of the gate occurred somewhere between the 1890s and the 1920s.[86] The reconstruction included the addition of at least two courses on both bastions, as well as the placing of the spout carved in the shape of a lion head that workers had found nearby. It had vanished by the time Maiuri began his work in the late 1920s. As a reminder of the reconstruction, the restored courses are composed of a brown tuff stone to serve as a contrast to the lower courses in yellow travertine. This observation seems trivial were it not for the fact that this

Figure 1.3 The rebuilt Porta Stabia with the lion spout on the top right. (After Cotugno et al. 2009, fig. 008)

restoration was never part of the literature on the gate. The modern masonry gives the erroneous impression that workers built the passageway first – a circumstance that misled even Maiuri when he analyzed the gate.[87] Instead, although the extent of the chronological separation remains difficult to establish, the results indicate the reverse where the exterior bastions precede the construction of the passageway through the *agger*.

Another investigation has targeted the Porta Vesuvio on the opposite end of the via Stabiana. The results confirmed Maiuri's hypothesis that the bastions built in travertine predate the gate's passageway walls built in brown tuff. The recovered materials further support Chiaramonte Treré's dating sequence, suggesting that the bastions and the passageway walls respectively date to the early third and mid-second centuries BCE.[88] The investigations also settled a longstanding debate whether an *opus incertum* vault ever existed on the city side of the gate. Its presence at the Porta Vesuvio was assumed based on similar vaults built elsewhere in the city gates, until construction of the aqueduct obliterated it in the Augustan period. Instead, the excavations have shown that the Porta Vesuvio exclusively featured a vault on the exterior field side of the gate. The results also indicate that the remains of the exterior vault date mostly to the late second century BCE rather than to the postearthquake period, as Sogliano and Maiuri had suggested previously.[89] Although some reconstruction did occur at the gate after the earthquake, including the vault, the development of the Porta Vesuvio largely falls in line with the phasing that Maiuri had postulated in the 1920s.

This traditional sequence has received an important challenge with the recent research campaign at the Porta Nocera. The published report does not confront the issue, but the results indicate that the outer fortification wall, built as part of the first Samnite enceinte, is the result of a single construction event. This development opposes Maiuri's substitution hypothesis that envisions the replacement of the upper travertine courses of curtain wall with tuff masonry during the expansion of the *agger* that occurred in the late third/early second centuries BCE. The excavations have uncovered a foundation trench for the interior reinforcement wall built in the second later phase, thereby indicating that the *agger* and therefore its exterior retaining wall must have been in place before its construction.[90] Rather than pointing to a substitution, this evidence indicates that both the travertine and tuff masonry present in the exterior facade of the fortifications wall are the product of a single construction event. Engineers built them into the first Samnite enceinte as part of an ornamental design that has remained largely unnoticed.

The most recent investigations have taken place on the western side of the city between the Temple of Venus and the House of Fabius Rufus (VII.16.22). Work in the precinct of the Temple of Venus has recovered a series of pappamonte blocks that probably doubled as a fortification and a terracing structure for the earliest two enceintes that protected Pompeii. Their presence indicates a possible early version of the temple in this location.[91] Farther north, investigations at the House of Fabius Rufus have confirmed the presence of a postern that gave access to this side of the city. The presence of a gate in this location has been the matter of a long debate because the ancient *decumanus* (the modern via delle Terme) ended here.[92]

This is the only postern known in the circuit. It is modest compared to the city gates, consisting of a narrow vault and a set of stairs. Evidence implies that it was in place since the first construction in the early third century BCE. In the following upgrade of the circuit in the early second century BCE, engineers equipped the postern with sewers to drain rainwater from the *decumanus* above.[93] The discovery completes the orthogonal layout of Pompeii, albeit with a smaller postern rather than a large-scale gate. Its reduced scale is probably the result of its location and function. It only served a small sanctuary and perhaps a suburb on the adjacent coast rather than opening onto a major regional road or the river harbor like the other Pompeian gates.

In addition to excavations, the fortifications have received restorations and minor interventions that are elusive. The example of the Porta Stabia previously highlighted reminds of restoration campaigns carried out throughout the twentieth century that have affected the current remains with the construction of large sections of modern masonry. Unfortunately, the records of these interventions are difficult to find. In some instances, such as the three towers on the north side of the city, a line of red bricks distinguishes between modern and ancient masonry. Presumably, this intervention occurred at some point after the major work that Maiuri conducted in the area. On other occasions, a plaque encased in the masonry highlights the date of a restored section. One example announces a repair to the curtain wall west of Tower X after a bomb demolished it during the Allied air raid of 1943. However, no plaque commemorates the damage on the Porta Sarno or the reconstruction of the outer facade of Porta Marina, obliterated during the same raid. Instead the restoration deliberately attempted to conceal the event, which infamously also destroyed the adjacent *Antiquarium* that functioned as the city's museum.[94] We can only assume that restorers accurately reconstructed the gate's masonry. Without markers, it is generally up to the viewer to distinguish between modern and ancient masonry whether by visual analysis, the image archives, or both.

A few gates have played important roles in the accessibility to the site and modern perceptions of the ruins, remaining on view since their discovery. The Porta Ercolano formed the main access to the ruins for much of the eighteenth and nineteenth centuries, greeting the European elite and artists engaged in the Grand Tour. The importance of the gate in the vision of Pompeii and antiquity is considerable. Francesco Piranesi produced the first engravings of the area, and the printed copies quickly spread throughout Europe, appearing with associated descriptions.[95] Early guidebooks and accounts invariably contain descriptions and illustrations of the fortifications. They provide precious insight into any embellishments that have disappeared and the slow deterioration of the fortifications.[96] The majority of these descriptions focus on the Porta Ercolano and the northern tract of the fortifications, whereas the Porta Nola, located farther off the beaten track, received somewhat less attention. As the excavations progressed, the fortifications slowly became psychological and physical barriers defining the city limit, just as Caroline Bonaparte had intended.

Once re-excavated, other gates functioned in similar roles. With the extension of the Napoli–Portici railroad in 1844, a new access to the ruins opened on the modern

Piazza Esedra, leading to the area of the Porta Marina. In the early 1900s, a new branch and station of the Circumvesuviana light rail built north of Pompeii led to the construction of the now abandoned entrance facility at the Porta Nola. It was the main access into the ruins until a new station, the current Pompeii/Villa dei Misteri stop on the Circumvesuviana line to Sorrento, opened in the 1930s. Only in 1948 did the new entrance near the amphitheater artificially bridge the city walls into the city.[97] Since then, the modern urban sprawl of the Bay of Naples has engulfed much of the surrounding territory. Throughout this process, the fortifications continued to act as the conceptual marker of the city. Unfortunately, this circumstance licensed the development of adjacent terrains with little supervision. Only recently has some research included the ancient territory of Pompeii. Today, the walls continue to form a backdrop for visitors, and, perhaps most importantly, they still help shape the concept of the limit of Pompeii as defined by the line of its defenses.

Notes

1 Pesando 2010a, 223; Coarelli and Pesando 2011, 37.
2 Mau 1902, 8; Fiorelli 1873, vii–xi, 78–86; Nissen 1877, 669.
3 Fiorelli 1873, viii; Sogliano 1918, 156.
4 Maiuri's excavations proved the early existence of the Porta Ercolano; Maiuri 1929, 120–139. Hans Eschebach revived Sogliano's idea in his later analysis of the Pompeiian urban environment; see Eschebach and Eschebach 1995, 62; Grimaldi 2014, 36–42.
5 Haverfield 1913, 63–66.
6 Von Gerkan 1940.
7 Eschebach 1970, 38.
8 Ward-Perkins and Claridge 1978, 37. Further criticism on Eschebach also in De Caro 1992b, 70.
9 De Caro 1985, 75–114.
10 De Caro 1992b, 70.
11 Pesando 2010a, 231–236.
12 Fiorelli 1873, 78–86; Mau 1902, 37–44.
13 Adam 2007, 99. Richardson 1988, 370 dismisses a Limestone period as fictitious. A debate also exists in print between Pesando and Amoroso. Pesando 2008a and b; Amoroso 2007.
14 For a recent discussion, see Brasse and Müth 2016, 75–100.
15 Romanelli 1817, 273. Mazois 1824a, 35; Fiorelli 1873, 89; Nissen 1877, 511; Dyer 1867, 58. Also Adams 1873, 49; Monnier 1865, 97.
16 Niccolini and Niccolini, Vol. 2 1862, 8; Overbeck and Mau 1884, 43.
17 CIL IV 5385; Sogliano 1898a, 68.
18 Mau 1902, 237–238.
19 On the development of the houses, see Noack and Lehmann-Hartleben 1936, 15.
20 Maiuri 1943, 242.
21 Fiorelli 1860–1864. The organization of the *Pompeianarum Antiquitatum Historia* is somewhat cumbersome. Volumes 1 and 2 each divide into three parts (1–3 and 4–6) with separate page numbers. Volume 3 is a single volume. Henceforth, the notes will contain the volume, part, page, and date. See also Laidlaw 2007, 620–621.
22 PAH 1 Pars Prima, 153.
23 Fiorelli 1875, 75.
24 For example, the entry for the start of the final excavation of the gate is actually July 22, 1769; PAH 1 Pars Prima, 234.
25 PAH 1 Addenda B, 114, #15, September 14, 1764.

26 PAH 1 Pars Prima, 233–236, July 29–November 11, 1769; Adams 1873, 50; Clark 1831, 73.
27 PAH 1 Pars Prima, 205, March 14, 1766, 211, October 10, 1766, 218, May 21, 1768, 225, November 19, 1768.
28 PAH 1 Pars Secunda, 14, January 2–January 9, 1783, describes work continuing on the city walls. PAH 1 Pars Secunda, 34 November 16–30, 1786, states work on the House of the Vestals (VI.1.6–8) slows down to uncover the city walls.
29 PAH 1 Pars Secunda, 13–14, September 12–December 26, briefly mentions the walls south of the portico associated with the Large Theater as excavated in 1782.
30 De Jorio 1828, 155.
31 PAH 1 Pars Secunda, 34, May 3, 1787.
32 For the Porta Nola, see García y García Rome 2006, 166. For the Porta Sarno, see PAH 1 Pars Tertia, 147, March 31, 1814. For the Porta Nocera, see De Jorio 1828, 157.
33 PAH 1 Addenda e Schedis Petri La Vega et Michele Arditi, 241. See Sakai 2004, 29.
34 PAH 1 Addenda e Schedis Petri La Vega et Michele Arditi, 275. See also De Clarac and Mori 1813, 2.
35 PAH 1 Pars Tertia, 101, December 26, 1812.
36 PAH 1 Pars Tertia, 98, November 21, 1812, 116, June 24, 1813, 130, September 27, 1813, 150, May 15, 1814.
37 De Jorio 1828, 47–48. See also Romanelli 1817, 34.
38 PAH 1 Pars Tertia, 69, October 26, 1811. Also PAH 1 Pars Tertia, 85, June 13, 1812.
39 PAH 1 Pars Tertia, 96, October 15, 1812.
40 PAH 1 Pars Tertia, 110, May 8, 1813. The full identification is a fact by May 29, 1813. See PAH 1 Addenda e Schedis Petri La Vega et Michele Arditi, 269. De Jorio 1828, 155 mentions May 1812.
41 PAH 1 Pars Tertia, 113, June 5, 1813.
42 PAH 1 Pars Tertia, 127, September 9, 1813.
43 De Jorio 1828, 155.
44 PAH 1 Pars Tertia, 149, May 5, 1814, 149. De Jorio 1828, 155 diverges a little, reporting the time span between May 1812 and June 1813.
45 De Jorio 1828, 155.
46 De Jorio 1828, 157.
47 PAH 1 Pars Tertia, 169, February 2–9, 1815.
48 PAH 1 Pars Tertia, 170–172, February 12, 16, March 8, 1815.
49 PAH 2, Pars Quarta, 1, January 16–23 1819.
50 PAH 1 Addenda e Schedis Petri La Vega et Michele Arditi, 275.
51 PAH 1 Pars Tertia, 148, April 17, 1814.
52 Maiuri 1943, 275.
53 Rainwater and ice caused large sections of fresco and plaster to collapse. See Mazois 1824a, 54.
54 PAH 2, Pars Sexta, 520. The PAH stops mentioning the excavations of the gate on March 15, 1852. Fiorelli states May 20, 1851–March 13, 1852, as the dates for the full excavation of the gate. See Fiorelli 1875, 27.
55 See Garrucci 1851, 21–38. Also Minervini 1851, 1–19; Quaranta 1851; Garrucci 1852, 81–84; Henzen 1852, 87–91; Bechi 1852, 21–22. For the full inscription, see Fiorelli 1875, 29.
56 Minervini 1853, 185–187.
57 Anonymous 1899, 406–407; Mau 1890, 283; Sogliano 1904, 301; Krischen 1941, pl. 5.
58 Mazois vol.1 1824a, pl. II and pl. X, fig. II; Mau 1902, 241; and Maiuri 1943, 284.
59 Maiuri 1960, 172, García y García 2006, 167.
60 Fiorelli 1861, 370.
61 Sogliano 1898b, 125; for Tower XII, see Sogliano 1901, 357–361.
62 CIL IV 5385; Sogliano 1898a, 68.
63 Sogliano 1918, 155 and 173; Maiuri 1929, 117.

64 Paribeni 1902a, 213; Paribeni 1902b, 564; best described in Paribeni 1903, 25–33.
65 Sogliano 1906, 97–107.
66 Spano 1910, 399–418.
67 Spano 1911, 377.
68 Maiuri 1929, 113–290.
69 Maiuri 1929, 218–219.
70 Maiuri 1939, 232–238.
71 A comparison between illustrations in the publications of the early 1800s and photographs published at the turn of the century reveal the extent of the reburial. Gell and Gandy 1833, pl. XVI, XVII; FitzGerald Marriot 1895, 28; Gusman 1900, 35–36.
72 Overbeck and Mau 1884, 52.
73 Maiuri 1943, 275–294.
74 Maiuri 1960, 166–179.
75 Nissen 1877, 457–466 describes the remains between the Porta Ercolano and Porta Stabia.
76 Maiuri 1960, 166–179.
77 Eschebach and Müller-Trollius 1993, 12.
78 De Caro 1985, 75–114.
79 Chiaramonte Treré 1986, 48.
80 The team has published a yearly excavation report, which the authors have recently brought together in a volume: Etani 2010.
81 García y García 1993, 55–70.
82 Sakai 2000/1, 94–96.
83 Etani 2010, 208.
84 Brands 1988, 191.
85 Devore and Ellis 2008, 13–15.
86 Anonymous 1899, 406–407; Mau 1890, 283; Sogliano 1904, 301; Krischen 1941, pl. 5.
87 Maiuri 1929, 205, 227. See Van der Graaff forthcoming.
88 Seiler et al. 2005, 224.
89 Maiuri 1929, 170 elaborates on Sogliano; Van Buren 1932, 37 suggests that the construction technique should make it contemporary to the other gate vaults in Pompeii.
90 See Gasparini and Uroz Sàez 2012, 9–67.
91 Curti 2009, 501–502.
92 For a discussion, see Eschebach and Eschebach 1995, 61–62 and note 212.
93 Grimaldi 2014, 36–42.
94 García y García 2006, 167–171.
95 Laidlaw 2007, 623; Goalen 1995, 181–202.
96 The authors include Romanelli, De Jorio, Mazois, Gell, Niccolini, Nissen, Overbeck, and Mau. Their debates on character of the gates provide details on now lost embellishments.
97 Longobardi 2002, 39–65.

2 Defense and the genesis of the community

As a testament to its viability as a settlement, the first occupation of the Pompeian plateau began in the Neolithic period and stretched into the late Bronze Age, when a community associated with the Palma Campania culture formed near Regio V and the Porta Nocera.[1] After an eruption of Vesuvius and a following hiatus in occupation, a new nucleus started to develop in the course of the seventh century BCE.[2] From this nucleus, Pompeii would begin to develop into a settlement through the sixth and into the fifth centuries BCE. During this relatively short period, the community built two main defensive circuits: the Pappamonte and the Orthostat walls. These events follow a broader pattern of synoecism in the Sarno valley, where a number of smaller hamlets united into larger settlements.[3] The shift in settlement pattern coincides with expanding trade routes to the eastern Mediterranean and the influence of the Etruscans spreading into Campania from the north. The fortifications and their construction would stimulate Pompeii's urban genesis and foster an early communal identity. Although much of the city's architecture of this early phase remains unknown, the materials used for construction, as well as Pompeii's placement in the landscape and the route taken by the early fortifications, would be critical factors shaping the early community and future city.

Pompeii lies on a strategic elevated plateau formed by the remains of a prehistoric volcanic edifice, or dome (see plate 3).[4] Steep cliffs naturally fortify the plateau on its southern, eastern, and western sides. Only the northern side, where a gentle slope comes down from Vesuvius, exposed Pompeii to direct attack. This weak spot would consistently receive the strongest fortifications throughout the history of the settlement. From this position, early Pompeii dominated the fertile plain that opens toward the east, as well as the estuary of the Sarno River that lay at the foot of the plateau. Strabo mentions that the river was an important communication route that made Pompeii the port city for the inland settlements of Nocera and Nola.[5] The settlement would also control an important land route mirrored today by the modern highway that passes a few hundred meters from the ruins to link Naples with southern Italy. A long topographic depression through the center of the Pompeian plateau accommodated a section of the main road that skirted Vesuvius to connect the Sarno River plain with Naples. Now known as the via Stabiana, the road would function as the main *cardo* of Pompeii throughout its history.

Early Pompeii also fulfilled the prerequisites of settlement location that Vitruvius and Aristotle prescribed. Its success depended primarily on location and the availability of natural resources.[6] Ideal sites required a healthy environment to sustain the population and ample arable land to retain a degree of self-sufficiency. The site needed a reliable source of water, preferably in the form of natural springs, nearby rivers, or human-made collection cisterns and wells reaching underground aquifers. A site also needed to lend itself to easy defense and provide refuge against natural or human-made threats. Ideally, approaches to the settlement should pass over difficult terrain to create strategic bottlenecks where small defensive units could fend off larger attacking forces. It also needed to be centrally located in relation to its territory as a strategic base to exercise rapid intervention over subjects and intruders and provide a convenient commercial exchange center.[7] Pompeii's powerful position in the landscape offered refuge to the surrounding population, who could seek shelter in the fortified plateau. It was a critical fortified node and regional psychological reference point that carried with it associations of safety and order for the inhabitants.

Given the expanding influence of Etruria upon the Italian peninsula at the time, a possibility exists that a group of Etruscans founded early Pompeii as a sanctuary in the landscape.[8] Similar foundations in Etruria would act as magnets attracting small settlements around them. Some signs of occupation dating to the ninth and eighth centuries come from the areas now occupied by the Doric Temple, the Temple of Venus, and the House of the Etruscan Column (VI.5.17), which was once the site of a sacred grove.[9] The Triangular Forum also hosts tangible signs of huts and wall foundations dating to the seventh century BCE.[10] This area, along with the Temple of Venus, is part of the zone that Von Gerkan recognized as the *Altstadt* – or old city (see plate 2). It coincides roughly with the farthest tip of the ancient lava spur that formed the Pompeian plateau and naturally fortified it with cliffs on three sides. The *Altstadt* still stands out in the urban layout of Pompeii because the irregular layout of its streets contrasts with the regular grid of the later Samnite city known as the *Neustadt* – or new city. Although the information is patchy, a possibility exists that the *Altstadt* began as a sacred *emporium* for the inland settlements of the Sarno valley. Similar contemporary examples existed in Etruria where the sanctuaries of Pyrgi and Gravisca functioned as the coastal emporia for the cities of Cerveteri and Tarquinia.[11] Such sanctuaries had strong political, religious, and economic associations with their communities. The southern side of the *Altstadt* would host the principle temples and the later Forum for the remainder of Pompeii's history. Any fortifications associated with this area, would continue to reflect the political as well as the religious and economic foundation of Pompeii.

The organized settlement began with the construction of the first walled circuit in the sixth century BCE. Known as the Pappamonte fortification, it receives its name after the tuff stone used to build it (see Figure 2.1). The circuit largely enclosed the area that would define Pompeii for centuries to come. Its route ran along the natural edge of the ancient flow that connected with Vesuvius on the north side. The visible remains of the Pappamonte fortification are scant: one or two courses of highly weathered blocks, some 40–50 centimeters high and 75–100 centimeters

Figure 2.1 Remains of the Pappamonte wall in the foreground with the Samnite enceinte in the background.

long, survive between the Porta Nocera and Tower III.[12] The natural erosion that occurred in the area must have exposed them to view in antiquity. Sections of the wall have emerged in Insula I.5. near Porta Stabia; buried in the later earthen embankment of the Porta Nocera; in the stretch between the Porta Vesuvio and Porta Ercolano; and most recently in the area of Tower IX.[13] Further remains have emerged in the precinct of the Temple of Venus Pompeiana and near the House of Umbricus Scaurus (VII.16.15), where a large stepped wall also functioned as a terracing structure along the steep ridge.[14] In some areas, the wall is still uncharted. Remains are notably missing near the Large Theater, where builders may have cut away part of the ancient ridge to accommodate the later building.[15] Vestiges are also absent near the Porta Nola, where the circuit may have taken a slightly different route or was simply absent along some parts of the ridge.[16]

The scant remains give rise to theories on the appearance of the enceinte and to debates whether it ever existed abound. A few hypotheses reconstructing the fortification envision a single freestanding wall, a low fortification with two facing facades, a low wall with a mudbrick or wooden superstructure, and a terracing wall bracing an earth embankment reaching some 3 meters high.[17] The problem is that pappamonte is particularly friable and weak, making it ill suited for large structures.[18] The limited flexibility of the stone and the shallow foundations of the remains indicate that the wall must have been small scale. Any pappamonte wall could not have exceeded six courses in height before collapsing under its own weight.[19] As opposed to the foundations of pappamonte walls uncovered elsewhere in the city, where blocks are arranged as headers with their long sides next to one another, those composing the city wall are laid out longitudinally as stretchers in an arrangement that implies a limited load-bearing capability.[20] The most probable reconstruction is a defensive line composed of a low wall supporting a small *agger*, or earth embankment.[21] It may very well be that this low wall also supported a superstructure in mudbrick or a wooden palisade, but any traces have disappeared. A pomerial street ran behind the wall to facilitate communications along the perimeter of the city.[22] This arrangement foreshadows the layout of the future fortifications, which would include a similar road until the late Republican period some 400 years later.

The proposed reconstructions almost invariably recognize the limited defensive capability of the wall, reinforcing the notion that it carried primarily symbolic rather than military associations. However, this assessment is rather narrow; the defenses included both symbolic and military components. In Italy, warfare at the time was predominantly a seasonal affair that consisted primarily of parties raiding enemy territory in the summer months.[23] Such warfare tends to exclude long sieges, which require standing armies able to conduct long-term operations and even occupy a site. Early siege warfare employed unsophisticated tactics. The strength of strongholds usually led armies to avoid direct attacks on cities. When they did occur, sieges featured the widespread use of manual weapons with a basic armory of spears, slingshots, and bows in both offensive and defensive operations.[24] If anything, offensive scenarios against cities or strongholds would be limited to small bands of soldiers attacking a sector of the fortifications. Any of the structures envisioned for the Pappamonte wall could respond to this pattern of warfare, which precluded the need for elaborate defenses. The pomerial street provided effective communications, and Pompeii already had strong natural defenses along its plateau.[25] Settlements often depended on their *nativa praesidia*, or natural topographical contours, for defense, which localized relatively simple fortifications, such as palisades or walls along vulnerable approach routes, could easily reinforce. Fortifications could cover a large expanse of land such as the Pompeian plateau to use the natural contours more effectively.[26] These open spaces offered further strategic advantage. They could accommodate the population and livestock of the surrounding countryside and host a measure of farming. In this framework, although unimpressive in height, Pompeii's first enceinte probably created a proper defensive line in tune with the military threats of the period.

The Pappamonte and Orthostat fortifications and the city

Fortifications carried a symbolic component that connected with the identity of a community and its image. The extensive use of construction materials implicitly tied a community with their source, which usually was located somewhere in a settlement's territory. The availability of local materials and their exploitation through technological innovation and communal organization were influential factors affecting the manner of defense and construction techniques. Vitruvius recommended the use of local materials for their construction and that engineers employ them in such a fashion to last as long as possible.[27] The presence of large forests, for example, encouraged the extensive use of wood, whereas an abundance of sedimentary soils encouraged the use of mudbrick. Importing construction materials was too expensive and logistically challenging. When employing stone, construction crews would actively seek out and use sources closest to the city walls as a measure of expediency. Quarries were often located just a few meters uphill from the actual course of the enceinte.[28] The Pappamonte enceinte is a typical example of this circumstance. Large outcrops of the material, which is present almost exclusively in the lower Sarno River valley, still exist on the southwestern portion of the Pompeian plateau.[29] Workers quarried it on site as a matter of expediency and economy.[30] Nevertheless, extracting and transporting stone for use in any given building remained a laborious task. The extensive use of pappamonte in any building must have carried an element of prestige for the patron(s) or community that financed it.

Geological factors and construction techniques could unite the settlements of entire regions into a distinct aesthetic. Ancient Etruria supplies an eloquent example. The relatively soft volcanic tuff abundantly present in southern Etruria could be cut into shape for easy quarrying, resulting in the use of regular isodomic ashlar masonry. In the northern region, the harder lime and sandstones would break along irregular natural fissures in the rock resulting in the use of polygonal techniques at sites such as Roselle, Cosa, and Populonia.[31] The development of better tools and building techniques had a significant effect on the construction of stone enceintes as masons progressively acquired skills that facilitated quarrying and stone dressing.[32] Well built fortifications, the techniques employed, and the materials used mirrored the skills and technological achievement of local craftsmen and communities. The permanence of stone as a material added an element of long-term memory to their accomplishment. These factors made the masonry in large-scale structures a reflection of the achievements and social cohesion of the community that built them. They were most prominent in fortifications because of their scale and their role as the first structure that viewers encountered when approaching a city.

The widespread use of pappamonte throughout early Pompeii must have created a unified image because building materials and techniques are the common denominators that unite the appearance of a settlement. At Pompeii, the Temples of Apollo and the Doric Temple of Athena would be among the earliest monumental sacred structures. A third cult dedicated to Mefitis Fysica probably existed in the precinct of the future Temples of Venus.[33] They exhibit a similar continuity of significance and place as the city walls. The function of these spaces and buildings as landmarks would shape the image and therefore the identity of the settlement right up to the eruption. The Temple of Apollo and the sanctuary to Mefitis both employed pappamonte as the foundation to support heavy superstructures, built in all probability using perishable materials. The character of the first Temple of Athena/Doric Temple is less clear, but it must have featured some sort of stone foundation as well. The use of pappamonte as a permanent construction material was a marker of prestige for these structures that were at the center of Pompeian religious and social identity.

Inevitably, the use of the same material in the fortifications and sanctuaries created a direct visual and symbolic relationship between the structures. This factor is evident in their spatial relationship on the southwestern side of the city where the Pappamonte wall acted as their terracing structure. The location of the Temples of Athena and Mefitis on the southern tip of the lava ridge acted as a catalyst to the creation of a community and the legitimization of power over the landscape. They served as beacons for those approaching the city from land and sea and acted as physical and religious reference points for the regional community. The fortifications and the sanctuaries complemented each other, with the defenses protecting the gods and the settlement, thereby stressing the strength of the city and its relationship with the territory. These aspects are more compelling considering the possible foundation of Pompeii as a sanctuary. Those financing the construction of both the sanctuaries and the fortifications projected a clear message legitimizing their power as an extension of their own benevolence to the community and their relationship to the gods. The significance of this relationship will remain a constant factor in the history of the city as the fortifications and the sanctuaries would receive almost concurrent upgrades.

The manner in which the course of the defenses anchored the road network at the gates united the city with its fortifications. When it comes to the major arteries, this layout is the result of defensive considerations associated with the placement of the main gates, which in turn exploited the natural defenses and topography. Early Pompeii had some sort of orthogonal plan that foreshadows at least in part the layout of the later Hellenistic city. The remains of two early gates have emerged: one sealed beneath Tower XI and the other at the current Porta Vesuvio. The via di Mercurio, which has a dead end on the tower, may be an urban fossil of what was once a main artery to the Forum.[34] At the Porta Vesuvio, substantial remains include a possible small forward bastion that served as a predecessor to the current plan of the gate.[35] The remains indicate that the via Stabiana behind the gate – and therefore the Porta Stabia in some form on its other end – has been part of Pompeii's urban layout throughout its history. The pappamonte foundations of contemporary houses have emerged from beneath the House of Marcus Lucretius (IX.3.5), the House of the Gladiators (V.5.3), and in Regio VI. Their orientation roughly follows the current street plan, suggesting that it shadows the layout of the early city.[36]

The Pappamonte enceinte and its relatively consistent course supply an insight into the character of the early city. Only wealthy communities could afford stone enceintes encompassing the entire settlement. Others relied on earthen ramparts and reinforced palisades to strengthen local weak spots in the natural defenses.[37] Stone fortifications or boundaries such as the Pappamonte circuit projected associations of power and wealth. This factor alone points to Pompeii as a relatively powerful settlement capable of mustering the resources to build an extensive boundary. It is hard to assess the image and urbanization of Pompeii any further. The known remains of the Archaic city are too fragmentary, whereas information on land use and the countryside remains buried and out of reach. Until the excavation in Regio VI, the ruling theory proposed that the plateau hosted little or no housing, painting a picture more or less consistent with contemporary Etruscan settlements of Veii and Tarquinia, where long enceintes covered dispersed settlements within in a broad plateau.[38] Such a layout could accommodate the population of the countryside in times of need and host a measure of farming to counter the effects of prolonged sieges. The real picture for Pompeii is probably somewhere in between, where a patchy urbanization aligned on a grid of streets and main arteries accessing the city.[39]

The Orthostat wall

In the fifth century BCE, Pompeii built a new circuit to replace the old Pappamonte fortification. In a dramatic shift from the low earthen *agger*-like construction of its predecessor, the new wall was freestanding. Engineers built it according to the typically Greek military construction technique known as *emplecton*, where they erected two facing retaining walls and filled the intervening space with earth.[40] At Pompeii, the walls are composed of a local yellow travertine dressed into vertical tall thin slabs (maximum 1.54 meters × 0.76 meters and between 0.26–0.47 meters thick), known as orthostats, which give the circuit its name. Substantial remains of the wall have emerged near the Porta Vesuvio (see Figure 2.2 and Plate 2), Porta

Figure 2.2 Remains of the Orthostat wall near the Porta Vesuvio. The inset shows its reconstruction as proposed by Krischen (1941, pl. 1).

Ercolano, Porta Stabia, and a long section between Towers III and IV.[41] A tract recovered near Tower IX is unique because it later changed function to a retaining wall for the subsequent Samnite fortifications.[42] Like its predecessor, the Orthostat wall is absent from the area of the Porta di Nola. It is also missing in the stretch between the Porta Ercolano and the Triangular Forum, where the pappamonte terracing may have continued to function for a while. The new enceinte largely shadowed the course of its predecessor and included some of its gates.

Reconstructing the appearance of the fortifications is not an easy task. Workers largely demolished the wall to make way for the later Samnite fortifications. The surviving remains indicate that the two retaining walls stood on average some 4.30 meters apart. At approximately every third slab, builders inserted a header into the earth fill to stabilize the wall face. Horizontal courses laid flat into the fill had a similar reinforcing role. The wall was pseudo-isodomic in an elegant yet somewhat irregular construction pattern of blocks quarried in slightly varying sizes. Each wall facade stood on an almost negligible foundation composed of a single course of horizontal slabs supporting a squat wall reaching 6–7 meters high, including the parapet.[43] So far, only Fritz Krischen, writing in 1941, has attempted to reconstruct the circuit, basing his drawings largely on Maiuri's data. Krischen certainly took a few liberties. He shows the wall with squat long merlons, although excavations have never recovered them. Other assumptions, such as the overall height of the fortifications, are probably correct. Access stairs recovered at the Porta Vesuvio were mostly narrow and steep, ending unusually close to or against the parapet. Due to the width of the wall, there must have been no more than ten steps.[44] On the exterior, a ditch, known as a *fossa*, ran in front of the wall to increase its relative height and defensive capabilities. On the city side, a pomerial street closely shadowed the fortifications, allowing for an easy communication route around the city.[45]

The shift in construction technique and materials from pappamonte to travertine is remarkable. A multitude of factors – each equally valid – can justify the switch: exhaustion of the pappamonte quarries, improved tools to extract the rock, or a simple recognition that travertine was a stronger and more durable building material. Other factors are an expanded territorial influence of Pompeii or a more developed social organization. As opposed to the pappamonte material retrieved near the city, the travertine quarries were located relatively far away near the modern town of Scafati and the higher reaches of the Sarno River valley. From there, the material arrived at Pompeii by means of river barges transporting it downstream.[46] Although the specific economic aspect is hard to gauge, the use of travertine in the fortifications required an increased socioeconomic organization to gather and transport the material. Once the structures were built, the material would have acquired an element of display, where the travertine implied a measure of Pompeii's territorial control and social organization.

Although the commissioner and builders of the wall remain unknown, the technical expertise required to design it hints at a Hellenic influence. Parallels for contemporary fortifications of this type exist in the Greek cities of Selinunte and Megara Hyblaea in Sicily and in Cuma and Neapolis across the Bay of Naples, suggesting close cultural ties.[47] At the time, Cuma was one of the most influential cities on the bay with its colony Neapolis as a subsidiary settlement. In the sixth century BCE, Cuma's wealth allowed it to build a series of enceintes along its perimeter. Right around 530 BCE, the tyrant Aristodemus commissioned a fortification of the *emplecton* type with orthostat retaining walls that is closest in design and layout to the Pompeian type.[48] As a testament to its defensive capabilities, this wall repelled a coalition composed of Dauni, Aurunci, and Etruscan troops who attacked the city in 524 BCE.[49] The battle marks a turning point against Etruscan influence in southern Italy. Fifty years later, a combined fleet of Syracusan and Cumean ships defeated an Etruscan fleet in the waters off Cuma in 474 BCE, thereby ending Etruscan ambitions in the area.[50]

The use of the *emplecton* technique, as opposed to the *agger*, may reflect a shift in the sociopolitical makeup or conquest of the city by an outside dominant group, Hellenic or even Samnite, intent on making a mark. However, associating the adoption of particular construction techniques as a measure of outside influence upon a community is problematic because the mechanisms of transmission related to military architecture are understudied and difficult to pinpoint. Given the strong Etruscan influence on early Pompeii, it would be easy to connect the Cumean military success with the construction of the new Orthostat wall. This situation induced Maiuri to identify its construction as the start of a Greek period for Pompeii.[51] Admittedly, the *agger* type of fortifications prevalent on the peninsula has an Italic origin. Yet such an approach is rather simplistic. Evidence for a direct military conquest of Pompeii is lacking, and the developments at Cuma are somewhat early for the Pompeian fortification. Another possibility is that the new enceinte is a Pompeian response to a threat arising from the hinterland. Samnite tribes descended from the Apennine Mountains during the course of the fifth century BCE, capturing Capua in 423 and Cuma in 421 BCE, according to Livy.[52] Only Neapolis would remain as a somewhat independent Greek settlement. Nevertheless, associating the construction of a new enceinte as a counter to a materializing threat requires a degree of foresight and planning, which remains difficult to ascertain and even justify. Assigning the construction of the Orthostat fortification to the subsequent Samnite occupation is equally difficult. The archaeological evidence does not indicate analogous architectural transformations elsewhere in the city.

Instead of assigning distinct Etruscan or Hellenic identities to Pompeii based on design, the shift between the Pappamonte and Orthostat enceintes must reflect the political and strategic considerations of an established polity. The impetus to build new defenses is the result of a combination of elements including political influence, military developments, technical expertise, and financial factors. An equally plausible scenario to those sketched above is the simple recognition of the Pappamonte wall as obsolete/inadequate to the social reality of the growing settlement. This narrative allows for the local elite to act as the commissioners of the fortifications as the result of both the development of the city and the events occurring in the region. Although, Strabo identifies the peoples who successively occupied Pompeii as the Oscans, Thyrrenians (Etruscans), Pelasgians, Samnites, and finally the Romans,[53] the archeological evidence offers little clarity on the dominance of a particular group.[54] Instead Pompeii was an independent community with a unique identity forged in ancient Campania – a crossroads of Etruscan, Greek, Punic, and Oscan influences exposed to diverse expertise and technical know-how.[55] The Orthostat and Pappamonte enceintes, irrespective of the population's original makeup, were defining architectural elements. The new fortifications, at twice the height and built in bright yellow travertine, were a dramatic contrast against the dark gray pappamonte used previously throughout the city. Such a contrast between enceintes and their respective patrons seems intentional. It conveyed a message of security to Pompeii's inhabitants and satellite communities that enhanced the standing of the commissioners. Whether it represents a generational divide, a cultural shift, a socioeconomic development, or a combination of the three is an open problem, which must resolve as a local solution.

Despite the considerable investment required in their construction, the new defenses herald a period of occupational hiatus and possibly implosion for Pompeii that began roughly in the aftermath of the Second Battle of Cuma in 474 BCE.[56] At Pompeii, these events led to large areas of the city languishing in abandonment and its gradual retreat into the area of the *Altstadt*.[57] The Temple of Apollo sees an abrupt halt in votive materials between the second half of the fifth and the first half of the fourth centuries BCE.[58] Throughout the city, a uniform abandonment stratum covers many Archaic buildings. The reasons for its formation are probably a combination of population decrease and natural phenomena, including flooding and mudflows that led to the further desertion of the city.[59] This period of regression inserts Pompeii into the wider trend of urban depopulation occurring in south-central Italy in the fifth century BCE.[60] A new Osco-Campanian society would arise because of the interaction between Samnite and coastal populations. It would not be until the late fourth century BCE that Pompeii began to expand back into the plateau after the construction of entirely new fortifications.

The Altstadt *as nucleus of Pompeii*

It is the *Altstadt*, then – whether as the product of an early nucleus, an urban contraction, or both – that was a central element to the final phase of early Pompeii. A question that has consistently accompanied its origin is whether the *Altstadt* had its own defensive circuit. A gate predating the current Porta Marina opening onto the Forum probably existed on the western edge of the plateau.[61] More remains beneath the House of Mercury (VII.2.35) have provided evidence of a defensive system, including traces of a ditch and wall foundations composed of both pappamonte and travertine blocks dating to the sixth century BCE.[62] Traces of a gate survive beneath the House of the Postumii (VIII.4.4–49) where it opened at an oblique angle along the via dell'Abbondanza, one of the main east–west arteries in the city. The presence of the gate substantiates how the early fortifications dictated the layout of the later city. The oblique angle in its layout suggests that the gate followed a typical *scaean* design. Named after the Homeric gate of Troy, *scaean* gates – rather than a distinct type – featured a characteristic layout. The oblique design gave troops on the walls a strategic advantage because it forced an attacking enemy to expose their right unshielded side to the defenders. Dating back to the Bronze Age with examples in the gates of Mycenae and Tiryns, the design is also common for the Archaic period in the gates of Roselle in Etruria and Selinunte in Sicily. The *scaean* gate and the stone revetment of the *agger* suggest that a fair amount of planning and resources went into the construction of the *Altstadt* perimeter.

A series of possibilities emerge regarding the relationship of the *Altstadt* wall with the Pappamonte and Orthostat enceintes. Its perimeter may have served as an interim between the Pappamonte and Orthostat defensive lines. It is also possible that the *Altstadt* fortifications are entirely separate, built before the Pappamonte wall, or are the result of the later sudden contraction of the city. It is difficult to be more specific because little information exists about contemporary secular buildings inside the *Altstadt*. Excavations under the Eumachia building in the Forum have recovered foundations built using both pappamonte and travertine blocks belonging to a row of shops dating to the middle of the fourth century BCE.[63] The construction technique is similar to the *Altstadt* walls uncovered nearby and seems to be unique to this period. Perhaps this mixed construction technique of pappamonte and travertine is the result of the reuse from other buildings in the abandoned plateau of Pompeii. A more probable scenario is that the *Altstadt* wall represents a secondary, inner, defensive line to either the Pappamonte or Orthostat walls or to both. In this role, the *Altstadt* functioned primarily as an urbanized citadel or sacred space in similar fashion to an Etruscan *arx* or Greek *acropolis*. These scenarios are more consistent with similar large settlements in Etruria and southern Italy.[64] Cities such as Atri, Volterra, and Veii – to name a few – had an outer perimeter and a sacred fortified *arx* that provided a defense in depth. The remaining plateau could have been open land, urbanized, or a combination of both until the construction of the new circuit in the late fourth century BCE.

Walls, gods, politics, and the early community

The association between fortifications and sacred spaces is a consistent element in the formation of communities. The growing importance of centralized sacred sites and the need to defend them contributed directly to the emergence of the city-state and consequently stimulated the cultural identity of a settlement.[65] The religious identity of a community rested on its identification with sacred spaces such as temples, shrines, altars, and localities. Their defense was of paramount importance. The idea of a deity or deities presiding over the well-being of a community, such as Athena was to Athens and Jupiter, Juno, and Minerva were to Rome, was a reciprocal process that required a communal effort to defend the tutelary gods. In Archaic Pompeii, this association first develops in the establishment of the three sanctuaries in the *Altstadt*. It developed further in the visual dialogue established between the temples on the southwestern portion of the city and the fortifications that created a twin military and divine message of protection projected over the landscape and the settlement.

The early Pompeiian sanctuaries display a significant cultural hybridity of Greek, Etruscan, and local elements that must have extended to the adjacent city walls. The sanctuaries and the fortifications would have a remarkable continuity of place, creating urban artifacts that would shape Pompeii for centuries to come. The design of the sanctuaries of Apollo, Athena, and Mefitis Fysica indicates the unique character of Pompeian identity. The first Temple of Apollo was an Italic-style building with a podium built with pappamonte blocks supporting a superstructure composed of perishable materials.[66] The roof carried Campanian *architectural terracottas*, that blended local and Etruscan elements.[67] Further east in the Triangular Forum, the Doric Temple dedicated to Athena/Hercules was Pompeii's most imposing landmark. Although the sanctuary had two previous phases in the sixth century, the best known version dates to the early fifth century BCE. The building had Doric column capitals and a peripteral layout according to Greek custom, yet it had the characteristic frontal orientation of Italic temples.[68] The result was a hybrid building with affinities to Temple B at Pyrgi and the Temple of Apollo Lycaeus at Metapontum.[69] The area of the future Temple of Venus is more enigmatic. A series of early foundation walls and collapsed mudbrick likely belong to an early sanctuary, dedicated to the Samnite goddess Mefitis Fysica.[70]

The connection between religion, politics, sacred spaces, and fortifications evident for Pompeii is common elsewhere in Greece and Etruria, where religious cults were a critical factor in the formation of a communal identity. In Greece, cults often found a place in the center of the city, creating a sense of partnership in the cultic beliefs and ceremonies related to the differing groups that formed the early state. By contrast, cults in Etruria initially found a place in the peripheral *arx*. Etruscan society placed much more emphasis on the celebration of power through tribal rituals and associated gentilicial cults than the Greek world. Ceremonies linked to the celebration of the individual *gens* reflected the feudal nature of Etruscan society. The investiture of power, or *imperium*, relied on the interpretation of divine will and the promotion of genealogies in myths and local legends. Consequently, the ritual of the *augurium*, or bird-watching, and the derived prognostic *auspicia* placed much significance on the hilltops where it occurred.[71] The legitimization of power through such ceremonies means that a direct relationship developed between religion, the elite, and the consequent peripheral location of the *arx*. At Pompeii, the presence of the *Altstadt* implies similar associations, further strengthened by the direct reference between the sanctuaries and the fortifications.

The Etruscan tradition implicitly tied the foundation of cities to their symbolic boundary according to a ritual known as the *sulcus primigenius*. An augur would trace the limit of the future city using a plow pulled by an ox and a bull. The ritual expelled otherwise uncontrollable forces of nature from the area, fixed its limits, and placed it under divine protection. The priest would lift the plow at the designated location of city gates to allow the passage of earthly things through the boundary.[72] The resulting sacred boundary was known as the *pomerium* that formed some sort of blueprint for the course of the urban defenses. The Etruscan ritual would enter Roman tradition where it was adopted for the foundation of colonies. In Rome itself, the *pomerium* marked the line within which no one could enter armed, no one could be buried, and the gathering of assemblies such as the *comitia centuriata* was forbidden. It also was part of the foundation myth of Rome when Romulus traced the boundary of the future city on the Palatine. Remus allegedly crossed the boundary illegally, prompting Romulus to murder his brother because of the sacrilege he committed against the gods.[73]

The notion of a strong Etruscan influence on the early city and the idea of its grand foundation have led to the proposal for a similar ritual and boundary for Pompeii.[74] It is at best unclear whether such a ritual accompanied the foundation of early Pompeii, or if the Pappamonte enceinte actually followed its hypothetical course. Part of the problem is that the *pomerium* is a rather vague notion with ancient and modern scholars alike arguing over its true meaning, extent, and course.[75] Past interpretations have connected the course of fortifications to the boundary defined during the *sulcus primigenius* ritual. However, it seems that city walls probably ran slightly in front of the established *pomerium* boundary, merely protecting rather than occupying its course.[76] At Pompeii, it is entirely possible that the line of the *Altstadt* actually formed the original *pomerium* of the city, considering that the area may have functioned as a separate sacred fortified Etruscan-style *arx*.[77] The presence of a sacred boundary would help explain its preservation in the street pattern long after urban development had surpassed it.

Francesco Vitale has assigned precise astronomical connotations to the spatial arrangement of Pompeii by correlating the course of the city walls to a perfect ellipse. Inside it, temples and gates align along precise linear patterns to create a protective religious umbrella resembling a pentagram over the city. Vitale notes that this layout follows precise astronomical calculations reflected in the street alignment with the solstices that formed a memory capsule of Pompeii's foundation for future generations. The layout is therefore primarily the result of the city's original foundation as a sanctuary evidenced by the plateau's relative isolation from readily available water sources. The sanctuary would have featured burgeoning sacred groves that subsequently grew to such an economic importance that they necessitated the construction of fortifications to keep out thieves and brigands.[78] Given Pompeii's possible foundation as a sanctuary, the approach is interesting. However, the arguments rest on coincidences and connections drawn from lines on a map that one can trace and interpret arbitrarily, creating endless relationships and interpretations.

Irrespective of cosmic alignments and foundation rituals, both the Pappamonte and Orthostat enceintes featured a strong religious association related to the fortified line acting as a liminal boundary between the urban space and the countryside. Such liminal boundaries included religious and cultural notions of self-determination, identity, and political authority important to those who financed their construction. The enclosures also protected sacred sanctuaries, which, in turn, symbolically protected the walls and those inside. The visual marriage between important sanctuaries and the fortifications carried further political and religious significance that legitimized the status of those who financed their construction. In this framework, the construction of fortifications and sanctuaries in their format as beacons dominating the landscape legitimized elite rule and control of the territory.

The process of building and financing those defenses acted as a political signal and a catalyst to the creation of a communal identity. It is unclear whether Pompeii could muster the resources to build the fortification on its own. Perhaps it received assistance from neighboring settlements such as Nocera or even Cuma. The makeup of the city's population in this period remains virtually unknown, but it is probable that the inhabitants of the town and surrounding countryside took part in the construction of the defenses, and, if not in the physical construction, then in the associated logistics of quarrying and transporting the stone. The division of labor on such large projects entailed crews of masons under the supervision a chief mason. Less skilled laborers, such as the farmers and population of the countryside, carried out the work of transportation and quarrying.[79] Such large-scale projects offered a measure of social control through work and distraction.[80] Although who commissioned the two enceintes at Pompeii is unknown, the subsequent construction process and the consistency of the design in both circuits indicate how each was part of single construction event that fostered ideas of community.

The role of the defenses in times of peace and war was equally symbolic. Given the patchy urbanization of the plateau and the length of the perimeter, the early settlement probably did not have enough manpower to defend the entire circuit. In a practice common in antiquity, authorities probably thinly manned the Pompeian defenses, if at all, during peacetime. Aineias the Tactician, writing on how to survive sieges, explains in detail how to organize the local population and man the walls, but only when a direct threat appears. In peacetime, commanders assigned any kind of nightly patrols to the smallest possible number of troops.[81] On a daily basis, then, the walls of Pompeii were mostly a symbolic and vacant boundary. Only the gates would have acted as toll points manned to control traffic in and out of the city. When threats did arise, much of the strength of the fortifications relied on the influx of manpower from the countryside. This combination of factors added to the role of the Pompeian plateau as an important communal psychological reference point offering security to the regional population.

The Pompeian fortifications of the period reflect an oligarchic model of government, one where the civic monuments and sanctuaries – that is, the fortified enceinte and the temples – were a direct manifestation of public display and munificence associated with its social hierarchy. Since political, economic, and social factors influenced settlement organization and defense, the nature of contemporary Campanian society can help explain the character of the two early Pompeian enceintes.[82] By the sixth century BCE, the Sarno valley experienced the culmination of an urbanization process spearheaded by aristocratic families organized in individual gentilicial clans. Southern Etruria forms a point of comparison. Here an oligarchic social model led to large settlements that occupied high expansive plateaus with natural defenses, including high natural cliffs and flanking rivers.[83] In both cases, the system rejected private ostentatious display. Instead, the elite built civic monuments and sanctuaries to express their wealth.[84]

The first two Pompeian enceintes were critical elements in the development of a local identity. Through their construction, manning, and day-to-day use, the fortifications performed as the embodiment of the social hierarchy they protected. The walls allowed a measure of dominance over the population and territory by commanding the land routes through the area. Their position next to the main sanctuaries created a direct association between the religious and secular protection of the city. The use of local construction materials served as a constant visual reminder connecting the settlement and its territory. The Pappamonte wall anchored the main arteries and substantial aspects of Pompeii's urban layout for the remainder of its history, thereby creating a sense of order for those living within the enclosure. They provided an indispensable measure of familiarity and security to the population of the city. The construction of the Orthostat wall in travertine symbolized both a distinct break and a measure of continuity for the settlement. The adoption of new construction techniques and materials along an almost identical course signaled a dichotomy of renewal and preservation for those experiencing the transitional period between the two structures. The adoption of Sarno travertine expressed a measure of territorial reach and social organization. Whether the construction of the wall is the result of a political transformation or a deliberate shift in military tactics or both is unclear, but the implications are the same. No other single act could further affirm the dominance on the immediate territory more than the construction of a new wall.

Notes

1 Nilsson 2008, 81–86; See also Robinson 2008, 125–138; for Neolithic materials, see Varone 2008, 354–359.
2 Thomas et al. 2013, 6; Di Maio 2014.
3 De Caro 1992b, 67; Cerchiai 1995, 100; Pesando and Guidobaldi 2006, 4.
4 Kastenmeier et al. 2010, 52.
5 Strabo *Geo.* V, 4, 8.
6 Vitruvius *De arch.* I. 4.; Aristotle *Pol.* 7.XI. See Pesando 2010a, 226; De Caro 1992b, 69; Cristofani 1988, 88; Winter 1971, 31.
7 Winter 1971, 3–55.
8 See Vitale 2000, 59; Carafa 2007, 65. For the sanctuary thesis, see Arthurs 1986, 38–39; for the earlier village hypothesis, see Wallace-Hadrill 2011, 422.
9 Carafa 1997, 26; Carafa 2007, 63–67; on the House of the Etruscan Column, see Bonghi Jovino 1984, 357–371.
10 Carafa 1997, 25.
11 Guzzo 2007, 41–43.
12 Lorenzoni et al. 2001, 36. De Caro associated the scant remains that Maiuri uncovered at Tower XI and Porta Vesuvio with the Pappamonte wall.
13 See Maiuri 1929, 151–163; Maiuri 1939, 233; De Caro 1985, 88–91, 104–106; Etani and Sakai 1998; Sakai 2000/2001, 92. Sakai and Iorio 2005, 328; Pesando 2010, 227.
14 Pesando 2010a, 227; Curti 2008, 50–52.
15 Guzzo 2007, 106–107.
16 Maiuri 1929, 206–218, Chiaramonte Treré 1986, 13–19. A complicating factor is that dating the pappamonte structures is challenging. Recent discoveries suggest that builders used it into the fourth century BCE; see Ellis et al. 2011, 7.
17 De Caro 1985, 89; Lorenzoni et al. 2001, 45–46; Sakai 2000/2001, 92; Pesando 2010a, 224; Pesando and Guidobaldi 2006, 29; Gasparini and Uroz Sàez 2012, 27.
18 Lorenzoni et al. 2001, 42.
19 Lorenzoni et al. 2001, 36.

20 E.g., the House of the Gladiators (V.5.3) in Esposito 2008, 77; also the Basilica in Maiuri 1973, 209–221.
21 Gasparini and Uroz Sàez 2012, 30–35.
22 De Caro 1992b, 89.
23 Cornell 1995, 122 and 130.
24 Lawrence 1979, 39.
25 De Caro 1985, 91–93.
26 Boëthius 1978, 67.
27 Vit. *De arch.* I.V.8.
28 Lugli 1968, 65–69.
29 Kastenmeier et al. 2010, 50.
30 Lorenzoni et al. 2001, 42.
31 Lugli 1968, 65–69, countered by Miller 1995, 78 and Becker 2007, 99–105.
32 Lugli 1968, 74.
33 De Caro 1986, 21; De Caro 1992a, 70–72; De Waele 1993, 113–114; Pesando 2010a, 226; Carafa 2011, 91–92. On Mefitis, see Coarelli 1998, 185–190; Curti 2008, 50.
34 Maiuri 1929, 168.
35 Seiler et al. 2005, 224.
36 See Wallace-Hadrill 2007, 281; Coarelli 2008, 174; Curti 2008, 50; Coarelli and Pesando 2011, 42–43; Pesando 2008a, 159; Pesando 2010a, 231–236. The occupation levels date to the sixth century BCE; see Castrén et al. 2008, 333; Esposito 2008, 73.
37 Lugli 1968, 65; Cornell 1995, 126–127.
38 Boëthius 1978, 67; De Caro 1992b, 72; Carafa 2007, 65.
39 Pesando 2010a, 234.
40 Maiuri 1929, 130 and 217; Chiaramonte 2007, 141–142.
41 Maiuri 1929, 155–158; Maiuri 1939, 232–233. More recently, De Caro 1985, 86–90; Gasparini and Uroz Sàez 2012, 35–39.
42 Etani 2010, 308.
43 Krischen 1941, 8.
44 Krischen 1941, fig. 3 and pl. 1.
45 The street covered the Pappamonte fortification, De Caro 1985, 89–90.
46 Kastenmeier et al. 2010, 56.
47 For a discussion, see D'Agostino et al. 2005; D'Agostino 2013, 220–224.
48 D'Agostino 2013, 220–224.
49 Dion. Hali. *Rom. Ant.* VII, 2–11.
50 Cristofani 1991, 16.
51 Maiuri 1929, 179–183, 222–224.
52 Livy *Ab ur. con.* IV, 44; De Caro 1991, 23.
53 Strabo *Geo.* V, 4, 8.
54 For a summary concerning the population of early Pompeii, see Lepore 1989, 147–175; De Waele 2001, 127–132.
55 Thiermann 2005, 158; Wallace-Hadrill 2011, 422.
56 Cristofani 1991, 16.
57 Coarelli 2008, 174; Curti 2009, 499; Pesando 2010a, 239; Giglio 2016, 42–44.
58 De Caro 1986, 20.
59 Pesando 2010a, 239.
60 The same trend occurs at Pontecagnano, the Sarno valley, and coastal Campania; see Cerchiai 1995, 178–194.
61 Arthurs 1986, 39.
62 For the House of the Postumii, see Dickmann and Pirson 2002, 298, 300–303; Dickmann and Pirson 2005, 158; Pirson 2005, 134. For the House of Mercury, see Pedroni 2011, 162–163; recently questioned by Giglio 2016, 21–24.
63 Maiuri 1973, 53–63.

64 Coarelli and Pesando 2011, 46.
65 Rowlands 1972, 448; Gat 2002, 131, 138–139; Torelli 2000, 195–199; Hansen 2000b, 168–169.
66 Remains of the wall are buried beneath the House of Tryptolemus (VII.7.5); see De Caro 1986, 21.
67 Thiermann 2005, 158.
68 Carafa 2011, 91–92.
69 De Waele 1993, 113–114; Pesando 2010a, 226.
70 Coarelli 1998, 185–190; Curti 2008, 50.
71 Menichetti 2000b, 588; Torelli 2000, 198; Menichetti 2000a, 206–207; Torelli 2000, 198.
72 Ovid *Fasti* IV, 819; Festus *De verb. sig.* 236.
73 Plutarch *Rom.* 10.1; Rykwert 1988, 29, 135–137.
74 Nissen 1877, 466–478; Della Corte 1913, 261–308.
75 Jacobelli 2001, 41–55; Camporeale 2008, 17–20.
76 Briquel 2008, 121–134.
77 Coarelli and Pesando 2011, 46.
78 Vitale 2000, 59, 89–94.
79 Diod. Sic. *Bib. hist.* 14.18; Lugli 1968, 68; Winter 1971, 61.
80 See Dion. Hal. *Rom. ant.* 4.44; in contemporary Rome, Tarquinius Superbus employed poor citizens on public projects to prevent uprisings.
81 See Aen. Tac. iii. 1 and xxii. 1, and xxii. 19, xxii. 26.
82 Rowlands 1972, 447–461.
83 Cristofani 1988, 88.
84 Cerchiai 1995, 99–100.

3   A new enceinte for a new city

In the course of the fifth century BCE, Samnite tribes descended from the Apennine Mountains to conquer Pompeii. Their arrival heralded the beginning of Samnite Pompeii, a new era of vigorous development that lasted from the fourth century BCE until the establishment of the Roman colony in 80 BCE. Pompeii expanded back out from the nucleus of the *Altstadt* in the southwest corner of the city into the plateau that now hosts the *Neustadt*. By the early second century BCE, Samnite Pompeii, a town still nominally independent from Rome, approached the apex of its development as a city with a sophisticated urban and architectural layout. A new city wall, at this point the third in the history of Pompeii, set the stage for the urban expansion in the late fourth century BCE.[1] Also known as the first Samnite wall, the circuit was a critical element in Pompeii's development that would anchor the urban layout for centuries to come. The new enceinte was the foundation of the future defenses. The final arrangement, with its twelve towers and seven gates, is the result of two subsequent upgrades on the primary circuit, which date to the late third and late second centuries BCE (see plate 1). The initial construction and following upgrades would transform the character and appearance of the defenses and the city.

The fortifications represented the largest public building of Pompeii. Although military precepts and strategic considerations went into the design of the enceinte, they were not the only factors governing the appearance of the walls. The city wall was also subject to aesthetic considerations as a reflection of its commissioners and builders. Given the size and scope of the structure, any ornamental effect worked on a similar scale. Factors including the building materials used, construction techniques employed, and the masonry quality indicate the care that builders took to present the circuit to viewers. These factors remain virtually unaccounted for in modern studies on the design of the Pompeian fortifications. Together with military functionality, these ornamental factors were the expression of a confident community in tune with martial and cultural developments occurring on the Italian peninsula.

Reconstructing the Samnite circuit

At over 3.2 kilometers in length and reaching an estimated 9 meters in height, the original circuit achieved a monumental scale. Engineers opted for a typically Italic design known as an *agger*, where an outer wall leans against a tall earthen embankment (see plate 1).[2] On the city side a small terracing wall contained the earth mound of the *agger*. An internal pomerial street hugged the circuit to facilitate communications for defensive troops as well as civilian traffic. The extent of the defenses on the exterior of the wall remains unexcavated. The most sophisticated contemporary defenses featured outworks on the field side of the walls, often in the form of elaborate systems of ditches and stockades designed to keep siege machines at a distance. The classic *agger* design, in its simplest form, consisted of an embankment composed of earth dug from a ditch in front of it creating an outwork of sorts. Traces of a defensive ditch have emerged near Tower IX, but the character of the outworks at Pompeii remain largely unknown.[3] Presumably, at Pompeii, as elsewhere in Italy, the breadth of such forward defenses was proportional to the natural strength of the topography, which the new Samnite wall followed and exploited. Between the Sarno gate and Tower VIII, it strayed from the path of previous enclosures, taking a wider course to exploit the ridgeline. Where the lava ridge reached its highest on the southwest between the Porta Marina and the Triangular Forum, the fortifications were largely terracing structures without the *agger*. By contrast, areas of low terrain, such as in front of the Porta Stabia or the gentle slope on the north side of the city, probably required an outer defensive ditch and even complex outworks as reinforcement.[4] These basic topographical elements and framework of the defenses would influence the design and appearance of the fortifications in all their successive forms.

Today the ruins of the fortifications present a medley of construction techniques and materials traditionally assigned to the successive building events on the original *agger* (see plate 4). According to this view, each construction event was a response to a military circumstance, such as the Second Punic War, evolving siege techniques, and battle tactics, or a combination thereof. Amedeo Maiuri, one of Pompeii's most influential superintendents, first proposed this chronology of upgrades based on excavations he conducted in the 1920s. He divided the construction sequence according to three distinct building materials. A yellow travertine known as Sarno limestone composed the first main circuit, erected in the late fourth/early third centuries BCE (see Figure 3.1). In a subsequent upgrade, at the time of the Second Punic War, engineers used a brown/gray tuff to erect a secondary internal wall and to raise the outer wall to accommodate a higher *agger*. The final intervention, dating to the late second century BCE, featured the construction of towers and vaults at the gates using *opus incertum*. With a few minor variations in the dates, scholars largely accept this tidy periodization, and recent investigations broadly confirm the construction sequence.[5] It radically altered the previous hypothesis that viewed the circuit largely as the product of a single construction event with the later addition of the towers and minor repairs.[6] However, projecting this succession of construction events on the circuit is a rather monolithic approach because it fails to take into account the localized interventions and repairs that probably occurred throughout the centuries. In addition, a distinct division in the application of construction materials is not clear-cut, nor is it possible to rule out a phased construction process and targeted interventions on individual wall sections or restorations.

Figure 3.1 Travertine wall section near Tower XI.

The most accepted hypothesis regarding the appearance of the first circuit comes from Maiuri, who based his reconstruction on just a few small sections of wall visible at the time – a circumstance that poses considerable interpretative challenges. He envisioned an *agger*-type fortification with neat, regularly stacked travertine ashlars leaning against an earthen embankment. One of these sections, roughly 22 meters long and 7 meters high, survives just west of Tower XI. It is without doubt the most regular travertine section of the circuit, presenting a well finished series of blocks, 40–50 centimeters high and 60–80 centimeters wide, stacked in a header-and-stretcher system. Internal piers periodically reinforce the wall, which leans at a slight angle to hold back the earthen mound behind it.[7] The travertine employed here and elsewhere throughout Pompeii is naturally coarse and therefore difficult to dress. Nevertheless, the blocks are highly finished; they are a testament to the great care that masons put into their work. An explanation for the blocks' presence and polished appearance may lie with their location, a place of high visibility next to an old city gate. The opening corresponded with Tower XI and the via di Mercurio, which was once an important road to the Forum. Engineers later shut the gate for military reasons because the gentle topography made the area especially vulnerable to attack.[8] This section of wall is somewhat of an anomaly in the circuit. Some of the blocks in this section date to the colonial period, and the entire tract may represent a much later repair.[9]

The few other predominantly travertine sections in the circuit display careless construction. They survive because of their function as a terracing structure on the western side of the city. Small tracts are still evident on either side of the Porta Marina and beneath the House of Umbricus Scaurus (VII.16.15–16), which preserves a section up to ten courses high that includes the only surviving original drainage spout. Another section, which now functions as a terrace for the House of M. Fabius Rufus (VII.16.21), is perhaps the least well built.[10] Plate 5 shows red lines indicating three construction seams that may represent section divisions assigned to individual building crews or later repairs. The seams signal a series of misaligned courses – a circumstance that does not imply any particular care for its final appearance. The owner of the later house that incorporated the wall hid the unsightly masonry beneath a plaster coat, remnants of which are still visible today. Such a careless display of disjointed masonry construction must be the result of its isolation out of sight from potential viewers. A nearby postern in the walls adjacent to the house indicates that little and mostly local traffic passed through this area. Similar examples of carelessly built as opposed to well finished tracts of masonry in places of high visibility are not limited to Pompeii, existing on the circuits of Norba and Cori, to name a few.[11] Considering that the circuit is over 3 kilometers long, these few tracts of roughly built masonry are a rarity that cannot justify Maiuri's vision of a purely travertine circuit.

Elsewhere, the fortifications have a radically different setup, displaying an average of only three to five preserved travertine courses, with surmounting tuff masonry. On the south side of the city, the section between Towers II and IV preserves a tall travertine framework reaching six to eight courses high. At first glance, travertine composed the entire facade, but a closer look tells a different story. The lowest four courses stand out because they are completely unworked as opposed to the smooth ashlars composing the facade above (see plate 6). The rough surface of these blocks has led some to identify them as a deliberate rustication, working as an ornamental element to the walls. However, similar unworked blocks, reaching up to six courses deep, functioned as a consistent foundation for the outer wall throughout the circuit.[12] Workers left them unfinished because they buried the blocks out of sight. Their exposure on the south side of the city is the result of soil erosion accelerated first by the cutting back of the cliff to make the wall more inaccessible and by a later lowering of the terrain to ease access through the Porta Nocera in the early colony.[13] Proof comes from the similarly exposed foundations of Tower III, the Porta Nocera, and Tower II, which still displays a high floating postern.[14] Just east of Tower II, a few tuff ashlars survive to form a row of blocks sitting above nine courses of travertine. They testify that the wall reached much higher than it does today. The remaining tuff courses have disappeared, presumably reused as construction material following the earthquakes of the 60s CE.[15] By subtracting the lowest four unworked courses as buried foundations, the number of travertine rows originally visible reduces to five, with surmounting tuff blocks, in accordance with the rest of the enceinte. The resulting layout points to a visual unity throughout the circuit that has remained unnoticed.

The next evident travertine/tuff juxtaposition runs between Tower V and the Porta Sarno, where the number of lower courses varies considerably, ranging between three and six (see plate 7). The presence of modern tool marks and cement holding the blocks together points to large-scale contemporary reconstructions, which may skew the original layout. Nevertheless, considering that the walls sit on a high ridge, minor variations in the number of courses were likely difficult to notice when viewed from below. The most regular travertine/tuff layout of the entire circuit survives between the Porta Sarno and the Porta Nola on the eastern side of the city. The number of upper tuff courses varies according to their survival rate. The lowest travertine blocks form a consistent level platform of three courses. This layout continues between the Porta Nola and Tower VIII. The first 27 meters west of the gate display four lower travertine courses followed by a section, some 23 meters long, that preserves five. The following tract resembles the curtain on the southern side of the city: it preserves three courses of travertine and no tuff because the blocks were spoliated later. The next 30 meters leading up to Tower VIII are more regular: two lower travertine courses support between four and ten blocks of surviving tuff.[16]

On the north side of the city, the section running from the Porta Vesuvio to the midmark between Towers X and XI displays an identical setup, where four lower travertine courses support a tuff framework (see Figure 1.1). Today, the juxtaposition is not immediately apparent because an earth fill dating to the early Imperial age and a subsequent dump of debris resulting from the 62 CE earthquake have buried the travertine.[17] The facade, perhaps the most intact of the circuit, is consistent. It displays gray tuff as the primary construction material, with a few lonely brown tuff blocks, and a rare travertine ashlar west of the tower.[18] The blocks are uniform, about 41–42 centimeters high, with slightly taller stones reaching 67 centimeters marking the parapet.[19] The wall face changes slightly 20 meters west of Tower X where the travertine socle increases to six courses and the parapet displays a few courses of later repairs. The curtain continues westward as a patchwork of tuff and *opus incertum* materials with a consistent travertine socle up to the completely rebuilt tuff wall that is considered an earlier phase of the current Porta Ercolano.

The west side of Pompeii features a prominent differentiation of construction techniques due, in part, to the role of the wall as a terrace functioning to equalize the terrain along the ridge defining the edge of the city. The consequent high stresses on the masonry caused by the weight of the earth fill behind the wall probably led to frequent collapses and substantial repairs.[20] A primarily travertine curtain is still evident on either side of the Porta Marina and the section beneath the House of Fabius Rufus. The neighboring House of Maius Castricius (VII.16.17), however, again displays a well finished tuff framework resting on a few courses of travertine. Beyond the Porta Marina, a single surviving tuff block encased in a later *opus incertum* framework still rests on a travertine socle in the Villa Imperiale.

Turning the corner, the southwest ridge of the city shows a unique set of circumstances. The remains of a tuff terracing/fortification wall, dating to the early the third century BCE, have emerged from beneath the Temple of Venus.[21] After a brief section of *opus incertum* masonry, the fortifications then disappear, embedded into later houses as the wall heads toward the Doric Temple. As is the case with the previous enceintes, the defenses on this side of the city functioned primarily as a terrace and lacked the *agger*. Here the fortifications would dictate the orientation of the houses and preserve their memory in the streets and outskirts of the neighborhood. A study of the area identified nine surviving sections of the exterior walls spaced out between Houses VIII.2.29–36. Terrace 20 in House VIII.2.29, in particular, displays two tuff blocks that still carry mason marks. They flank a lonely travertine block still in situ. Terrace 19 in House VIII.2.30 displays a similar superimposition of construction techniques, with travertine blocks partially covered in later plaster and incorporated into *opus incertum* masonry.[22] Finally, a small tract of wall composed of two tuff courses surmounting a travertine socle survives as a terrace for the Triangular Forum.

Achieving an ornamental effect

The pattern that emerges is that the fortifications display a consistent socle of travertine supporting a framework of tuff above it. Maiuri hypothesized that the tuff masonry represents an upgrade of the fortifications that raised the previous travertine wall from about 9 to 11 meters.[23] However, considering that the standard travertine block is about 40 centimeters high and the average height of the travertine is four to five surviving courses, the first circuit would have barely reached 2 meters in elevation. The low height of the proposed first travertine wall has remained virtually unnoticed and largely ignored. Maiuri explained it away by proposing that engineers substituted large sections of earlier travertine masonry with tuff.[24] However, this hypothesis, which would require the demolition and reconstruction of most of the circuit, does not make economic sense and seems impractical. If anything, the reasons for this layout would have been aesthetic. Tuff was a prestige material surpassed only by the marble copiously introduced in the later colony. Its use on buildings throughout the city was a means to flaunt wealth. There is no reason to believe that tuff, as a construction material, carried any strategic advantage compared with travertine. The presence of tuff on the fortifications was a means for the community to flaunt its own assets to viewers approaching the city.

So what does this observation say about the Samnite first circuit? Perhaps it functioned primarily as a boundary marker rather than as a fortification, or it never reached full completion. Results from the work at Tower IX hint at a short time interval between the travertine and tuff curtains.[25] Another possibility is that the travertine acted as a socle for a superstructure built with perishable materials later substituted with tuff. Stone socles were a common remedy to protect mudbrick walls from rainwater damage.[26] Vitruvius offers a similarly practical solution for the application of a travertine socle, specifically recommending its use as a foundation stone because of its load-bearing capacity.[27] These scenarios can explain the relative regularity of the lower courses. They also help to clarify the presence of large sections of travertine on the western side of the city, where mud-brick would have been entirely inadequate as a terracing structure. Nevertheless, there is no tangible evidence for such a superstructure. Instead, the uniform size in the blocks of both materials on the outer curtain implies that they are part of the same construction event.[28] Moreover, the defenses at the Porta Nocera indicate the presence of an extensive *agger* connected with the first wall akin to the 9-meter-tall embankment projected for the north side of the city.[29] Such a structure does not seem to be the kind that a mudbrick or wooden framework could hold back effectively.

The most probable hypothesis is that builders deliberately applied both travertine and tuff when they first built the exterior wall to achieve a distinct aesthetic effect.[30] The application of these two construction materials, one bright yellow and the other brown-gray, results in a distinct contrasting effect – more evident when rainwater accentuates their colors (see plate 8). Even if the tuff does represent a subsequent construction event, its differentiation against the travertine signals the extent of the intervention and therefore the investment of those financing the refurbishment. This architectural play with materials is a mainstay of the walls. An example comes in the traces of the same effect that survive in the internal supporting wall spanning the via di Mercurio on the city side of the fortifications. Partially demolished and buried in the later expansion of the *agger*, the wall survives in a fragmentary state. Large travertine blocks compose the majority of the terrace. They are identical to those on the outer facade, but a small section built in tuff spans the width of the street, where it marks a repair or closure of a gate soon after the construction of the first Samnite wall.[31] The later dark brown tuff blocks contrast markedly with the yellow travertine. They also preserve a slight rustication that, along with their smaller size, served to emphasize visually the limit of the *agger* at the end of the street. The effect contributed to the overall grandeur of the via di Mercurio, which accommodated a concentration of elite housing.

Gates and towers in the first Samnite circuit

With the construction of the first Samnite wall, the gates and towers received their definitive placement, thereby anchoring the main thoroughfares into and out of the city for centuries to come. Although relatively little remains of the gates in this early stage, their plan must have foreshadowed their current layout due to the *agger* design of the fortifications. They took on the shape of a forecourt gate, a design that included two forward bastions flush with the curtain wall opening onto a gate court through the *agger*. Today, only the bastions are evident, surviving at the Porta Stabia, Nola, Sarno, Vesuvio, and Marina. The remains at the Porta Vesuvio and Stabia have a secure construction date to the late fourth/early third centuries BCE, which makes them contemporaneous with the outer curtain wall.[32] The twin bastions at each opening were powerful bulwarks of solid masonry protecting the outer edge of the gates. Each is composed of travertine ashlars with distinct grooves either facing the inner corners, such as the Porta Vesuvio and Stabia, or looking outward, as at the Porta Nola. They once accommodated the doors of a simple two-leafed wooden gate that closed the opening. A further gate also marked the threshold on the city side of the gate.[33] The overall layout is simple, designed to allow passage through the deep embankment of the *agger*. It also carried a defensive advantage to trap any attackers in the court of the gate, allowing troops to pelt the enemy from higher ground.

As the natural candidates for ornamentation because of their visibility, the gates of Pompeii display aesthetic considerations in their masonry composition that are equally important to their design. The outer bastions dialogued architecturally with the outer curtain to add emphasis to the gates (see plate 9). Masons further highlighted the bright yellow color of the travertine by polishing each block to a smooth surface despite the material's natural coarseness. The result presented viewers with solid towering facades on each side of the passageway. The adjacent curtain, composed of a lower yellow travertine socle and upper brown tuff courses, created a contrast with the bastions, highlighting the gates and the passageway into the city. The resulting effect must have created a look of military impenetrability to the bastions, which were otherwise flush with the wall and did not stand out in any other way. Since the outer curtain wall and the bastions would remain unchanged at most gates, this primary ornamental emphasis would remain a staple for Pompeii.

The Porta Stabia – considered the oldest gate in the city because it is composed of primarily travertine masonry – displays the basic layout despite its later additions (see Figure 3.2). It is a typical forecourt gate, with two forward bastions, a long internal passageway through the *agger*, and an internal threshold into the city. The ruins have generated rather intense discussions concerning its original appearance. Much of the debate centers on whether the exterior bastions supported an arch, whether the steps leading to the parapet featured an intermediate landing, the precise length of the inner passageway, the exact position and number of gates closing the passageway, and whether the *opus incertum* arch replaced an earlier tuff version.[34] All of these discussion points relate to the original design of the gate as a forecourt gate, which came in many variants but usually featured some sort of overhead passage linking the external bastions and internal threshold. The chronology that Maiuri outlined for the gate is somewhat divergent compared to the broader development of the circuit. The thin vertical orthostat blocks that compose the gate court are the would-be relics of the first phase upon which workers later added the two outer bastions. They function as part of the same type of Greek *emplecton* construction technique that engineers had used in the previous Orthostat circuit.[35] However, this proposed sequence is doubtful. Instead, the court walls and the technique are at least contemporaneous with, if not later than, the bastions built in the third century BCE.[36]

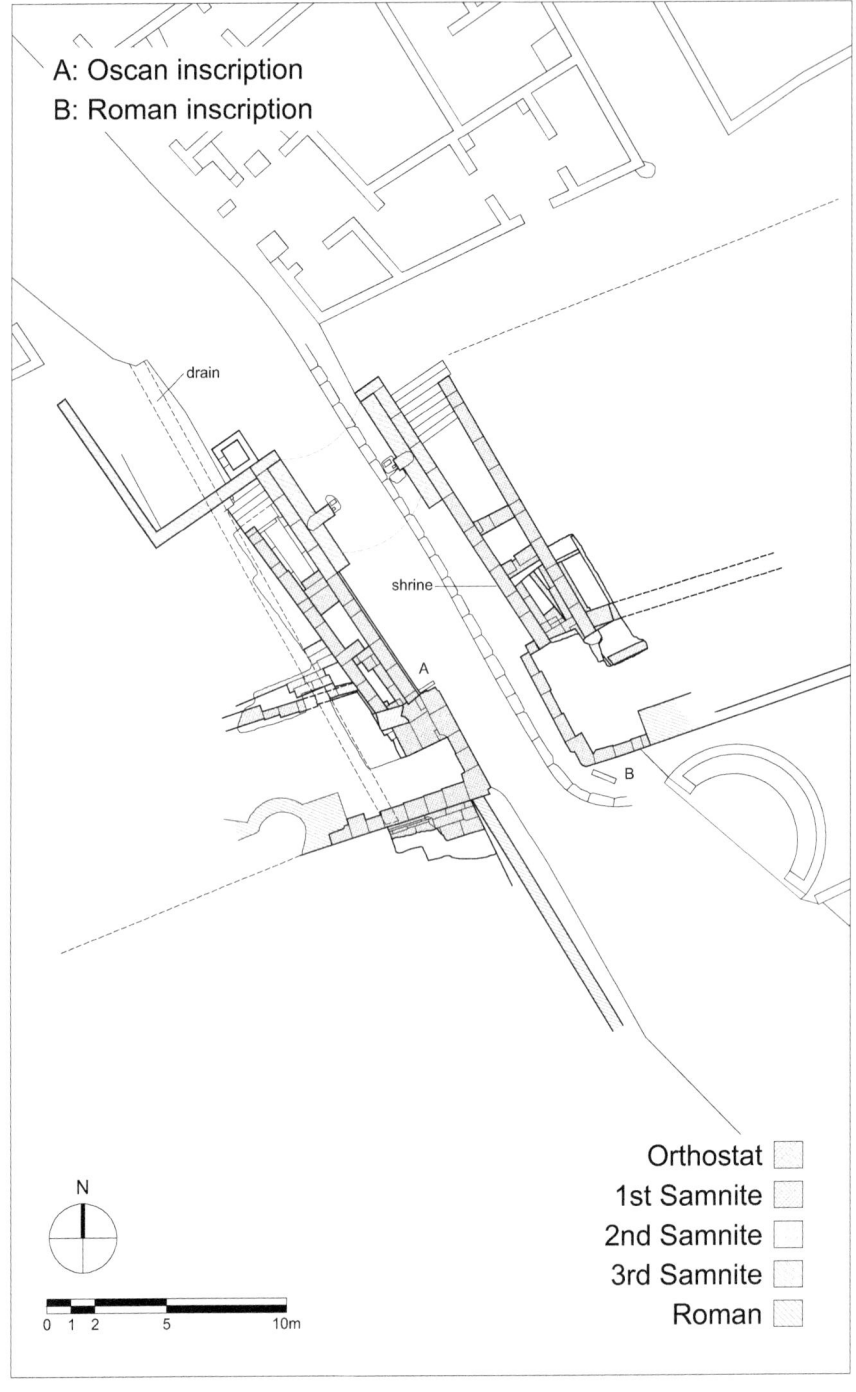

A: Oscan inscription
B: Roman inscription

drain

shrine

A

B

N

0 1 2   5   10m

Orthostat
1st Samnite
2nd Samnite
3rd Samnite
Roman

Porta Stabia   1:300

Figure 3.2 Plan of the Porta Stabia. The key highlights the phases. (Drawing by T. Liddell)

Figure 3.3 View toward the exterior of the city from the court of the Porta Stabia. Note the contrast between the masonry of the court walls and the outer bastions.

This construction sequence indicates how builders and patrons envisioned the ornamentation of the gate where the differing building techniques highlighted a distinct tripartite division of the passageway. The bastions, composed of solid horizontal ashlar masonry, indicated the most powerful exterior line of the fortifications[37] (see Figure 3.3).This must have been a formidable sight. The roadway that first passed through the gate sat a meter lower than today, thereby imparting a tall cavernous appearance to the gate.[38] Two closing mechanisms recovered on either side of the gate probably belonged to double leafed door closing both thresholds.[39] The twin gateways on either end functioned to further mark the entry and exit into the passageway. Once inside the court, the vertical orthostat blocks signaled the transitional space between the exterior and the interior of the city. Admittedly, the use of the *emplecton* technique may simply be the result of a further reinforcement to hold back the *agger*. Yet the choice to employ orthostats rather than more stable ashlars is an arrangement that deliberately highlights a viewer's transition through the gate. Together with the bastions and the threshold, each element created a clear stage marker for any traffic passing through it.

Anthropological studies on rites of passage help explain the presence of this type of ornamentation at the Pompeian gates. Victor Turner has outlined how rites of passage between spaces, in our case between countryside and city, are crucial to notions of inclusion and exclusion from a community.[40] In Turner's view, fortifications expressed a taboo against accessing the other side, whereas gates represented magico-religious liminal areas in between spaces. Arnold Van Gennep identified three stages for such territorial passages: the preliminal, the liminal, and the postliminal. The construction techniques at the Porta Stabia marked the passage through the *agger*. The ashlars of the bastions and the closing gates on each end of the passageway emphasized the pre- and postliminal phases, whereas the orthostats in the gate court marked liminal space.[41]

Pompeians had evident concerns with such liminal spaces and a desire to highlight them at the city gates, where religious attributes still survive. In antiquity, the dangers and insecurity posed by liminal spaces required special safeguard; city gates and fortifications were no exception. They often received religious protection or carried apotropaic devices to protect the community.[42] At Pompeii, the Porta Stabia and Marina still display niches designed to host tutelary divinities; the Porta Vesuvio includes two altars, whereas the keystone of Porta Nola features a bust depicting Minerva. These attributes of the relationship between the community and the city walls are the subject of chapter 8. For now, it suffices to say that both the religious and secular protective roles of the fortifications came through in the careful display of construction materials at the gates and on the walls.

The ornamental emphasis placed on the earliest gates finds a slight variation at the Porta Vesuvio because of its design (see Figure 3.4). Its bastions are set slightly aback from the outer curtain. This layout created a double bastion on either side of the entrance – a circumstance undoubtedly related to its location in a weak area of the defenses. The gate was particularly vulnerable to attack because of its position on a natural downward slope in the terrain. In an effort to strengthen the opening, engineers transformed the northwest tip of the gate into a tower. Today its remains are not immediately evident: The southern and western facades are buried in the later *agger*, whereas the northern and eastern flanks are flush with the curtain wall. Steps recovered on its southern side offered access to the building in its first phase. The remains represent the only trace of a tower in the first Samnite circuit. Workers probably demolished it after the construction of Tower X slightly further west.

The presence of a tower in this early phase attests to the sophisticated design of the Pompeian enceinte, in tune with the latest military developments in the region. Defensive towers are common in contemporary circuits; similar examples existed in nearby Cuma across the bay and Syracuse.[43] Its presence at Pompeii has led to a debate whether further examples existed in the early circuit. One suggestion sees the eleven towers that are currently visible as replacements for earlier versions, but any archaeological evidence of this setup remains elusive.[44] Considering the terrain and the vulnerable position of the gate, the reason for the presence of a tower at the Porta Vesuvio is immediately evident. The tower essentially transformed the gate into a *scaean* type, forcing any attacker to expose their right, unshielded side to the defenders as they approached the city. What is striking for Pompeii is that – with the exception of the Porta Vesuvio – none of the gates actually consider this principle. Instead, as exemplified by the Porta Ercolano and Nola, the natural terrain of sharp cliffs tends to force the opposite, where routes approach the gates from the left. The Porta Vesuvio, then, is an exception that addresses the weakness of its position.

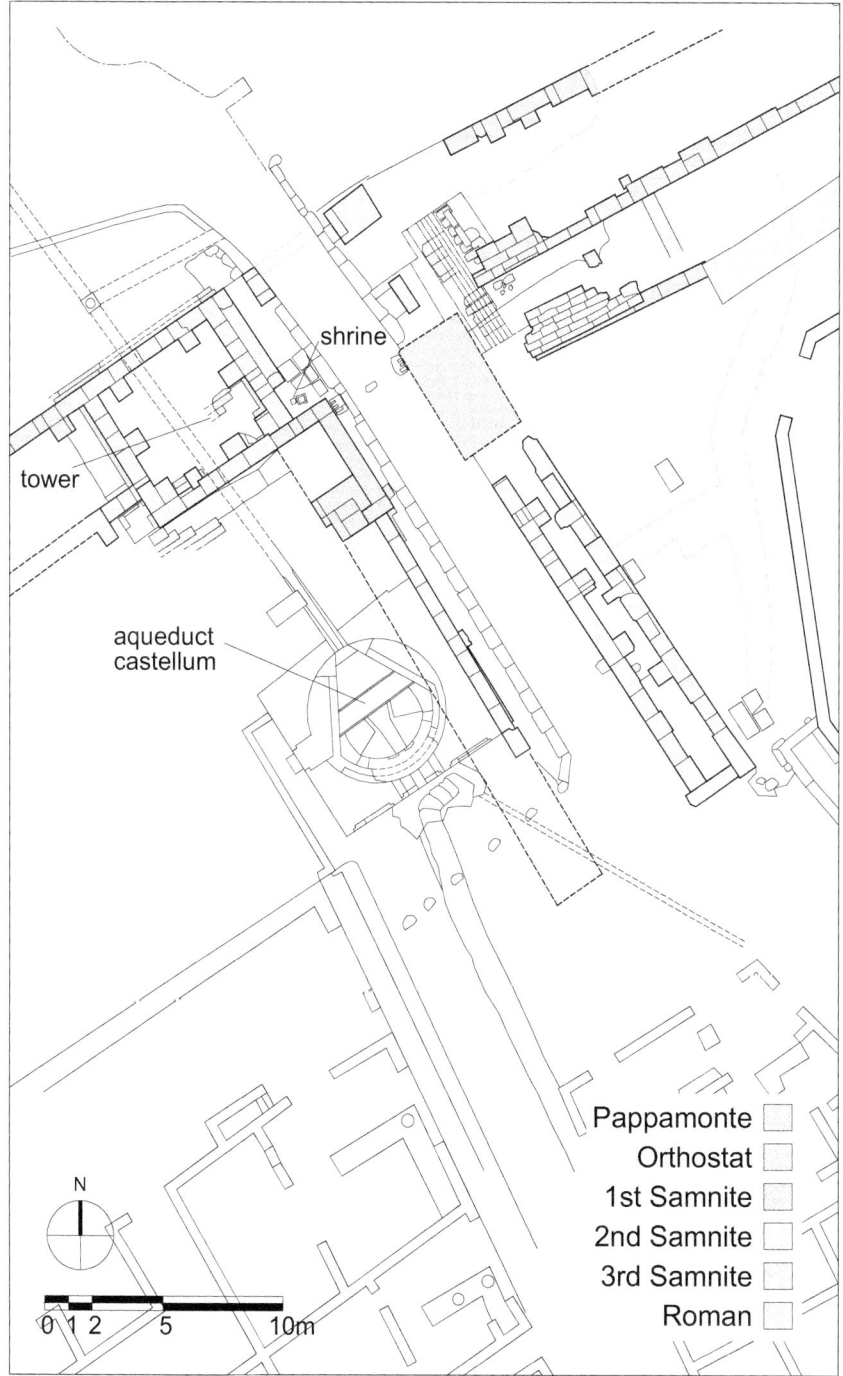

shrine

tower

aqueduct
castellum

N

0 1 2   5   10m

Pappamonte
Orthostat
1st Samnite
2nd Samnite
3rd Samnite
Roman

Porta Vesuvio   1:300

Figure 3.4 Plan of the Porta Vesuvio. The key highlights the phases. (Drawing by T. Liddell.)

In addition to strategic considerations, the tower displays the same ornamental effects that are present in the outer curtain. Excavations have revealed its surviving framework. On the city side, the tower displayed six courses of tuff blocks resting on a socle of eight travertine courses. On the exterior, the travertine socle reduces to four blocks.[45] On the western flank, the tuff blocks display a subtle decorative rustication. It has disappeared elsewhere due to later modifications. As with the curtain wall, the traditional view assigns the applied construction materials of travertine and tuff to subsequent construction events.[46] However, just as is the case with the adjacent curtain wall, this juxtaposition of materials is probably the result of a single construction event. The tower obeyed the same aesthetic principle where the masonry displayed a prominent yellow socle with surmounting tuff blocks. The rustication present on the tuff ashlars added extra ornamental emphasis. Considering the location of the Porta Vesuvio at the highest point in the ancient city, these elements transformed the tower and the gate into a landmark visible for miles in the landscape. This ornamental arrangement essentially prefigures in stone the white First Style ensembles applied to the later towers and gates of the circuit that used stucco to emphasize a derivative socle of orthostats with surmounting ashlars. It is a remarkable testament to the continuity of the ornamental display present on the circuit.

Upgrading the defenses

After the establishment of the first Samnite circuit, the defenses witnessed a radical transformation with the addition of an internal wall and the expansion in both depth and height of the *agger* (see plate 1). The intervention on the walls is rather difficult to date precisely because excavations have failed to produce definitive datable evidence. The consensus generally views it as a response to the Second Punic War. Hannibal brought the threat of destruction very close to Pompeii after having laid waste to the neighboring town of Nocera in 216 BCE.[47] Nevertheless, the triggers that caused the upgrade – the gathering war clouds, the conflict itself, or a simple recognition that weak fortifications needed reinforcement – remain difficult to pinpoint.[48] Despite these uncertainties, the expansion of the *agger* and the addition of an internal wall is an event attested on numerous occasions.[49] Moreover, the blocks composing the internal and external walls display divergent types in the surviving quarry marks, suggesting that construction occurred in separate events.[50]

The area between the Porta Ercolano and Vesuvio highlights the main elements and reasons behind this massive reinforcement. The internal wall performed both structural and defensive roles. The new structure supported a taller *agger* behind it, increasing the height of the embankment from about 8 to 11 meters. A series of regularly spaced internal piers mirroring those on the external curtain further stabilized the structure. The new wall functioned to relieve the outer curtain from the pressure exerted by the *agger*, which probably caused previous collapses. It seems unlikely that engineers needed to raise the exterior curtain since builders piled the taller *agger* behind the inner wall, which was more than sufficient to support it (see Figure 3.5).

The internal wall is composed almost entirely of tuff. Where exposed, the blocks still carry prominent quarry marks attesting to a well organized production and

Figure 3.5 Photo taken from Tower XI toward Tower X showing the development of the
fortifications.

construction process administering the operation. A few interspersed travertine
blocks derive from the partial demolition of the terracing wall on the city side of
the old *agger*.[51] As part of the overhaul, builders added a new containment wall
on the city side of the defenses to accommodate the newly expanded *agger*. The
operation included the construction of a new pomerial street that hugged the base
of the embankment. The result was a dramatic improvement of the overall defensive capabilities of the fortifications. The increased height of the *agger* provided
cover for neighboring houses, as well as for troops and supplies moving along the
new street.[52] The combined external curtain and the new wall created a double
parapet that provided an in-depth defensive line against attackers. It also allowed
for the widening and equalization of the wall-walk, which, in turn, led to improved
communications and more space for troop movements.[53]

Although engineers expanded the *agger* throughout the circuit, the construction
of the internal wall is not uniform. It is notably missing at the Porta Stabia and
between the Porta Vesuvio and Tower IX, where a remnant of the Orthostat wall,
later replaced by an *opus incertum* terracing structure, supported a smaller *agger*.[54]
A large tuff stairway offering access to the wall-walk replaces the wall between the
Porta Ercolano and Tower XII. The internal wall is also missing in the southeastern
quarter of the city where construction of the later amphitheater demolished it. In
the southwestern quarter of the city, the wall remains unidentified between the
House of Fabius Rufus and the Doric Temple, where it may be missing because
of the natural strength of the Pompeian plateau. The desire to maintain the view

of the landscape from the Doric Temple and the Temple of Venus probably also precluded the presence of the full defensive system.[55]

The battlements of both walls have completely disappeared, but the double defensive line must have been a formidable sight. The earliest archaeological reports of the nineteenth century describe how each wall included L-shaped merlons extending onto the internal piers reinforcing the facade. This shape best covered the front and left sides of the defenders and liberated their right arms to pelt the enemy with projectiles.[56] At present, only three capping stones still lie near the battlements: one on the wall-walk west of Tower XII and two just outside the Porta Nocera. All three bear signs of a joint on the top right, indicating that they originally capped the inner pier. An extended cornice, lined with spouts to drain rainwater from the wall-walk, marked the transition between the curtain and the parapet. This simple functional marker in the wall added to the overall ornamentation. Recent research has questioned the presence of drainage spouts for this phase because most survive in much later masonry. However, this is likely a circumstance of reuse – reconstructions have otherwise invariably placed them in this phase of the walls.[57] The spout carved to resemble a lion's head recovered outside the Porta Stabia is exceptional (see Figure 1.3).[58] After identifying it as a spout, August Mau projected it as a type that adorned the entire circuit. Basing such a projection on a single surviving example is a little far-fetched. Reconstructions have otherwise tended to ignore it. Instead, a more plausible assumption is that this spout was a feature unique to the gates and perhaps the Porta Stabia alone.[59] Such subtle details added to the general presence and ornamentation of the gates, thereby enhancing their representational role for the city behind them.

Emphasizing the transition through the fortifications

The completion of the new *agger* required a broad revision of the gates to accommodate the added height and depth of the embankment. Given the complexity and extent of the fortifications, establishing exactly when and how each gate changed its appearance remains problematic. Any date is approximate to the first half of the second century BCE, as attested only for the gate court at the Porta Vesuvio.[60] According to the traditional view, each gate received new court walls composed of tuff as a substitute of a previous travertine framework, as well as a vault marking the threshold into the city. The Porta Stabia, with its surviving travertine court walls, would be the exception. The tuff vaults would be the predecessors of the current ones in *opus incertum*.[61] The Porta Nola, the best preserved example, formed the basis of this chronology (see Figure 3.6). In its current layout, the gate includes outer travertine bastions, a tuff gate court, and a vault built in *opus incertum*. Maiuri proposed that the tuff blocks embedded in the *opus incertum* masonry, along with the protome of Minerva that acts as the keystone of the arch, are the relics of the earlier vault. The protome in particular is a would-be example of reuse for purposes of cult continuity or nostalgia.[62] When it comes to the vault, much of this theory is highly debatable; any evidence for the substitution of a previous vault is completely lacking.[63] However, the dramatic expansion of the *agger* must have incurred some sort of intervention at the gates in order to contain the extended earthen embankment.

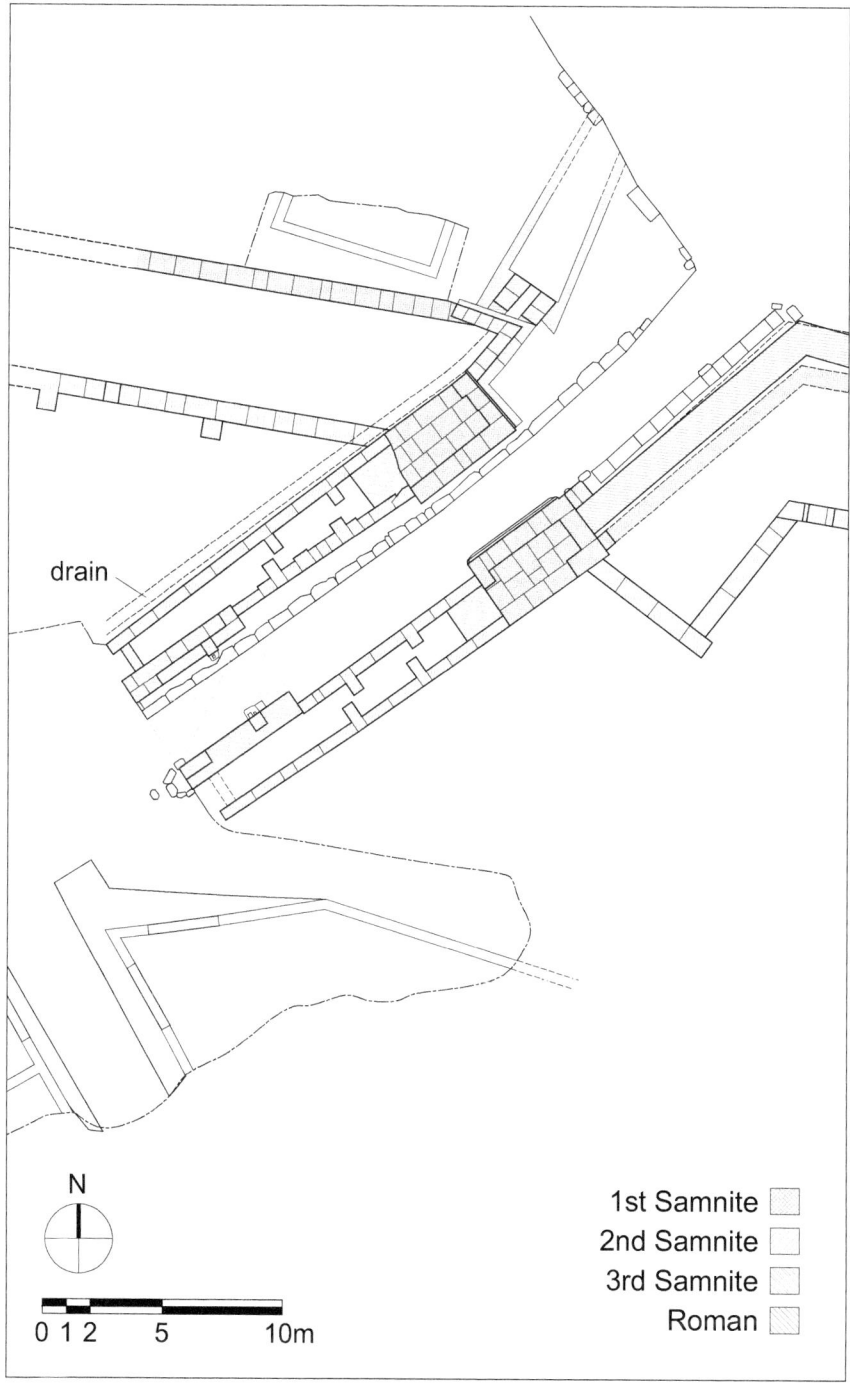

drain

N

0 1 2   5   10m

1st Samnite
2nd Samnite
3rd Samnite
Roman

Porta Nola   1:300

Figure 3.6 Plan of the Porta Nola. The key highlights the phases. (Drawing by T. Liddell.)

The gates that received an overhaul as part of this upgrade of the defenses respond to strict aesthetic rules concerning their appearance. Four of the best surviving examples, the Porta Nocera, Sarno, Nola, and Vesuvio, display an almost identical layout to the Porta Stabia but with one significant difference: orthostats in brown tuff mark the court (see plate 10).[64] The result is a deliberate effect of contrast between the yellow travertine bastions and tuff masonry of the gate court. The subtle design of the Porta Stabia now gives way to a clear differentiation of the gate's functional elements. Combined with the effect of the polychrome masonry on the outer wall, the new gates created an ornamental effect encompassing the city. Such a development is in tune with the contemporary trend of vigorous urbanization occurring on the Italian peninsula.[65] The monuments that cities would build, including temples, fortifications, baths, and theaters, would intertwine with notions on the image and idea of cities. The ornamentation on the Pompeian fortifications is a striking statement of aesthetic unity on an otherwise military structure. It served to emphasize the fortifications as a civic monument built by a confident community.

Porta Nocera

The Porta Nocera displays an ornamental factor governing the design of the gates and its long-term continuity (see Figure 3.7). Its earliest phase is contemporaneous with the first Samnite enceinte. Its later development as compared to the main interventions on the circuit are less clear.[66] The current structure repeats the familiar layout: outer travertine bastions, a tuff gate court, and a concrete vault. In the traditional view, each element should represent a separate construction event. The remains tell a different story. The foundations of the gate are visible because engineers lowered the road passing through it after the establishment of the Roman colony.[67] Proof of this arrangement comes from simple observation. The sidewalk leading up to the gate, as well as the struts that once supported the gate leaves, are unduly high in comparison to the road. The lowest four courses of the masonry composing the gate court display quarry marks and are laid down as ashlars. By contrast, the upper courses stand vertically as orthostats and are dressed to a smooth surface. All of the masonry of the gate court stands on a thick *opus incertum* foundation. This concrete base displays a seamless transition with the masonry of the vault behind it (see plate 11). Unless the gate court walls were lifted and put back into place – an event that seems unlikely – the current passageway and vault of the Porta Nocera are the result of a single construction event using both concrete and tuff.[68] This sequence leaves three options concerning the gate's appearance: its current layout is a late addition, the gate received a late refurbishment restoring the tuff masonry, or the current gates are relatively late buildings built together with the towers. Although a definitive answer remains elusive, each scenario implies the presence of a universal design or at least the consideration of a uniform aesthetic in the presentation of the gates.

Given the uncertainties concerning the date of the first introduction of *opus incertum* at Pompeii, the Porta Nocera is important evidence for its early use. So far, its circumstances in the context of the Pompeian fortifications are unique.

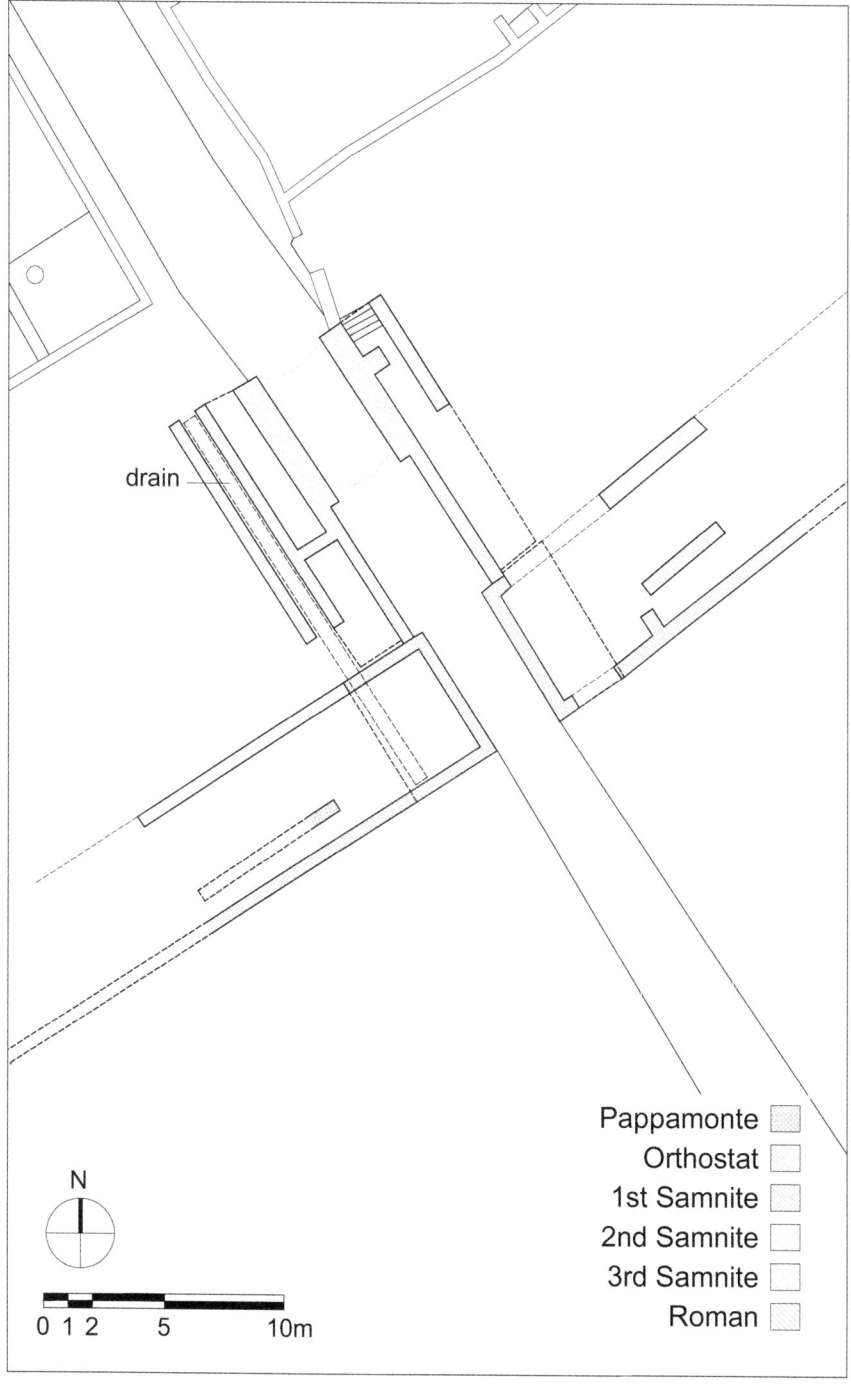

drain

Pappamonte
Orthostat
1st Samnite
2nd Samnite
3rd Samnite
Roman

N

0 1 2   5   10m

Porta Nocera

1:300

Figure 3.7 Plan of the Porta Nocera. The key highlights the phases. (Drawing by T. Liddell.)

The Porta Stabia and Vesuvio do not preserve a similar use of *opus incertum* in their foundations,[69] whereas the vaults at the Porta Nola and Sarno abut the gate court walls, indicating that they are later additions. Nevertheless, given the massive expansion of the *agger*, the current layout of the Porta Nocera must be contemporaneous with the upgrade of the adjacent fortifications that occurred in the second century BCE.[70] This date makes the Porta Nocera a laboratory of sorts for the application of *opus incertum* as a new construction technique.

Porta Sarno

The Porta Sarno is perhaps the most damaged of the standing gates (see Figure 3.8). The entire northern flank is missing as a result of the Allied bombing in WWII. Only the lowest course, parts of the *opus incertum* vault, and traces of a drain remain. The southern side is in a somewhat better state. The outer bastion is heavily damaged, rising to a maximum of six courses. Its rear half features a series of tuff courses, which, due to the tool marks and very good state of preservation, must be modern. They resemble similar blocks applied as restorations on the adjacent curtain wall. The gate court features large tuff orthostats on top of a foundation composed of travertine ashlars. Occasional crowbar marks suggest that workers removed some the original masonry for reuse elsewhere.[71] A lowering of the roadway that occurred in the Augustan period led to the exposure of the foundations visible today. A number of tuff blocks are inserted headfirst into the *agger*; given their similarity with the blocks on the outer bastion, these must also represent a modern reconstruction. Their different orientation probably represents an effort to highlight the restoration. Behind the gate court are the remains of the *opus incertum* vault. A hole in the concrete masonry reveals that part of the gate court masonry continues behind the vault, indicating that the arch is a later addition. The development of the gate remains elusive. However, given its position on the via dell'Abbondanza, it must have been part of the original Samnite circuit. The masonry indicates that the Porta Sarno followed a similar development to the other gates, including the completion of the monumental tripartite layout.

Porta Ercolano

The Porta Ercolano was the most monumental structure among the Pompeian gates (see Figure 3.9). The ruins exposed today represent its Flavian version, built as a monumental arch crowning the reconstruction effort following the earthquakes of the 60s CE.[72] Its form is the result of the gate's special status because of its position at the head of the via Consolare, a busy regional road connecting Pompeii with Naples and, through the via Domitiana, on to Rome.[73] In its previous Samnite version, the gate repeated the layout and masonry of the other gates, except that its northern flank was at a slight angle because of the terrain. The northern bastion originally sat on a small natural terrace that not only dictated the orientation of the gate but also added height and therefore an effect of impenetrability to the fortifications.[74]

drain inlet

N

Orthostat
1st Samnite
2nd Samnite
3rd Samnite
Roman

0 1 2   5   10m

Porta Sarno   1:300

Figure 3.8 Plan of the Porta Sarno. The key highlights the phases. (Drawing by T. Liddell.)

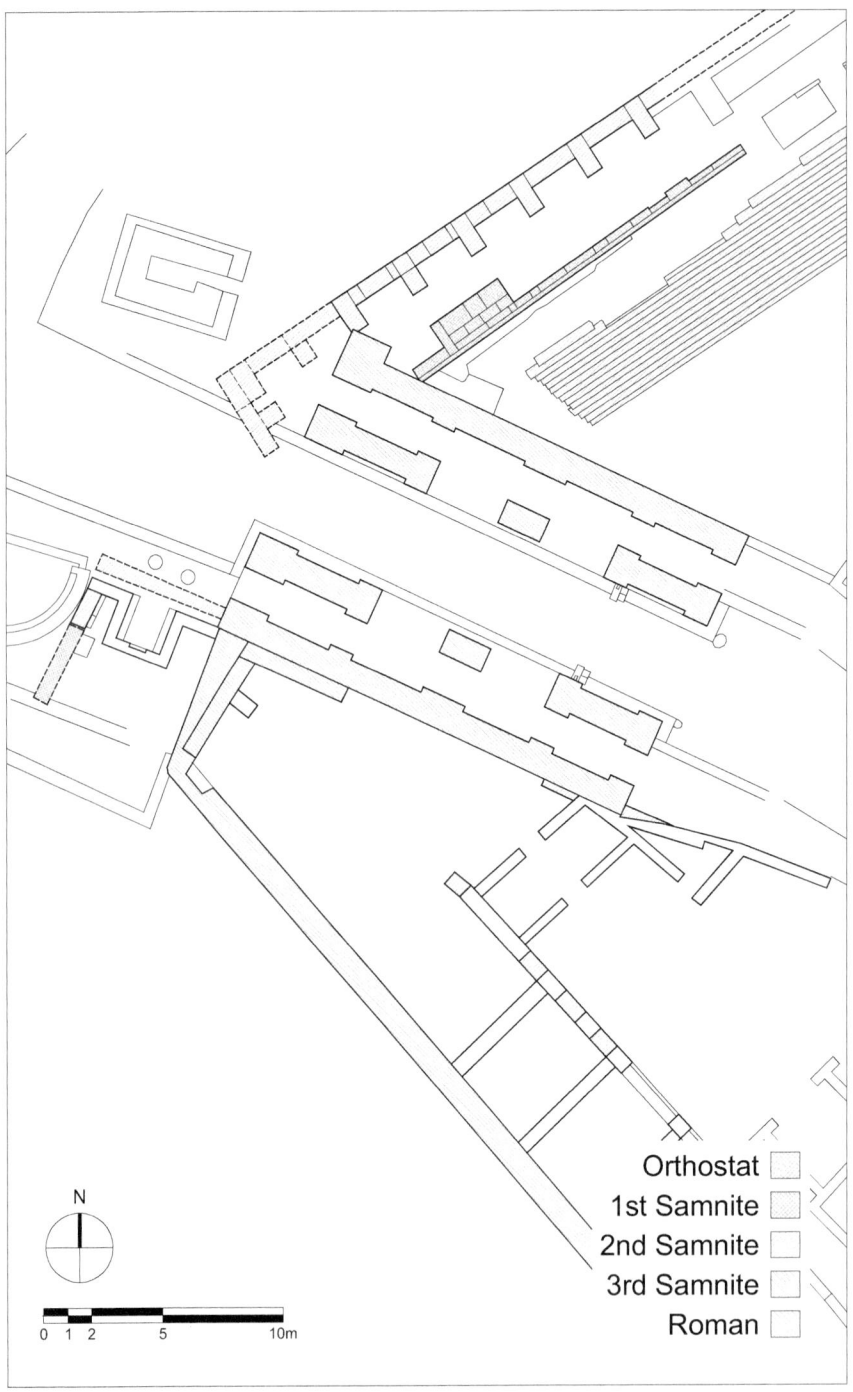

Orthostat ☐
1st Samnite ☐
2nd Samnite ☐
3rd Samnite ☐
Roman ☐

N

0 1 2   5   10m

Porta Ercolano   1:300

Figure 3.9 Plan of the Porta Ercolano. The key highlights the phases. (Drawing by T. Liddell.)

Figure 3.10 Stretch of wall next to the Porta Ercolano in tuff masonry.

Despite the design variation, the Samnite Porta Ercolano included the same differentiation of materials explicitly highlighting the entrance into the city. Today the only stretch of wall that survives once ended in its northern bastion (see Figure 3.10). Littered with scars from the Sullan siege, the fourteen-course section still displays some of the highest-quality finished masonry of the entire circuit.[75] It is also unique as the only stretch of fortifications composed entirely of tuff – a circumstance reflecting its prominent position next to the gate. Builders used it as a display material, paying special attention to the arrangement of construction materials to contrast the travertine of the bastions against the tuff of the curtain and gate court. The resulting effect set the stage for approaching viewers of the architectural character of the city behind the gate.

Porta Vesuvio

Neighboring Porta Vesuvio presents a unique variation of the basic plan (see Figure 3.4). The gate responded to its vulnerable position in the defenses with its travertine bastions set back behind the curtain and the addition of a forward tower. With the second upgrade of the *agger*, this layout remained essentially unchanged. The added height of the embankment led to the construction of the tuff gate court somewhere between the late third and mid-second centuries BCE. The resulting layout transformed the gate into a formidable defensive outpost with double tuff and travertine bastions that created a deadly cul-de-sac. This strategic consideration explains why the Porta Vesuvio featured a closing mechanism on the exterior side of the travertine bastions; defenders could isolate attackers more effectively. At a later stage, in a clear effort to monumentalize the gate, engineers built an arch using *opus incertum* on the exterior of the passageway. Its placement on the exterior, as opposed to those on the city side on the other gates, is a natural continuation of the outward emphasis that engineers placed on the Vesuvian gate. It is also the result of the smaller *agger* on the eastern side of the gate, which, in this area of the defenses, was shorter than elsewhere. Whether another vault once existed on the city side of the gate remains unclear due to the heavy damage it sustained during the earthquake.[76] Despite the alterations, the gate again staged a powerful three-step entry into the city with one exception: The curtain rather than the travertine bastions announced the passage into the city (see plate 12).

Considering the collective evidence from the gates, as well as the recently suggested date for the introduction of concrete at Pompeii, a challenge emerges to the established chronology identified for the various phases of the Samnite circuit.[77] It indicates a possibility that the upgrade of the fortifications occurred in the mid-second century BCE, as opposed to the traditionally assumed late third century. Further investigations are needed to sort out the chronology. Nevertheless, the variety of building techniques employed for the foundations of the gate courts also suggests that their construction occurred separately rather than in a single event. This scenario makes the presence of the ornamental effect in the fortifications even more compelling – each gate had to match a clear aesthetic standard. It means that a vision existed of what the fortifications were supposed to look like as a proper monument representing the city and its urban decor, and that it was maintained throughout various construction events.

Reaching new heights: *opus incertum* and the next phase of construction

The ornamental emphasis on the fortifications found its greatest expression in the final upgrade of the Samnite period when Pompeii achieved its developmental peak before the establishment of the colony in 80 BCE. The public buildings erected in this period allowed Pompeii to develop an autonomous architectural identity with the fortifications as an integral part of an image that reflected the community they protected. The current view equates most of the works on the fortifications as a response to the gathering clouds of the Social War, but it is hard to assess this kind of foresight. The upgrade may also represent a reaction to the threat posed by Rome's war with the Cimbri and the Teutones that included battles in the Po valley. At the same time, the threat of conflict may not have constituted a factor at all. Pompeii fits into a general trend of the late second and first centuries BCE when cities throughout Italy increasingly upgraded or built new defenses.[78] This activity is part of a wider trend of urbanization where Italian settlements added the essential architectural elements such as theaters, baths, temples, and fortifications in order to complete their image as a city.[79]

The building campaign carried out at the turn of the first century BCE included the construction of twelve towers inserted into the earlier curtain and the vaults on the gates (see plate 13). Engineers resorted to *opus incertum* to build the additions. The material is a cheap form of cement faced with irregular, fist-sized stones.

Builders usually covered it with plaster to mask its crudity. Concrete carried a distinct set of advantages over regular ashlar masonry. It was adaptable to molds and allowed for rapid construction. Relatively unskilled builders could participate in the construction process because the material did not require expert masons to the dress masonry. Architects introduced concrete on a widespread scale at Pompeii starting in the mid-second century BCE,[80] but it is unclear when they were confident enough to start using it in elaborate tall structures. Architects would eventually apply it to extensive vaulting, which, along with the inherent strength of the material, allowed them to achieve taller and more complex buildings at a cheaper cost than was possible using ashlar masonry.

The Pompeian circuit preserves large sections of *opus incertum* as part of its curtain wall. Their similar building technique as the towers and gate vaults has tended to lump the construction of these long stretches into a single event. The longest sections still in place are south of the amphitheater, a bastion of Porta Nola, east of the Porta Ercolano, either side of the Porta Stabia, and the terracing wall supporting the Temple of Venus. It is without question that these extensive tracts of wall indeed represent repairs. However, given the inherent ambiguities surrounding the dating of masonry, assigning them all to this single construction event is problematic. It would mean that no construction activity occurred on the fortifications in the 170 years after this intervention.[81] I consider the majority of these sections as colonial, either related to repairs or isolated building campaigns such as the Amphitheater or the Temple of Venus.

Instead of viewing concrete as the sole means to repair the fortifications, several distinctions in the ashlar masonry point to numerous refurbishments in the curtain wall conducted in the same travertine/tuff framework that is typical of the Samnite enceinte. They point to a conscious desire to preserve the appearance of the walls, indicating, once again, that commissioners sought to highlight an ornamental aspect to the walls. A small section of curtain some 47 meters west of Tower XI presents less finished and homogeneous blocks that implies a later repair.[82] Similarly, a 23-meter stretch of wall located 27 meters west of the Porta Nola displays slightly smaller travertine blocks that likely represent a later refurbishment (see plate 14).[83] Variations in the block size in the curtains flanking Towers VII and VIII also indicate further repairs that likely occurred during their construction.[84] The striking aspect of these refurbishments is that they imitated the previous *opus quadratum* masonry layout, including the travertine/tuff sequence.

Construction of the towers significantly changed the role and appearance of the battlements. Their insertion into the curtain wall required engineers to dismantle parts of the masonry and lower the adjacent *agger* to allow access into the new buildings. As a result, the secondary inner parapet largely disappeared, moving the line forward and pinning defense on the dominating height and firepower of the towers.[85] This process revealed much of the undressed masonry and the quarry marks on the blocks of the internal wall.[86] The new layout required workers to pierce new access points to the wall-walk through the internal wall. A clear example, walled up at an unspecified later date, still exists between Towers X and XI, where workers gave the *agger* a gentle slope that reached a small postern cut into the tuff blocks (see Figure 3.11).

Figure 3.11 Reconstruction of the Tower X as seen from the city side. (Drawing by L. Kukler.)

The traditional reconstruction of the fortifications in this phase proposes an unroofed wall-walk as a matter of continuity with the previous layout of the battlements. A key to understanding this arrangement are the drainage spouts that are still evident today, designed to drain rainwater from the wall-walk. A recent reinterpretation has cast doubt onto this reconstruction. Rather than seeing the inner tuff wall as the relic of a secondary parapet, the hypothesis proposes that it supported the rear of a roof sloped toward the field side of the walls. The roof subsequently vanished because of a demilitarization effort that included the demolition of the battlements carried out by victorious Sullan troops after the siege. The series of drainage spouts that survive in the current walls would then be part of a later reconstruction effort conducted in the early colony that did not restore the roof. The fact that most spouts survive in *opus incertum* masonry would constitute further proof because the masonry type represents the extent of the repairs.[87] The theory does carry some weight. In fact, a striking aspect of the Pompeian fortifications is the derelict state of the battlements. However, this circumstance is more likely a result of much later events, including their spoliation for building materials after the earthquake.

The demolition of wall crowns was a common military tactic aimed at prohibiting further use of the defenses. The victor did not always fully demolish the defenses – an extensive, expensive, and laborious task – opting instead to demilitarize the structures in such a way to prohibit their effective use in military operations. This act was also symbolic, signaling the loss of independence of a community by crippling their ability to defend themselves.[88] What remained was essentially a symbolic fragment of previous independence. Covered battlements carried the distinct advantage of protecting defending troops from the elements and enemy missiles.[89] A main problem with the roofed battlements theory is the complete lack of any further supporting archaeological evidence; there are no discernible beam holes in any of the surviving masonry that allude to a wooden superstructure. Moreover, a curtain section located 15 meters east of Tower XII displays a top course of vertical travertine orthostats with alternating tuff spouts set in ashlar masonry. Unless this section also represents a later repair to the battlements, their presence disproves the theory of a previously covered wall-walk. Moreover, the interior wall is missing in sections of the defenses. Although the precise layout of the battlements is elusive, the projection of a roof covering the wall-walk for the entire circuit in this phase seems unlikely.

Given the strategic importance of the battlements, perhaps it is not without coincidence that the only truly intact part of the wall top that survives is primarily ornamental. It stretches along the edge of a cliff bordering the street known as the vicolo dei Soprastanti, where it functioned to emphasize the city's edge. The section is composed of six travertine Doric columns engaged in an *opus incertum* curtain. Stylistic analysis of the capitals suggests a late second century BCE date for the wall, making its construction roughly contemporaneous with the towers.[90] With the exception of a small triangular window designed as a viewing port or firing slit, the wall has no real military purpose (see Figure 3.12). Although it has

Figure 3.12 Section of wall in the vicolo dei Soprastanti. Arrow points to triangular window.

now completely disappeared, plaster very similar to that covering the towers probably decorated this stretch of walls. No other part of the circuit better highlights
the desire to embellish the fortifications than this ornamental marker of the city's
edge. Its presence here is hardly accidental. This area preserves some of the most
opulent houses in the city.

New elements to old gates

Together with the construction of the towers, the city gates would receive new
vaults built in *opus incertum* to mark the threshold into the city. The vaults at the
Porta Vesuvio, Sarno, Nola, and Stabia clearly abut the ashlars that compose the
court, indicating that they are the last major architectural addition to the gates. A
clear date comes from the Porta Stabia, where construction of the vault occurred in
the final decades of the second century BCE.[91] The vault of the Porta Vesuvio is the
only example built on the exterior side of the gate. It can have found a place there
only after the construction of Tower X farther west, which made the earlier tower
flanking the gate redundant. The construction dates and sequence of the vaults
at the gates remain debatable, but the evidence indicates that they are roughly
contemporaneous with the erection of the towers.

The function of the arches is primarily ornamental (see plate 15). The *opus
incertum* framework that composes them originally featured some sort of covering in plaster to disguise the masonry and protect it from the elements. With the
exception of a few slivers of plaster on the inside of the vaults, any decoration
has almost completely vanished due to its exposure. By means of prints and written accounts, the various monographs dedicated to Pompeii published over the
centuries offer a few clues to their original appearance. Only two written accounts
describe the plaster in any detail. Unfortunately, they supply little information on
any possible decorative schemes and even less on any potential colors employed.
Bechi, writing in 1851 shortly after the excavation of the Porta Stabia, described
the vault as collapsed but decorated in the same white First Style stucco as the
current Porta Ercolano.[92] Predictably, considering his instrumental role in defining
the four Roman painting Styles, the only author who actually mentions any color
on the structures is August Mau. He described the interior of the vaults at the Porta
Stabia and Porta Nola as carrying a yellow elevated socle receding into a simple
smooth white plane above.[93] Such a scheme is relatively straightforward, yet, as
Mau explicitly details, the plaster itself achieves the highest possible quality.[94] It
is tempting to associate the yellow socle within the vaults with the natural color of
the travertine present in the bastions and the lower courses of the exterior curtain.
However, it is the white color, as opposed to any other, that composed the main
ornamental scheme of the vaults.

The earliest depictions of the gates offer further clues, although their schematic
nature and the monochromatic limitations of the black-and-white print lessens
their usefulness. Many gates were in an abandoned state or reburied – a circumstance that forced authors and artists to come up with sometimes imaginative
reconstructions or rely on previous publications.[95] French architect Mazois, who

worked in the decades immediately after the walls were uncovered in the early 1800s, produced some of the earliest and most accurate renditions. Subsequent authors and guidebooks often copied his prints. In his cross section of the Porta Nola, Mazois carefully depicts the state of the gate including a vanished election notice painted onto the tuff of the passageway. This is an important detail. It implies that plaster did not cover the masonry of the court or the outer bastion. The rendition confirms how the masonry and materials were an integral part of the adornment of the gates.

An engraving by William Gell, a contemporary of Mazois, reveals a far greater surviving extent of the plaster on the vault of the Nolan gate, with sections of smooth plaster still in situ. He also illustrates the keystone depicting Minerva with an adjacent inscription.[96] Neither author emphasizes Mau's yellow socle, which, although discolored, is still distinguishable today. There is no trace of the imitative ashlars typical of the First Style found on the towers and the later Porta Ercolano, but with so much of the evidence gone, it is hard to assess the embellishments on the vaults with any greater certainty. It may very well be that the plaster on different gates featured slight variations, responding to differing construction or restoration events, the wishes of different patrons, or even their placement and association with particular districts of the city. Complicating the matter further is the large gap between their first construction and the eruption, a period in which the individual structures may have changed their appearance.

Despite the insecurities on the details, the vaults were the first arched monumental structures to appear in a city otherwise built almost entirely using the post and lintel system. As an architectural novelty, they emphasized the tripartite passageway into the city. The materials composing the Pompeian gates now plainly marked each step through the *agger*: The travertine and vaults mark the pre- and postliminal phases, whereas the tuff emphasizes the liminal.[97] This arrangement marks the processional elements of the gates in an architectural tradition first established at the Porta Stabia some two centuries earlier. The commissioners of the arches were clearly referencing and reinforcing a conventional ornamental emphasis on the gates to mark their role as protectors of the community. The ornamentation at the gates began a trend in the expanding city where architects increasingly used arches in a monumental fashion in the developing urban decor.

The towers

The twelve towers built into the Pompeian circuit must have formed its most imposing elements. Today they are in ruins, their tall structures toppled during the earthquakes of the 60s CE and the eruption of Vesuvius. Engineers inserted the towers into the curtain by demolishing the two walls composing the fortifications. The buildings straddled the wall-walk and extended slightly beyond each side of the two parapets. Each tower consisted of three floors supported by barrel vaults and interconnected by means of rear internal stairways. A door in the back provided access to the city side, and one on either flank opened on the wall-walk.

A small postern in the ground floor usually opened onto the exterior side of the defenses (see plate 16).

Many uncertainties exist on the precise date and sequence of the construction of the towers. Most publications treat the towers only cursively, and none has recognized the subtle differences in layout, spacing and construction techniques that each building adopts. With the exception of later reconstructions, these differences seem to be the result of military and geographical factors rather than any major chronological separation in construction events. The cursive treatment of the towers in publications means that particular hypotheses and their reconstructions became entrenched without further questioning. An example comes from the debate concerning the existence of Tower XIII. It was a matter of much contention until the excavations of the 1950s and 1980s finally uncovered enough of the circuit to prove its existence unlikely.[98]

A similar unresolved question concerns the placement of Tower I. The surviving *eituns* inscriptions indicate the existence of at least twelve structures and their counterclockwise numbering in Roman numerals. The inscriptions, a group of six painted in red throughout the city, pointed the way to individual sectors of the fortifications for use by the *eituns*, or Samnite troops, in defending the city during the Sullan siege.[99] Two of them describe Towers X and XII as the buildings east of Porta Vesuvio and west of Porta Ercolano. With the current number of unearthed towers at eleven and the counterclockwise count, Tower I should be located somewhere between Tower II and the Porta Marina. A common consensus places it at the tip of the Triangular Forum, where it must have created a stunning visual relationship with the Doric Temple nearby.[100]

The reconstructions of the towers divide chronologically along the interpretations provided first by Mazois and later by Maiuri. Each recognized a three-story building but fundamentally differed on the final reconstruction. Mazois believed that the top floor was unroofed. He proposed a small covered stairway in the rear decorated with a Doric frieze after excavations recovered a few pieces of stucco. Mazois based his reconstruction on an amalgam of the ruined remains of Towers VII, X, XI, and XII (see Figure 3.13). With their elevation gone, he projected the three window openings surviving on the ground floor onto those above.[101] In a separate plate, Mazois depicted the ruins of Tower VII as the most complete surviving example. At the time, much of the plaster was still in place, but it has since disappeared. Mazois noted a slight variation of the plan, but did not explain the left rear door of Tower VII in direct opposition to the right door present on the other three towers he saw.[102] This small detail radically changes the arrangement of the stairways connecting the floors. The drawings that Mazois published remained the standard in subsequent publications until Maiuri uncovered substantial pieces of Tower X buried in the volcanic fill in front of it. The pieces allowed him to reconstruct a gabled roof ornamented with a Doric frieze, and four windows on the second and third floors. Maiuri also based his reconstruction on the fresco of the brawl in the amphitheater where the towers clearly show regular roofs.[103] For reasons that remain unclear, the currently rebuilt remains of Tower X feature crenellations rather than a gabled roof. Perhaps this reconstruction follows Mau's

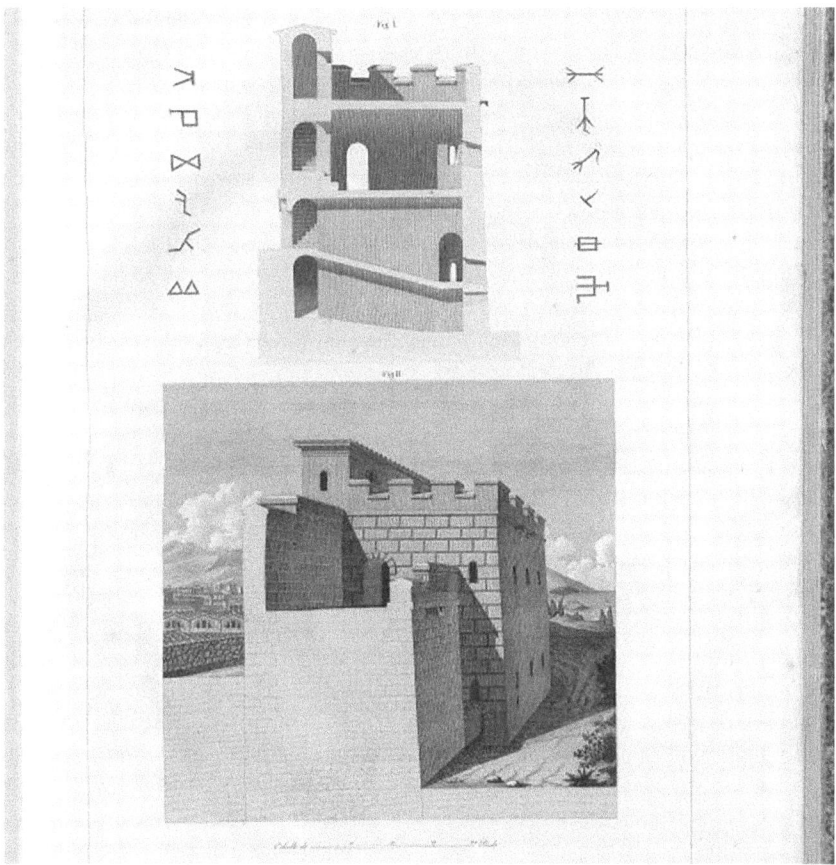

Figure 3.13 Reconstruction of the towers as projected by Mazois (1824a, pl. 13).

suggestion that the pediments were added to the towers when they were demilitarized.[104] The most recent reconstruction by Russo and Russo shows the crenellations supporting a sloping roof, but it remains unclear how these small masonry merlons could support such a heavy load.[105]

Types

A few significant differences in the internal design of the towers have remained unnoticed. Most publications consider the layout of the towers to be uniform, basing their observations largely on the structurally similar Towers X, XI, and XII in the northern sector of the defenses.[106] This is hardly surprising since this portion of the fortifications remained visible after its first exposure, whereas other sections were reburied over time. The main problem with the towers is that an accurate survey and plan of each structure is still lacking, making them the least under-stood buildings of the city despite their massive presence and visibility. Yet even a cursory observation points out different types, which I divide into four groups.

Group one, described by various authors due to their visibility throughout the modern history of Pompeii, includes Towers X, XI, and XII (see Figure 3.11). They feature an entrance on the rear right side of the building opening up on two sets of stairs: one heads straight down to the first floor, the other ascends to the second floor on the left. This setup allows for the presence of arrow slits at the corner of the staircases that provided both frontal and flanking fire to the towers. The first floor is a wide-open barrel-vaulted chamber. A door in the back opens on a rear corridor heading to the postern placed on the right side of the building. On the second floor, two doors opened onto the wall-walk, whereas another staircase in the back gave access to the top third floor.

Group two, which includes Towers II, III, IV, and VIII, is slightly different. The entry door from the *agger* is also on the right side, but it opens directly onto the second floor; a staircase immediately on the left heads to the first floor below. The group features a faux corridor in the rear of the ground floor, mirror-ing the functional one accessing the postern in group one. Instead, the posterns of group two, present in Towers II, IV, and VIII, open directly onto the ground floor chamber. The staircase layout precludes a flanking arrow slit set directly above the postern, leaving the tower in a weaker position. The type also features a slight variation in the decorative scheme. Whereas the other towers, with the exception of VII, all display raised voussoirs defining the postern vaults, the openings of group two are flush with the masonry. Tower II features a high floating postern, implying a later lowering of the ground level around the build-ing (see Figure 3.14). Tower III, perhaps because of its elevated position and inaccessibility, lacks a postern entirely, whereas Tower IV uniquely features a door opening on its left side. As opposed to the other examples, the position of the postern in Tower IV is counterintuitive to military tactics. An effective defensive strategy included sending out small parties of infantry to harass enemy positions with surprise sorties.[107] The position of the posterns on the right flank of the towers offered sortieing infantry a measure of protection behind the shield they held in their left arm. The postern of Tower IV does the opposite, forcing troops to expose their unprotected right side to the enemy. Perhaps this postern purposefully functioned as a reentry point for sortie parties who could protect their left flank when approaching the tower.[108]

Group three includes Towers V, VI, and VII, where the access door from the *agger* is on the left side (see plate 17). The door opens up directly onto the second floor; a staircase immediately on the right leads down to the ground floor chamber. This type features the same corridor as group one, winding behind the main cham-ber to the postern. The layout of the staircase heading down to the vaulted chamber, however, lowers the ceiling of the corridor. This arrangement cuts the arrow slit at the end of the corridor in half, rendering it ineffective. Instead, the arrow slit is purely ornamental or even deceptive: It gives the towers in group three the same external appearance as those in group one.

Figure 3.14 Tower II viewed from below. Note that the terrain around the tower has lowered significantly since antiquity. One can access the postern only on the left
by walking on unexcavated volcanic material. The smaller vault on the right
was covered with a stairway leading to the second floor.

Finally, group four includes only Tower IX. Its appearance is radically different
because the main building collapsed in an event associated with the Sullan siege.
The ruins present no trace of staircases, posterns, or windows; the main chamber
on the ground floor preserves only the faint outline of a barrel vault that once
covered the space. A door in the middle of the back wall offered access to the main
chamber from a street on the city side. This is an unusual situation for Pompeian
standards because it also implies the absence of the *agger*. This version of the
building, however, is a much later reconstruction dating to the colonial period. It
probably served as a storage area rather than as a defensive structure. The layout
of the previous version is largely unknown. Excavations have recovered pieces
of masonry covered in white First Style stucco, proving that the earlier version
featured the same ornamental features present on the other towers.[109]
    Despite the differences, the layouts of the groups are essentially very simi-
lar, displaying variations only in the positions of the rear doors, stair access, and
placement of the posterns. The types concentrate in specific areas of the circuit,
indicating that the differences in design are a matter of military tactics, different
construction events, or both. For instance, the towers composing group one concentrate on the north side of the city where the terrain sloping gently into the city

gives attackers a strategic advantage. Presumably, this is also an area that featured outworks designed to keep siege machines at a distance. Harassing the enemy with effective sorties from the towers would provide an extra measure of defense – a circumstance which explains the presence of posterns. The towers of group two, with the exception of Tower VIII, are located on the low, steep lava ridge on the south side of the city. The topography limits the possibilities for direct enemy attack, but it still allowed defensive troops to mount sorties against the enemy.[110] This relative inaccessibility explains why the posterns directly accessed the main chamber as opposed to those in group one, where the corridor provided a further measure of defense in case of a breakthrough. The buildings composing group three are concentrated near the amphitheater on the southeastern side of the city, where the steep lava ridge diminished the necessity for short-range weapons on the lower floors – a circumstance that explains the blocked firing slits on the ground floor. It is difficult to pinpoint the reasons for the differences more precisely, but they probably relate to the surrounding terrain and the outworks in front of them, which remain largely unknown.

Another possibility is that the divergent groups represent phased construction events or the piecemeal replacement of earlier towers, as financed by individual patrons. Such a gradual approach to understanding their construction can explain the slight variations in building techniques. For instance, whereas most of the towers show an almost uniform application of *opus incertum*, Towers II and V have prominent toothed quoins composed of fired brick. Tower VII features traces of similar quoins composed of yellow tuff bricks, whereas Tower III displays the *opus vittatum mixtum* technique where tuff and fired bricks alternate in layers. The presence of these techniques, almost invariably dated to the colonial city, is somewhat of a conundrum for structures built toward the end of the second century BCE.[111] Thomas Frölich has convincingly isolated the use of *opus vittatum mixtum* to the postearthquake period after 62 CE.[112] None of these studies has considered the towers, and this particular aspect deserves further scrutiny. The presence of these techniques on the buildings is likely the result of refurbishments carried out after the establishment of the colony. The similarities in plan and decoration of the towers point to their construction occurring in close succession or even as part of a guiding design or template.

A striking aspect of the towers is their unequal spacing and general indifference to military precepts. Vitruvius and Philon of Byzantium both prescribe that towers should be regularly spaced within an arrow's flight of one another to offer mutual cover against attackers.[113] Their framework should protrude considerably from the curtain and their design made round or polygonal to better resist bombardment and avoid weak corners vulnerable to battering rams. Furthermore, the masonry composing the towers should be separate from the curtain to prevent large sections on the wall collapsing with it if a building went down during the fighting.[114] At Pompeii, the ruins display the opposite, where the square towers are almost flush with the wall and are concentrated in the vulnerable northern and southeastern sectors. This layout almost invariably relates to Pompeian topography. With the exception of the northern sector, the proximity of the walls to the natural tuff ridge inevitably forced the towers into the curtain.[115] This was the result of circumstance. Engineers inserted the towers into a preexisting circuit, but it was not a weakness. The ridge carried the distinct advantages of inaccessibility and the look of great height, thereby imparting a formidable appearance of impenetrability to the circuit. The placement of Tower V at the tip of the amphitheater particularly illustrates the point. Here the cliff ridge turns north at a ninety-degree angle. The tower certainly worked in a defensive context but also carried decorative and monumental elements, giving the entire area the appearance of a fortress.[116]

Although at first glance the siting and concentration of the towers might appear primarily related to weaknesses in the circuit, their location elsewhere is not entirely arbitrary. For instance, the location of Tower X essentially transformed the Porta Vesuvio into a massive *scaean* gate by forcing attackers to expose their right unshielded side to direct flanking fire. Towers VI, VII, and VIII are spaced exactly halfway between neighboring gates or at drastic angles in the circuit. Although isolated from the other towers, each of these buildings essentially acted as a small fortress defending its sector more effectively and forming a last line of defense if any enemy troops managed to overcome the walls.[117] The towers also invariably appear at the end of streets, commanding the internal axes that facilitated communications toward them. This aspect suggests that the towers replaced earlier posterns in the circuit – a circumstance that also explains why such openings are otherwise curiously absent.[118] Due to their prominent placement, the towers were a constant visual presence in the city. As a result, they projected a sense not only of security but also one of authority onto the urban area. Fortifications, in addition to their military character, also functioned as effective policing structures that restricted population movement and provided expedient tax barriers at the gates.[119] Through their imposing visual presence, the towers provided a constant reference to the authority and social hierarchy they protected.

Despite the internal plan variations, the towers all appeared equal from a distance. Although most of their decoration has now disappeared, each structure featured a coating of white plaster arranged according to the design principles of the First Style. In the domestic sphere, First Style schemes usually featured a formulaic design consisting sequentially of a socle, followed by orthostats, ashlars, and topped with a frieze. Its arrangement on the towers had a slightly simpler variation consisting of an elevated socle, imitation ashlar blocks, and a Doric frieze. The application of a uniform ornamentation must partly be the result of a military consideration to disguise any internal disparities that, if detected, might expose strategic weaknesses. The First Style must also have carried symbolic connotations. Scholars have associated various meanings to the style, connecting it to notions of *pietas* as understood in terms of service to the state, citizenship, and urban decor. In this context, both towers and gates received nonmilitary and purely aesthetic decorative ensembles as a reflection of the city they enclosed. The style also marked a prominent intervention on the walls. Each white tower was a landmark visible from miles away that collectively served to crown the city visually. On a broader scale, the towers, curtain-wall, and gates worked together in a type of large-scale ornamentation that befitted the size of the building.

Aesthetic and military considerations worked together in the design of the city walls. To this end, builders applied construction materials to achieve a distinct polychrome effect. The result created a monumental statement that not only emphasized the defenses but that, by extension, also highlighted the relationship between the community and its territory. At Pompeii, patrons carefully simultaneously preserved, appropriated, changed, and enhanced this effect. This is most apparent at the gates, where builders elevated the processional emphasis through the passageway with construction materials, establishing a tradition after a first experiment at the Porta Stabia. The towers show how military tactics influenced their design and placement, whereas the ornamentation on each building completed the image of the fortifications as a monument protecting the community. At this point, the circuit reached its symbolic apex in the history of Pompeii, dialoguing both socially and architecturally with the community they protected.

Notes

1 See Von Gerkan 1940, 9–14.
2 Krischen 1941, 11.
3 Etani and Sakai 2003, 133–137.
4 Maiuri 1929, 197.
5 Gasparini and Uroz Sàez 2012, 9–68; Brasse 2014, 22–28.
6 Overbeck 1854, 40–41; Richardson 1988, 45.
7 Maiuri 1943, 281–283; Sakai 2000/2001, 94.
8 Maiuri believed in the closure of the gate at this stage; see Maiuri 1929, 155–156. Krischen contends this view; Krischen 1941, 9–11. Pesando and Guidobaldi, suggest the presence of a postern rather than a gate; Pesando and Guidobaldi 2006, 29. Guzzo believes that the tower's construction led to the closure the gate in the late second century BCE; Guzzo 2000, 108.
9 See Hori 2010, 288–289.
10 Cassetta and Costantino 2008, 198; Pappalardo et al. 2008, 300.
11 See chapter 7 of this volume.
12 De Caro 1985, 106.
13 For the cliff, see Maiuri 1959, 108; for the lowering, see D'Ambrosio and De Caro 1983, 29; Conticello de' Spagnolis 1994, 19.
14 Tower III preserves parts of the original level kept in place by the remains of the previous circuits.
15 Maiuri 1939, 233; De Caro 1985, 79.
16 Chiaramonte Treré 1986, 25.
17 Maiuri 1943, 279.
18 They have a diverse erosion rate; the gray type appears to be more resistant to the elements. Despite the divergence, the difference in the appearance is actually very subtle.
19 A recent article notes that the courses on either side of the tower are misaligned; see Hori 2010, 286. See also Hori et al. 2007, 1–5; the article, however, ignores the Allied aircraft bomb that destroyed the curtain west of Tower X in 1943 and its subsequent reconstruction; see García y García 2006, 164.
20 Maiuri 1929, 159–162.
21 Curti 2008, 53.
22 Noack and Lehmann-Hartleben 1936, 5–15.
23 Maiuri 1929, 159–163, 218.
24 Maiuri 1929, 218.

25 Sakai and Iorio 2005, 328.
26 Winter 1971, 62; Sakai 2000/2001, 94.
27 Vitr. *De arch.* II.7.2.
28 Brasse 2014, 27.
29 The internal wall is indeed a secondary event, see Gasparini and Uroz Sàez 2012, 43–48.
30 Heinrich Nissen described an ornamental socle on the fortifications near Tower VII; he was likely referring to the lower travertine courses; Nissen 1877, 465.
31 Maiuri 1929, 160. On the gate closure, see Maiuri 1929, 168; Guzzo 2007, 67; Hori 2010, 288; each sees this event occurring with the construction of Tower XI. See also note 8 supra.
32 Seiler et al. 2005, 232. See Van der Graaff forthcoming.
33 Brands 1988, 8–29; Van der Graaff forthcoming.
34 For a summary of the debates, see Brands 1988, 177–191.
35 Maiuri 1929, 227. The two walls creating the *emplecton* in the court would also host the stairs that led up to the parapet.
36 See chapter 1 of this volume; Van der Graaff forthcoming.
37 Fiorelli 1873, 80.
38 Maiuri 1929, 191.
39 See Van der Graaff forthcoming.
40 Turner 1977, 94–98.
41 Van Gennep 1961, 18–25.
42 Turner 1977, 94–98.
43 Karlsson 1992, 106–107; D'Agostino 2013, 216.
44 See De Caro 1992b, 78; Chiaramonte 2007, 143 for the replacement theory. Brands argues for its unique status; Brands 1988, 183–184.
45 The southern facade differs slightly, displaying a few stray travertine blocks in the tuff courses.
46 Maiuri 1929, 188.
47 Livy *Ab ur. con.* XXIII.15.1–6; XXVIII.3.6–9.
48 Chiaramonte 2007, 142. For a summary of the debate, see Sakai 2000/2001, 90–92; Brands 1988, 174–176.
49 Gasparini and Uroz Sàez 2012, 43–48.
50 Brasse 2014, 27.
51 Maiuri 1929, 159–162.
52 Richardson 1988, 45.
53 Adam 1982, 41.
54 Sakai and Iorio 2005, 328. For the tuff wall, see Seiler et al. 2005, fig. 11. For the *opus incertum* wall, see Maiuri 1929, 184, who demolished it during his excavations.
55 Noack and Lehmann-Hartleben 1936, 167 project a questionable internal wall in Regio VIII.
56 Mazois 1824a, 36 and plate XII; Clark 1831, 68.
57 Mazois 1824a, pl. II and pl. X, fig. II. See Mau 1902, 241; Maiuri 1943, 284.
58 See chapter 1. See Anonymous 1899, 406–407; Mau 1890, 283; Sogliano 1904, 301.
59 Krischen 1941, pl. 5.
60 As also dated by the evidence recovered at the Porta Vesuvio; see Seiler et al. 2005, 232.
61 Maiuri 1929, 218.
62 Maiuri 1929, 211–213. If we follow this line of reasoning, the tuff vault should therefore date to after Second Punic War. Maiuri acknowledges a more complicated periodization of the walls than he first delineates.
63 Pesando and Guidobaldi 2006, 33.
64 Given the uncertainty regarding the construction dates at the Porta Stabia, this circumstance suggests that, like the other gates, it may have received the court walls during this overhaul of the defenses.

65 Gabba 1972; Jouffroy 1986.
66 A series of buried steps that once led up to the parapet of the first Samnite wall imply a similar layout.
67 D'Ambrosio and De Caro 1983, 24.
68 Pesando and Guidobaldi 2006, 34.
69 Seiler et al. 2005, 224; Devore and Ellis 2008, 13–15.
70 D'Ambrosio and De Caro 1983, 24.
71 Brands 1988, 191.
72 Fröhlich 1995, 153–159.
73 De Caro and Giampaola 2008, fig.14.
74 The oblique angle has led to speculation that the previous road aligned with the via Superior currently passing in front of the Villa of the Mysteries; Pesando and Guidobaldi 2006, 31. On the topography, see Maiuri 1929, 132–133; Maiuri 1943, 284–285; Eschebach and Eschebach 1995, 74–76.
75 Maiuri 1943, 284–285.
76 Seiler et al. 2005, 219–224 and 230–233.
77 On the date for concrete, see Mogetta 2016, 71.
78 Gabba 1972, 108–110; Jouffroy 1986, 25.
79 Gabba 1972, 93.
80 Mogetta 2016, 71.
81 As explained further in chapter 4.
82 Maiuri 1943, 283; Hori 2010, 288.
83 Chiaramonte Treré 1986, 25.
84 Hori 2010, 293.
85 Adam 1982, 46–76.
86 Maiuri 1929, 120.
87 Russo and Russo 2005, 71–75; Hori 2010, 286.
88 Lawrence 1979, 115; McK. Camp II 2000, 48.
89 The walls of Athens are a prominent example; Adam 1982, 39.
90 Cassetta and Costantino 2008, 200; Aoyagi and Pappalardo 2006, 19–22.
91 See Van der Graaff forthcoming.
92 Bechi 1851, 42; also Mazois describes a white plaster coating on the vaults; see Mazois 1824a, 36.
93 Mau 1879, 236.
94 Mau 1882, 58; Gell 1832, 90.
95 For example, Overbeck and Mau refer the reader to Mazois for the buried sections of the Porta Nola; Overbeck and Mau 1884, 52.
96 I will address its significance in later chapters; see Mazois 1824a, 53 and pl. XXXVI figs. 1 and 2; Gell 1832, pl. 29.
97 Van Gennep 1961, 18–25.
98 Maiuri 1948, 31.
99 Antonini 2004, 279.
100 Van der Poel 1981, 88; Bonnet speculated that the tower was west of the Porta Stabia, where it would have covered the gate with flanking fire; Bonnet 1980, 329.
101 Mazois 1824a, pl. XII and pl. XIII.
102 Although Mazois claims to show the plan of Tower XII, he clearly rendered Tower VII. He also depicts names carved in the wall section between the tower and Porta Nola; Mazois 1824a, pl. XII, fig. 4. Overbeck and Mau 1884, 42, fig. 11 copy the plate in Mazois.
103 Maiuri 1943, figs. 8 and 9.
104 Mau 1902, 241. Mau uses Mazois to justify the presence of the pediments in the Riot at the Amphitheater fresco.
105 Russo and Russo 2005, 59.

106 Only Reinicke and Van der Poel point out some discrepancies with the published plans. Although mistaken, Reinicke pointed out that Towers II and III are mirror images of X, XI, and XII; Reinicke 1896, 83. Van der Poel points out discrepancies in Mazois but confuses the tower depicted in Reinicke, which is actually VII, with number VIII; Van der Poel 1981, 88–90.

107 Phil. *Polior*. I.33–34.

108 Overbeck and Mau 1884, 44–45 comment on the lack of such posterns in the wall circuit at Pompeii, but Tower IV was buried at the time. See also Mau 1902, 241.

109 Etani 2010, 307–308.

110 The ancient terrain in front of Tower VIII remains unknown; Chiaramonte Treré 1986, 10.

111 Wallat 1993, 353–382; Pesando and Guidobaldi 2006, 72; Adam 2007, 108.

112 Fröhlich 1995, 153–159.

113 Vitr. *De arch.* I.5.4.; Phil. *Polior*. I.20–24. For an English translation, see Lawrence 1979, 67–107.

114 Phil. *Polior*. I.62.

115 Johannowsky 1994, 133.

116 Maiuri 1960, 179.

117 De Caro 1992b, 72 and 77–79. The recent excavation of Tower IX confirms this concern with placement.

118 De Caro 1992b, 78; Chiaramonte 2007, 143 for the replacement theory.

119 Pinder 2011, 74.

4 Establishing an image for Samnite Pompeii

The construction of the Samnite enceinte in the late fourth century BCE heralded the start of the Samnite period of Pompeii that would last until the early first century BCE. The city would flourish under both Roman and Hellenic influences, all the while maintaining a nominal independence from Rome until its fall to Sulla during the Social War.[1] The Samnite period began in a time of conflict that may have stimulated the construction of the new fortifications. The Sarno valley was a theater for hostilities during the Second Samnite War.[2] Roman troops landed near Pompeii to sack the inland territory of Nuceria (Nocera) in 310 BCE.[3] It is unclear whether the new Pompeian enclosure existed at the time, but the war party bypassed the settlement, either because of its unimportance or because it possessed a well defended enclosure. A peasant uprising, launched perhaps from a stronghold such as Pompeii, eventually ambushed the Roman forces as they made their way back to their ships, suggesting that the Pompeian territory hosted a typically Samnite decentralized occupation.[4] This picture would soon change after Pompeii entered an alliance with Rome as a *civitas foederata* at the end of the war.[5] The city would develop outward from its *Altstadt* core and back into the plateau of the *Neustadt*. The construction of and each subsequent intervention on the city walls were transformational events for the image of Pompeii. The architectural muscle of the fortifications – their presence, scale, function, and meaning – would resonate throughout the city to imbue a civic identity.

The first Samnite enceinte was a sophisticated structure complete with decorative embellishments that only a well organized society was capable of financing and building. The chronic lack of knowledge for the social makeup of the city makes answering the question concerning who or what acted as a catalyst for the construction of the new circuit a difficult task. On the eastern side of the Forum, new shops sprang up, attesting to the rise of a new merchant social class, but it is unclear whether they were capable of financing the defenses.[6] The neighboring city of Nuceria (Nocera) underwent a similar urban transformation at the same time, suggesting that this was a wider process in the Sarno valley. At Pompeii, the sheer scale of the operation, however, indicates the involvement of some outside element, such as neighboring Nuceria or Rome, for military expertise and financial backing if we consider the scattered occupation of the hinterland. The alliance with Rome resulted in a wave of wealth and commerce with the eastern Mediterranean and

Italy. This new relationship included extensive social and religious alliances that would influence Pompeii for the remainder of its history.[7]

Calibrating the image of the expanding city

The very act of building fortifications fostered the creation of a communal identity and set the stage for a new city. Given the size and scope of the building, we can only imagine the spectacle associated with the construction of the new enceinte. Hundreds of carts and workers must have toiled to carry earth and stone to build the fortifications. Greek historian Diodorus Siculus gives a glimpse of the wall construction process and its effect on a community in his account concerning the fortifications of Syracuse, which were famous for their length and scale. In about 400 BCE, Dionysius I, the tyrant of the city, expanded the enceinte to include the strategic Epiploae plateau in anticipation of a war with Carthage he was instigating. Dionysius mustered over 60,000 free peasants and 6,000 yokes of oxen from the surrounding countryside to build a section of wall some 30 stades long – about 5.4 kilometers. He put a master builder in charge of each stade and assigned six parties, each 200 men strong, to build a plethron length of wall, or roughly 30 meters, under the supervision of a mason. The remaining workforce quarried and transported the stone. As the story goes, Dionysius incited the men with rewards and led by example, supervising and participating in the construction process. He inspired such zeal and competition in the builders that crews finished the entire fortification, including the towers and six gates, in just twenty days.[8] The historian describes how onlookers watched in wonder at the fervor and labor of so many people on a single project. It is unclear whether such concerted efforts were typical of the time. Diodorus probably recounts this particular episode because of the notable and out-of-the-ordinary achievement. The numbers Diodorus furnishes are unreliable and probably exaggerated since he wrote about the event roughly 350 years after it happened. Nevertheless, the episode tells us something about the labor organization on such massive projects and the division of construction into distinct sections. It is a vivid example of how the building process and its spectacle could foster a sense of community and legitimize the power of the commissioner.

Any sort of similar calculations for Pompeii runs into a few issues: Part of the circuit remains unknown and buried, whereas the Samnite wall has disappeared behind much later housing in the southwestern portion of the city. An attempt to calculate the number of blocks in the Samnite wall has to make a few assumptions including their average size. Although the blocks have not been properly measured, I estimate that on the average they are 45 centimeters high, 100 centimeters long, and 75 centimeters wide (or about 0.34 square meters).[9] The curtain had an average height of 9 meters, to which one can add at least six courses of buried foundation blocks, which brings the elevation of the masonry to about 12 meters. When multiplied with the 3,200-meter length of the enceinte, then the volume of the stone used comes in at about 28,800 square meters. If we then count the internal piers as a single line of vertical blocks spaced on average 3 meters apart reaching 9 meters high, then we must add 1,066 (piers) $\times$ 9 $\times$ 0.34 = 3,261 square meters, which brings our total to 32,061 square meters of stone, or 94,297 blocks. The numbers concerning quarrying rates for tuff and travertine in the Pompeii area are unknown. The closest parallel comes from Rome, where Cifani has calculated that it took a team of three men – one mason and two workers – about 10 hours, or one workday, to quarry about 2 square meters of tuff.[10] When applied to Pompeii, this rate, under ideal conditions, suggests that it would have taken a team of 300 men about 160 days to quarry the stone for the first Samnite enceinte. Certainly, there are many issues with these calculations: They do not include the time needed to dress the stone or the number of stretchers vs. header blocks; they treat the curtain wall as a uniform solid; and the distance the stone had to travel remains elusive. Nevertheless, the estimated numbers for the stone alone give a sense of the work needed.

At Pompeii, aside from envisioning chief masons supervising small groups of workers and individual sections, it is hard to assess who exactly built the wall.[11] Whether the builders of the Pompeian fortifications were its citizens, as was the case at Syracuse, remains a matter of speculation. Nevertheless, accounts of workforces employed on large-scale projects suggest that they offered a measure of social control by providing ample work and distraction. For example, Tarquinius Superbus, the last king of Rome, actively employed poor citizens on public projects in order to keep them distracted from the affairs of state and to prevent uprisings.[12] Such a deliberate exploitation of the workforce probably also occurred for the construction of the Archaic walls built to protect Rome in the sixth century BCE.[13] Setting the local population at work on construction projects was probably a common occurrence in the early Republic. At least a portion of the population in the Sarno valley took part in the construction of the Pompeian fortifications. Given the scale of the construction process, this local effort would have fostered a strong sense of community and identification for the population who depended on the region's resources for survival.

Building materials and the image of Pompeii

The materials used to build a city are a key component of its urban image. Where these materials came from and how builders and patrons employed them throughout the urban landscape is a measure of how Pompeii's citizens envisioned their city. It seems evident that the use of local building materials would tie a community explicitly to its territory. Each material had specific characteristics that made it apt for a particular use or gave it an aesthetic value. The fortifications of Pompeii are unique because, throughout their history, they constituted a kind of laboratory for the application of construction materials. They employed each known type at some point, including earth, travertine, tuff, and *opus incertum*. The techniques once experimented and applied on the enceinte would then find resonance throughout the city.

The earth composing the *agger* is undoubtedly the most abundant material used in the fortifications. Any attempt to calculate its volume runs into almost unsurmountable obstacles since the depth of the earth fill on the cliff edge on the south side of the city is unknown, as is the extent of the *agger* in the buried areas of the circuit. Hundreds if not thousands of carts and laborers must have been necessary to transport the earth during construction in a spectacle of organization and logistics. The origin of the earth remains elusive, and it is frustratingly clean of artifacts, which would allow a more precise dating of the fortifications. The lack of artifacts indicates that it came from an uncontaminated extra-urban setting, perhaps an exterior ditch dug in front of the defenses. Some of it may stem from efforts to level out the plateau, or the reorganization of the countryside carried out to support the expanding city.[14] It is arguably the least glamorous of the materials present in the fortifications. Nevertheless, its transport, excavation, and amassing was part of considerable operation that projected a strong association among the fortifications, the town, and its hinterland.

The travertine used in the fortifications was a relatively cheap and plentiful resource that even today lies in banks along the Sarno River valley. Sarno travertine is soft, allowing workers to saw it into blocks but making the stone difficult to carve beyond basic geometric shapes.[15] Its application seems to be a more practical, rather than an aesthetic choice. Travertine quickly strengthened after contact with air, and its prolonged exposure to rain led to precipitation that sealed the joints between blocks and created exceptionally strong masonry.[16] Unfortunately, its precise quarries remain unknown beyond a general approximation of the area around modern Scafati and the lower Sarno River valley.[17] Travertine is constantly forming in the Pompeii area as lime suspended in the springs of the Apennines naturally precipitates in the Sarno plain. For this reason, quarrying the stone prevented entire areas from transforming into sterile swamps unfit for cultivation. Any construction event that employed this material also reclaimed land to expand farming activities to allow the city to grow.[18] The large-scale use of travertine in buildings such as the fortifications undoubtedly expressed a strong symbolic statement of domination over the landscape.

Tuff is a relatively soft stone, easy to quarry and sculpt. More versatile than travertine, its compact character also allowed masons to dress it to a uniform smoothness, which is ideal for the application of plaster veneers designed to imitate marble. A plaster veneer could protect the most porous types of tuff from water infiltration.[19] The Pompeian types seem exceptionally well suited against weathering. Two hundred years after its re-excavation from the ruins of the eruption and over 2000 since its quarrying, some masonry still presents a smooth facing despite its exposure to the elements. Tuff also lent itself to elaborate carving. Sculptors used it to make life-size statues. These properties ensured that tuff remained in use as a building material from the late third century BCE up to the final days of the city.[20] Tuff would be a prestige construction material surpassed only by the marble introduced during the Imperial period. Patrons stipulated its use in highly visible places, such as colonnades, stylobates, facades, and impluvium linings, as a means to flaunt their wealth. For instance, when the owner of the House of Pansa (VI.6.1) built his peristyle in the second century BCE, he left the tuff columns naked to display the material; only in the Imperial period were they covered with a plaster coating.[21] Many tombs outside the Porta Nocera, built in tuff during the

Roman period, exhibit graffiti on the masonry suggesting that they did not carry a plaster veneer. As commemorative structures, they attest to the recognition of tuff as a viable display material.

Architects first used tuff on a large scale on the city walls of Pompeii, making them a test site for the material. The exact quarries that fed the construction of the fortifications are still unknown, but miners extracted the stone from areas on the Sorrento peninsula, the Monti Lattari, and the neighboring town of Nuceria (Nocera), transporting it to town by ship, river barge, and carts.[22] Such diversity in the quarries explains the brown to gray color range of the tuff employed throughout the city, but the fortifications tend to present a consistent gray/brown type, indicating its origin from a common location. Together with travertine, the abundant use of tuff tied the fortifications with the city to signal its territorial hegemony in the Sarno River valley.

After its introduction, *opus incertum* would become the cheapest and most readily available construction material in Pompeii. It consists of a concrete of lime mixed with pozzolana (volcanic ash) and aggregate that workers faced with fist-sized stones, often basalt or lava quarried directly from the Pompeian plateau.[23] This crude material lacked any type of aesthetic appeal. Builders usually applied a veneer to embellish and protect it from exterior moisture, often in the form of plaster, but also with tuff or marble slabs on more prestigious projects. *Opus incertum* gained popularity in the middle of the second century BCE, when it facilitated construction of private houses and monuments. It had two great advantages over traditional masonry: Workers could put it into forms, and relatively unskilled laborers could work with it as opposed to the training needed to dress quarried stone. *Opus incertum* led to a radical departure from traditional building techniques, encouraging the use of the arches and vaults used in the towers and gates of the city.

Samnite Pompeii defines its image

After the construction of the enceinte, Pompeii's expansion back into the plateau would gain increased momentum between the late fourth and the early first centuries BCE. The redevelopment began at the sanctuaries: Votive offerings resumed at the Temple to Apollo, and the Doric Temple received a refurbishment around 325 BCE.[24] Once again, the fortifications connect intimately with the temple, functioning as a new terrace and reasserting the dual martial and divine protective elements upon the city. The construction of housing was a central element to the developing city. It is undeniably difficult to ascertain how many surviving houses date to the early Samnite period. However, many of the earliest dwellings are similar in design and layout, pointing to the presence of an overall unifying concept to their construction. They include the House of the Scientist (VI.14.43) (see Figure 4.1), the House of Amarantus (I.9.11–12),[25] the House of the Surgeon (VI.1.10), and the House of the Naviglio (VI.10.11),[26] the House of the Centaur (VI.9.5), and Houses VI.9.1 and VI.14.39, all spanning the third century BCE.[27] These spacious dwellings included wide-open Tuscan atria, indicating that they belonged to the higher levels of Pompeian society. Other, more modest houses (I.11), the so-called *case a schiera*, were row houses of comparable design and size.[28] Most houses employed a similar construction technique, using travertine *opus quadratum* for their facades and *opus africanum* (loose mortar laced with large travertine blocks) for the interior. The very regularity of these houses and their construction technique has induced past Pompeianists, such as Fiorelli, Nissen, and Mau, to dub early Samnite Pompeii as the Limestone period. They then placed the construction of the circuit into the Limestone period because of the use of travertine as the primary construction material.[29] The practice of assigning specific parameters to the use of a particular construction material is somewhat suspect since significant overlap can exist among the applied materials. As evidenced previously, the isolated use of travertine in the first enceinte is unlikely. Instead, the fortifications pioneered the use of travertine and tuff in a way that would resonate throughout the city in the following centuries.

Although one must use caution with easy generalizations, the evidence suggests that many of these early houses abundantly used travertine, where it carried much more meaning and symbolic significance for the city. These houses employed *spolia* coming from the previous Orthostat wall that workers had demolished to accommodate the new enceinte.[30] The blocks resulting from the dismantling, too thin and unsuitable for the new fortification, became the facades and foundations of the new city. Clear examples still exist in the House of the Surgeon (VI.1.10)

Figure 4.1 Facade of the House of the Scientist.

and the House of the Naviglio (VI.10.11). The reuse of the blocks was undoubtedly a matter of expediency. Nevertheless, the event also created a layer of cultural memory. A small part of the earlier fortifications would remain visible near Tower IX, where they would have created a vivid reminder of Pompeii's past.[31] Throughout Pompeii, the reused blocks would shape the new image of the city through the demolition of the old. To those who witnessed the construction event, the old blocks were a clear reference to the past. They created an important dual message of renewal and preservation in the newly expanding city.

The surviving facades highlight how the *spolia* functioned to create a uniform image to the city. Most are actually uniform facades only occasionally pierced with small window slits. To a viewer, they appear as rectangular solids in a layout that resembles the shapes of the outer bastions in the city gates.[32] Although extensive First Style ornamental programs must have embellished the interiors of the houses, no indications exist for something similar on their exterior.[33] The wide swaths of plaster covering the masonry in places such as the via di Mercurio are the result of the much later aesthetic choices of the Roman period to disguise earthquake damage repair. The very survival of the travertine facades up to the eruption, even after the interiors saw radical alterations, is an indicator of their importance. Given their extent and the materials employed, these were houses that belonged to the upper classes, and their exteriors carried a representational element related to the status of the owner. The reuse of the blocks may represent a measure of antiquarianism, nostalgia, ancestral continuity, or conquest and renewal of the city. The presence of the travertine along the regular gridded streets of the *Neustadt* offered a further contrast to the irregular layout of the *Altstadt*. It highlighted a dichotomy between past urban retraction and renewed expansion. The fortifications, as the largest and newest public building, further expressed this message.

Fortifications and the urban network

The expansion of building activity at Pompeii included the basic arrangement of its street network, which the main gates in earlier enceintes had already partially anchored. The Pompeian walls foreshadow Vitruvian prescriptions where the general plan of the walls, as well as the placement of other public buildings such as temples and marketplaces should regulate the construction of a town.[34] This factor is more pronounced with the new enceinte where other entry points, in the form of posterns or towers, aligned along minor north–south and east–west streets dividing the plan.[35] At Pompeii, the design originates with the commensurate via Stabiana (the main north–south *cardo*) and the via di Nola (the east–west *decumanus* of the city (see plate 4). The via dell'Abbondanza acts as a second *decumanus*; together with the via di Nola, it divides the via Stabiana into three equal sections. The third section also returns in the length of the via di Mercurio and the distance between the Porta Sarno and Nola, whereas the gap between via di Nocera and the via Stabiana is exactly half the length of the *cardo*. The city subsequently expanded in a tiered development in precisely parceled plots set out along the main axes. As in the archaic period, the enclosed area never achieved full urbanization. Regiones I and II remained thinly occupied with agricultural buildings up to the eruption of Vesuvius.[36] The wide area encompassed by the walls could accommodate the population of the *chora* and established a measure of self-sufficiency with extensive protected farmland within the enclosure. Although no direct evidence exists for a refoundation of the city, the coherence of the grid layout is the product of an initial single plan, indicating a desire to monumentalize the city. Remarkably, future architects and planners would retain and respect this orthogonal plan in the following centuries.

The grid layout has generated much discussion concerning any potential meaning it may carry. Any such significance implicitly refers to the fortifications since their layout ultimately dictated the arrangement of the street network. The discussion divides broadly between two camps: a purely secular vs. a heavily symbolic view that has recently also ventured into archaeoastronomy. The problem revolves around identifying a predominantly Greek, Etruscan, Samnite, or Roman character of Pompeii. Orthogonal planning was primarily a secular endeavor for the Greeks. In Greek colonies, religious, residential, and artisanal areas were often separated according to function. Aristotle explains how roads needed to be oriented according to topographic or hygienic considerations, requiring that a city was open to the winds for a healthy environment – a view that Vitruvius strongly discouraged.[37] By contrast, a more symbolic approach considers a primarily Italic aspect to town foundations. Etruscan orthogonal plans carried far more symbolism with the designation of the celestial *templum* onto the city, which relates primarily to the orientation of temples within a city but which also resonates in urban layouts.[38] Two axes divided the heavenly sphere into four quadrants. Ideally, each quadrant found an earthly correlation in the notion of the circular city that had a center point – a *mundus* – at the intersection of *cardo* and the *decumanus*. These two axes inscribed the quadrants of the *urbs quadrata*, or square city.[39]

The arguments for Pompeii primarily concern the history of theories on the city's development, that is, the grand single foundation design envisioned by early scholarship vs. the *Altstadt/Neustadt* theory and its later adaptations. This circumstance led to an early consensus equating the street pattern   thus the associated *Neustadt* – with the first foundation of the city and any associated rites such as the *sulcus primigenius* ritual.[40] With a possible Etruscan foundation of Pompeii, the crossing of the via di Nola as the *cardo* and the via Stabiana as the *decumanus* would be the *mundus*, and the original starting point for the layout of the city. Pompeii's *pomerium* would then follow the present line of the Samnite walls. A more recent hypothesis suggests that if the *Altstadt* represents the original nucleus of the city, then the Forum would be the *mundus* at the intersection of a *cardo* marked by the via di Mercurio/via del Foro/via delle Scuole, whereas the via Marina/via dell'Abbondanza would be the *decumanus*.[41] The *pomerium* would thus follow the line of the *Altstadt* fortifications, whereas the later *Neustadt* would instead follow a secular Greek design.[42] A further hypothesis traces an axis from the via di Nola east toward the Monte Torrenone, where the Samnite federal sanctuary of Foce a Sarno marks a source of the Sarno River as part of a deliberate orientation mirrored in the contemporary layout of neighboring Nocera.[43] These alignments would then symbolize a wider political union and rebirth after the insertion of the two cities into the Roman sphere of influence.[44] From here, the cities and the division of the territory along the main axes would come under divine protection from the federal sanctuary.[45]

At this point, projecting only Etruscan and Roman foundational rites on Pompeii seems premature in light of the region's rich cultural diversity. However, there does seem to be somewhat of a deliberate orientation following the principles of Roman *agrimensores* (land surveyors) who would use the sun for orientation. *Agrimensores* borrowed heavily from Etruscan ideals but transformed them into secular notions. When dividing the countryside into equal plots, in a process known as centuriation, surveyors used the solstices to orient the land division – presumably also as a timekeeping device for the agricultural cycle.[46] At Pompeii, the orientation of the via dell'Abbondanza and the via di Nola with the summer solstice, as well as the via Stabiana with the winter solstice, suggests that the layout of *Neustadt* follows the principles of the Roman *agrimensores*.[47] This is a far more pragmatic approach that allows for the inclusion of topographical elements such as the natural depression that accommodates the via Stabiana.[48] It also fits better with a militaristic scheme where the walls and street network interacted to offer the greatest defensive capabilities.[49] In this layout, the minor streets facilitated communications toward the individual sections of the enceinte and the later towers.

Stepping back from this array of theories, the important point is that the fortifications fall into a broader discourse concerning the urban layout, the landscape, and its parcellation. The gates are the structures that anchored both urban and rural roads along which surveyors laid out fields in equal plots. Although much of the Pompeian *chora* remains buried, a similar organized landscape must have supported the city. Hints of such a land parcellation have emerged extending beyond the walls on axis with the via di Mercurio to the north and south toward the Sarno River along the via di Nocera.[50] As is the case with the urban layout, the division of the countryside imposed an artificial organization on the landscape. The fortifications acted both as distinguishing and unifying elements for the urban and nonurban territory of Pompeii. As such, they also were a filter point for movements between the city and its territory. Although a clear symbolic foundation rite for Pompeii is elusive, the defenses would cast a religious resonance on the landscape through their visual association with the *Altstadt* sanctuaries. Such an organized landscape is something that communities actively developed throughout Republican Italy. It would become synonymous with an urban center and its identity.[51] The walls were a symbolic nexus highlighting the new organization, simultaneously unifying and distinguishing the city and its countryside. This framework explains the drive to apply the polychrome ornamentation on the Pompeian fortifications. It is the sign of a community that is cultivating its identity and urban ambition. Whatever modifications followed on the fortifications, patrons clearly intended to cultivate this image.

The picture for Pompeii is one where the city lays the groundwork for its future appearance. This massive operation included laying out streets, designating public spaces, regularizing the countryside, and constructing the walls. These elements and the position of walls in particular will remain virtually unchanged as urban relics for the remainder of the city's history, creating a clear factor of continuity for its population.[52] The fortifications would be a constant visual background, whereas the streets formed visual axes focusing on the *agger* and the later towers.

These elements continuously projected a sense of security on the population. The defenses also conveyed a measure of social hierarchy and internal power because they were capable of controlling the population. These notions of security and any associations with the patrons who built the fortifications permeated the rest of the city's history, adding a measure of cultural memory that consciously or unconsciously affected those living inside the protection of its walls. This image would not remain static. It would be altered in subsequent interventions on the city walls that would prove transformational for the image of the city.

The image of the high Samnite city

It is once again an intervention on the fortifications – this time through the addition of an internal wall and the widening of the *agger* – that marks a pivotal event for the history of the city (see plate 1). After the end of the Second Punic War and the neutralization of Carthage, Pompeii tapped into the new Roman trade routes with the east. The city was still nominally independent from Rome. The Samnite elite controlled much of the export and centralized landownership. Slaves increasingly worked the land, producing a massive influx of population from the countryside into the city.[53] This shift eventually caused the stresses that led to the crisis of the Social War and the installation of a Sullan colony in Pompeii. The city would also participate in Rome's conquest of the east. Lucius Mummius, the sacker of Corinth, dedicated a statue of Apollo at the god's temple in gratitude for Pompeian support.[54] The rich spoils of war, as with other centers in Italy, swept Pompeii into the currents of feverish construction activity occurring on the peninsula.[55]

The second century BCE would be the age when Pompeii developed the necessary architecture to call itself a city. A list of the public buildings involved highlights the dramatic shift of its image. In the Forum, the center of public life, patrons would reconstruct the Temple of Apollo as well as build the sanctuary to Jupiter, the Basilica, the Comitium, and a new two-story portico. The square also received a new pavement in *opus caementicium*, a crude cement, framed with tuff slabs. To the south and east, the Large Theatre went up along with the colonnades and tuff pavement of the Triangular Forum, the Samnite Palaestra, and the Stabian Baths.[56] In some measure, each of these projects would employ travertine and, increasingly, tuff and concrete. The result would be the creation of a unified civic image modeled largely on the construction materials used.

The two sanctuaries on the southwestern side of the city would see a vigorous redevelopment. The construction of the colonnades in the Triangular Forum would include travertine foundations and tuff columns in a technique of contrasting materials similar to the city walls, suggesting that the practice found a ubiquitous application. Farther west, a redevelopment of the sanctuary to Mefitis Fisica would occur as a predecessor of the current Temple of Venus.[57] The temple had two phases in the Samnite period: the first coinciding roughly with the first enceinte and the second with the late second century BCE when the towers and vaults of the gates were built. A wall built primarily with tuff acted as a terracing structure for both phases. If the first phase is elusive, the second temple was a grand concept imitating on a small scale the great terraced sanctuaries of Praeneste, Tivoli, and Terracina. A double terrace supported three porticoes surrounding a central temple with an open view toward the sea and the Monti Lattari.[58] The complex towered some 30 meters above the riverine port below, dominating the landscape for miles around and forming a landmark for ships at sea. From a distance, the tuff fortification wall was an enormous podium for the sanctuary, creating, as in the previous phases, a direct visual connection between the walls, a protective goddess of the city, and the temple itself, one of Pompeii's most ambitious architectural projects. Such symbolic visual statements were critical to the Pompeian elite who financed the building. It signaled their ambition to join the architectural developments occurring on the Italian peninsula and legitimized their relationship with the deities and fortifications protecting the city (see Figure 4.2.)

Figure 4.2 Entrance to the House of Pansa.

The tuff employed in an ornamental fashion on the city walls found extensive use in houses of the Pompeian elite. It would resonate with the fortifications to create a distinct architectural image for the city. The rapid development of Pompeii included the large-scale construction of houses that display similar, perhaps unambiguously designated specifications.[59] The wealthiest families built on a lavish scale. Their dwellings, sometimes spanning the area of a full city block, emulated the opulent Hellenistic palaces of the east. Craftsmen applied and painted plaster in the manner of the First Style in interior spaces to imitate the marble used in palaces. On their exterior, tuff blocks feature prominently on the facades of the most opulent houses. In the earliest examples, the tuff ashlars display rather sober carvings with plain rustication. They eventually gave way to elaborate carved pilasters and canonically proportioned entablatures with carved Corinthian and Doric capitals framing the doorways.[60] The increased sophistication in these facades indicates an element of competition among the homeowners vying to flaunt their wealth.[61] The design of the facades dialogues with the First Style, which applied similar heavy cornices and imitative ashlars. Where patrons employed the First Style to imitate high-quality marble masonry on the interior of the buildings, the tuff mimicked this concept through the application of a local prestige material on the exterior.

A number of elements indicate that patrons left the masonry on the facades naked to underscore the material. The expense incurred to build and carve the tuff probably led the owner to want it left uncovered. Today a few of the surviving examples, such as the House of the Faun (VI.12.2), carry a thin plaster veneer, but its presence here is a much later colonial phenomenon, or even a postearthquake, effort, aimed at masking divergent repair masonry.[62] Instead, this facade and others still carry *programmata*, graffiti, and inscriptions, including the famous *eituns* inscriptions. They owe their preservation to a plaster veneer applied later in antiquity that fell off the facades after modern excavation.[63] In other instances, such as the House of Pansa (VI.6.1) and the House of the Little Fountain (VI.8.23), red *programmata* and graffiti painted directly onto the tuff masonry were on view at the time of the eruption.[64] These examples indicate a desire to preserve the facades through successive generations, implying the continuity of a local identity despite the subsequent installation of the Roman colony.[65] Notably, the construction of tuff facades ceased abruptly after the Sullan conquest, and many elite homes would remain the property of the old Samnite elite.[66] Instead, plaster found its most common application as a veneer applied to *opus incertum*, which inherently carried less prestige as a construction material. As representational elements promoting the owner's status,[67] the presence of tuff facades throughout Pompeii would connect to concepts of proper civic and elite representation associated with the First Style.

The extensive remains of tuff facades throughout Pompeii indicate how they must have established a dramatic visual presence in the city.[68] The western tract of the via dell'Abbondanza, running between the via Stabiana and the Forum, originally featured a coherent lining of tuff facades on both public and private buildings (see Figure 4.3). The earthquake demolished much of the original masonry, but enough remains, including the Stabian Baths, to imagine the stateliness of the street. To emphasize the point, a tuff gateway on the east side of the Forum

Figure 4.3 Looking down the via dell'Abbondanza from the Forum.

announced its beginning. The unified architectural appearance functioned as a purposeful ceremonial background for religious processions between the Forum and the Doric Temple.[69] A series of facades lining the via di Nola and the via degli Augustali to the north indicate that the idea of a unified civic image permeated further throughout the city. A concentration of these buildings exists in Regio VI, whereas others scattered throughout the city invariably face the main connecting arteries.[70] The layout of the facades along primary roads reflects a conscious shaping of the architectural image of the city.

It seems reasonable to assume that this vision was also present in the city walls as the largest structure in the urban framework. The use of both travertine and tuff on the outer curtain of the fortifications is a pioneering technique that would be a standard in the city. The facades of elite houses sometimes rest on travertine foundations that occasionally still peek through the raised sidewalks.[71] Perhaps it is here that one finds Vitruvius's prescription noted earlier on the qualities of travertine as a foundation stone. Admittedly, it is difficult to trace the technique to every house since later sidewalks partially bury most exteriors. Nevertheless, the travertine/tuff technique on the fortifications is so marked and exaggerated that it must have carried an intended aesthetic effect. This is true even if the divergent materials denote different construction events as Maiuri envisioned in the past. It essentially marked the ambitions of the community and of the elite to define Pompeii as a proper city.

Although patrons considered tuff a valued material, the presence of plaster and the application of the First Style decoration were similar factors shaping Pompeii's image. The style featured prominently on the towers and gates of the city where it performed an official public role. With the increased use of *opus incertum* and its later derivatives, the First Style would become ubiquitous on public and private buildings in the colony, where it stayed in use up to the eruption.[72] The style was popular in the homes of wealthy Pompeians between 200 and 80 BCE. As a derivative of actual masonry, the style was often formulaic, featuring a lower socle with a surmounting middle zone of drafted imitative orthostat blocks and an upper third zone with mock ashlars and surmounting cornices. Widely popular throughout the Greek world, the First Style spread to Italy in the third century BCE through trade centers such as Delos.[73] The brightly painted stucco panels imitated the prestigious marbles found in the great palaces of the eastern Mediterranean. A plainer white version often found a place on the facades of houses and tombs, as well as Pompeian public buildings including the Basilica and the Temples of Apollo and Jupiter in the Forum, to name a few. The style would thus carry aulic associations reflective of civic institutions, the eastern kings, and, by association, the connotations of power and the state.

The towers were perhaps the most visible buildings to display the style in the (extra) urban landscape. Until the discovery of Towers VIII and X, some doubt existed about the coverage of the plaster on the towers, with some authors only recognizing imitative ashlars on the tower flanks and smooth plaster on the facades.[74] Despite the internal plan variations, all buildings appeared equal from a distance, displaying a coherent image. This was perhaps also a strategic element designed to present attackers with unknown tower layouts. Anne Laidlaw has highlighted how the embellishments on the towers included a formulaic First Style white decorative scheme featuring an elevated socle, imitation ashlar blocks, and a Doric frieze.[75] (see Figure 4.4). It differs slightly from the traditional scheme because the imitative orthostats above the socle are absent. There are also a few modest variations, such as the slightly lower height of the socle on Tower VIII, but these are noticeable only up close. Rather than a deliberate break from the desire to achieve a coherent appearance for the towers, these details indicate a later remodeling or that a different workshop applied the stucco.

The application of the First Style decoration on public buildings such as towers and gates invariably connected to themes of patronage, civic image, and the state. Laidlaw even suggests that it links with *pietas* – the idea of duty toward the city and the state – but this hypothesis remains debatable since the notion seems to be more a code of conduct rather than an implicit ideal.[76] On public buildings, the First Style emulated the marble and stone masonry of monumental civic buildings and temples in ancient Greece. Structures such as the Propylaea of Mnesicles and the Parthenon on the Athenian Acropolis were clear antecedents of the style. The First Style as displayed on public buildings would thus connect with the proper representation of public space, the state, and civic duty, and on temples it carried religious associations.[77] At Pompeii, the widespread use of the style throughout the Samnite period and the colony reproduces similar ideals. During the Augustan

Figure 4.4 Surviving decorations of Tower X. The inset shows a close-up of the Doric frieze.

period, it would come to reflect a symbolic return to the values and social *mores* of the Republic after the chaos of the first century BCE.[78] On the towers, the First Style undoubtedly reflected their part in protecting Pompeii.[79] It would connect to the aulic connotations of the style and the role of the fortifications in the projection of the proper image associated with a Hellenistic city. The style would thus reflect the euergetism of those who commissioned the towers as leaders and protectors of the community, thereby visibly legitimizing the leaders of the community. The application of the First Style highlighted the message of dominance and order that the fortifications imparted over the city to foreigners and Pompeians alike. The towers and the defenses – observable for miles in the countryside and at the terminus of almost every Pompeian street – formed a constant architectural background that acted to convey notions of dominance and order upon the city.

The specific use of the Doric order on the towers seems to have carried further meaning. Traditionally associated with mainland Greece, its use was a distinguishing element that had once carried implicit ethnic and political associations. By the late Roman Republic, such strict political associations were largely obsolete as a distinguishing element in architecture.[80] Nevertheless, architects continued to recognize that the squat proportions of the Doric order imparted notions of strength and impenetrability upon viewers. Vitruvius notes that the strong proportions of the Doric order reflect the martial strength of Mars, Minerva, and Hercules, specifically recommending its use on temples dedicated to Minerva.[81] Her bust still decorates the keystone of the vault at the Porta Nola. The Doric order on the towers and, by extension, the fortifications of Pompeii may represent an oblique allusion to Minerva. Such a religious factor implies a complex interaction that expressed the relationship between the individual commissioners, the state, and the gods, each contributing to the safety and well-being of the community.

Patronage and display at the walls

Fortifications were the object of elite patronage just like any other civic monument. Many of the *municipia* and *coloniae* in Republican Italy built their fortifications through the euergetism of wealthy patrons. Every new construction event on the defenses signaled projections of strength and independence (see Figure 4.5). Surviving inscriptions attest how, in the early decades of the first century BCE, local magistrates of some twenty cities on the Italian peninsula built or refurbished urban fortifications with the number rising to thirty-five at its close.[82] The funding often translated to the construction of individual sections, gates, or towers. Single groups of patrons financing an entire circuit, such as Gaius Quinctius Valgus, Marcus Magius Syrus, and Aulus Patlacius at Aeclanum, are rare, indicating that only wealthy centers or outside cities such as Rome could finance complete enceintes, as would be the case with the foundation of a colony.[83] By the late Republic, magistrates dedicating city walls in Italy and the Roman west often did so as a symbolic act reflecting the (re)foundation of the city or its achievement of municipal or colonial status under Rome. The intervention on the Pompeian walls in the early colony is an example of this process. However, it is often unclear what such interventions or dedications entailed. The postsiege reconstruction effort at Pompeii must have involved far less resources than building a completely new circuit. Eventually, wall circuits would fall under imperial patronage in the Augustan program of civic renewal in Italy and the Roman west as a reflection of the new political order. It would lead to the great city gates of the Antonine age and thereafter endowed at Attaleia, Timgad, and Bizya.[84] The gates

Figure 4.5 Surviving wall section in Telesia seen from the city side. Note the concave
    section of walls and the towers protruding toward the exterior.

unequivocally symbolized the protection and benevolence of the emperor and the
state toward the community and its reciprocal allegiance to Rome.

The euergetism on single towers or gates that was common practice in the
late second and first centuries BCE potentially complicates the traditional view
of a single upgrade of the city walls at Pompeii into a multitude of events. The
towns of Telesia and Grumentum, both extensively rebuilt after the Social War,
show how successive magistrates built individual towers or wall sections. An
inscription from Telesia indicates how the magistrates Lucius Mummius and
Gaius Manlius were fulfilling the *summa honoraria* – the money that they paid
to the community as part of their office – by building two towers rather holding
public games.[85] The circuit of Telesia, with its regularly spaced round towers, is
remarkably uniform despite these different interventions. Each new construction
event followed a single preexisting plan, which successive patrons did not alter.[86]
The successive magistrates from each of these towns were building a monument
to the idea of the city.

The archaeology at Pompeii reveals little about whether the gates were built separately or as part of a single construction event that included the towers. Only the
Porta Stabia has supplied a secure late-second-century BCE date for the construction
of its vault, making it, according to the traditional chronology, roughly contemporary with the towers.[87] The design of those surviving at Porta Nocera, Sarno,
and Nola suggests that they are concomitant. An inscription at the Porta Nola
implies that some sort of individual euergetism went into the construction of the

vaults. Here the patron prominently associated himself with its reconstruction by means of a text written in Oscan to address the local audience:

v. půpidiis v.
med tův
aamanaffed
isidu
pruphatted[88]

Aggressively translated into English, it reads:

Vibius Popidius (the son of) Vibius
meddix tuticus (magistrate)
commissioned
and
dedicated (the gate).[89]

Vibius was a rich wine trader belonging to the powerful Popidii clan. He actively built public structures around the city. A stray inscription indicates that he financed the tuff colonnade on the south side of the Forum.[90] Clearly, Vibius valued both projects as worthy enough of his euergetism. Each equally contributed to the image of the city and his standing in the community. The single addition of the vault indicates that it was part of overall embellishments and obeyed a grander monumental design repeated at each gate.

In terms of a viewer's passage into and out of the city, each new vault completed the tripartite passage through the *agger* in a layout that stretched back to the very first construction of the Pompeian fortifications. The Porta Nola, the best surviving example of the type, highlights how the juxtaposition of travertine bastions, tuff corridor, and a plastered vault accentuated the progression through the *agger*. Each material demarcates, in a monumental form, a viewer's exclusion, transition, and inclusion from the countryside into the city and vice versa. Architects and patrons had already marked out this effect in the first phase of the Porta Stabia. They now accentuated it further in a monumental fashion that was part of a coherent design. Its adaptation over 200 years later indicates that the local population recognized the role of the fortifications as a monumental as well as a military structure.

Although the arch was a well-established architectural form at the time, monumental vaults were a new addition to Pompeii that must have added novelty and prestige to the gates in a city built primarily using post-and-lintel architecture. Use of the arch in military architecture was nothing new, stretching back to the Canaanite gate of Ashkelon dating to 1850 BCE. In the Hellenistic east, the late-fourth-century BCE main (east) gate of Priene and the posterns at Cnidus are contemporary with the first Samnite enceinte.[91] In Italy, the arch had come to merge with gates in a monumental form at the Etruscan cities of Volterra and Perugia in the third and second centuries BCE.[92] By the late second century BCE, the introduction of concrete allowed architects to exploit arches on a massive scale. Vaults would become a quintessentially Roman monumental form.[93] At Pompeii, vaults would find one of their earliest applications in the fortifications. Even if Maiuri's hypothesis that the *opus incertum* vaults replaced earlier ones built in tuff is true – a claim there is no evidence for – the gates would still be the place where architects first applied the form. The arches at the gates allowed patrons to display themselves as the protectors and benefactors of the community and, perhaps more importantly, put the city on par with the latest architectural developments.[94] The monumental emphasis at the gates as well as in the barrel vaults in the towers put the fortifications at the forefront of architectural innovation at Pompeii. They would find clear resonance in some of its largest monuments, including first the amphitheater and then the Large Theater in its later Augustan reconstruction.

Only two more inscriptions come from the Pompeian gates. Both appear at the Porta Stabia. One, written entirely in Oscan, sits along the western passageway wall facing those leaving the city. It reads:

.siuttiis m, n půntiis m
aidilis ekak viam
terem[na| a] tens. ant půnttram staf(i)anam viu te(r)emnatust per.
x· iussu via půmpaiiana ter
emnattens perek III ant ka.
la iůveis meelikiieis.[95]

An intense debate has surrounded the exact translation. August Mau translates it loosely as:

The aediles M. Sittius and N. Pontius improved the street heading out of the (Stabian) Gate as far as the Stabian Bridge and the Via Pompeiana as far as the temple to Jupiter Milichius; these streets as well as the Via Jovia and . . . placed [them] in perfect repair.

It announces how the aediles M. Sittius and N. Pontius repaired the road from the gate up to the *pons stabianus* crossing the Sarno River and the *via Pumpaiiana* up to the Temple of Jupiter Meilichios as well as the Via Jovia.[96]

The other inscription, located on the exterior of the eastern bastion, is in Latin and dates to the Augustan period (see Figure 4.6). It reads:

L · AVIANIVS · L · F · MEN ·
FLACCVS · PONTIATVS ·
Q · SPEDIVS · Q · F · MEN ·
FIRMVS · II · VIR · I · D · VIAM ·
A · MILLIARIO · AD · CISIARIOS ·
QVA · TERRITORIVM · EST ·
POMPEIANORVM · SVA ·
PEC· MVNIERVNT ·[97]

It states that the duumvirs L. Avianius Flaccus and Q. Spedius Firmus repaved the road from the gate to the station of the *cisarii*, the drivers of the *cissium*, a light two-wheeled cart, at the limits of Pompeian territory, at their own expense.[98] The fact that these are the only other inscriptions known from the Pompeian gates indicates that the Porta Stabia had a particular prominence. It was a place of high visibility, offering access to the port for the inhabitants and to the new theater and its adjacent buildings for crowds from the surrounding countryside. For the Oscan inscription, the road renovations seem to associate rather closely with the construction of the vault on the city side of the gate. The road improvements indicate how the two aediles conceived of the gate as a monument upon which to advertise their achievement. In both cases, the inscriptions relate to road works. Their presence at the gate stressed its role as the conceptual beginning and end of the urban matrix.

Instead of being outliers in the urban space, the fortifications were at the center of the social and political developments of the city. In addition to official inscriptions, the gates would function as dynamic billboards upon which the population would paint notices and graffiti. Many have now completely faded away, but the Porta Ercolano, Nola, and Marina once preserved election notices and announcements of gladiatorial games painted onto the plaster and on occasion the tuff masonry of the passageway walls. At the Porta Marina, an inscription mentions the harlot Attica and her price.[99] The record preserves those painted at the time of the eruption, but we can assume that their presence stretches back considerably

Figure 4.6 The inscription of Avianus Flaccus outside the Porta Stabia.

before that. Over time, authorities covered the notices with thin layers of paint to keep the gates presentable. Rather than static structures, the gates were highly dynamic visual buildings in the social context of the city. Their appearance would subtly change with every new announcement of gladiatorial games and elections.

The evidence for inscriptions and individual euergetism becomes scarcer for the towers. The three types of layout present in the buildings suggest a more complex development of phased or individual construction events associated with individual acts of benevolence. A few fragments of letters survive, carved into the plaster and painted black near Tower VIII.[100] They were part of a monumental Oscan inscription that probably faced viewers approaching the exterior field side of the tower:

T V D VI T·

The surviving fragments have occasioned a number of interpretations, including that they belonged to a dedicatory inscription where they spelled out the Oscan word *svddit* (tower) as mentioned in the *eituns* or that they simply marked the tower with its numeral VIII.[101] Whatever the correct interpretation, the fragments indicate how the towers carried markings to follow a civic military arrangement or displayed dedicatory inscriptions related to the patron(s) who financed them. In either case, such a symbolic inscription referenced the city and its institutions. Taken together, the inscriptions on the towers and gates found a place here because they were places of high visibility. The position of the inscription on the exterior of the towers highlights the notion that viewers should conceptualize the walls from the exterior side of the city.

The fortifications thus announced to onlookers the civic and social organization that they enclosed. The city walls provided the conceptual setup for the *postmurum*, or the city beyond the wall. They were the stage that declared the independence and political identity of the city.[102] The gates and walls implemented a structured entrance and exit to the hierarchy of public monuments – temples, theaters, baths, and the like, as well as to private dwellings.[103] At Pompeii, the clear association between the walls and the city resonated through construction materials and ornamental schemes. The fortifications explicitly connected with the ambitions of the local elite through the appearance of the defenses, the facades of their houses, and the public monuments. The patrons who financed the defenses further emphasized this message as part of their own legitimization of power. The enceinte not only announced and conceptually visualized the city, it also reflected the independence, social hierarchy, wealth, and status of a fully developed urban matrix.

The fortifications and gates held such a prominent role in the social and urban layout of the city because they acted as filter points for everything entering and exiting the city. On a daily basis, gates functioned as tax barriers and collection points, where gatekeepers and tax collectors stopped people and goods passing through. At night, authorities could shut the gates to prevent the uncontrolled passage of people and goods into the city.[104] As a result, the figure of the gatekeeper, as the holder of the gate keys, was particularly important. Aeneias the Tactician warns specifically to keep a close eye on gatekeepers since their bribery or differing political allegiance had led cities to fall.[105] The fact that the Pompeian gates had closing mechanisms indicates that some sort of gatekeeper force must have existed for the city, but their organization and the system that was in place to police the gates are elusive. The presence of lodges at the gates to shelter the keepers is uncertain. One such shelter at first identified for the Porta Ercolano turned out to be the Tomb of Marcus Cerrinius Restitutus, whereas a small structure indicated on Mau's plan for the Porta Stabia has disappeared.[106] Nevertheless, given the closing mechanisms, the Pompeian gates must have acted as policing barriers. The ornamentation of city gates and the neighboring curtain wall would therefore have functioned as a stage symbolizing the city and its authority. The architectural correlations between the city walls, the gates, and the urban architecture would come together into a unified urban image.

As Pompeii developed, the design of its gates had to take into account the increased demographic and economic pressure. The narrow passageways present in the earliest gates became points of congestion. As commerce and internal stability increased throughout the empire, the design of the gates also changed to include separate pedestrian and cart entrances designed to ease traffic.[107] This layout was a practical one related to the gates' role as tax barriers. Carts could now be stopped, whereas pedestrians could move more easily through the gates. The gates at Pompeii would respond to trends. The design of the Porta Marina is essentially a large barrel vault separated into a small entrance for pedestrians and a large arch for carts. The Porta Stabia would receive a separated sidewalk, whereas engineers would lower the roadway passing through the Porta Nola, Sarno, and Nocera to ease access for carts. The trend and its incorporation with elements of civic representation and decor would culminate in the Porta Ercolano, where we find two pedestrian entrances framing a central roadway in a monumental design reminiscent of a triumphal arch.

As is the case of goods, animals, and people, city walls also functioned in relation to a proper water supply and the drainage of sewage and rainwater from the streets. Their management was equally important to civic decorum. As access points through the circuit, city gates were a natural venue for such infrastructure.[108] From a military perspective, drains and aqueducts were weak holes in any circuit. Their presence near the gates made them more defensible.[109] The fortifications were also on public land that did not require costly expropriations for infrastructure projects. With the exception of the Porta Ercolano, the Pompeian gates all include some sort of water feature. The most visible is the *castellum* at the Porta Vesuvio (see plate 12). At the highest point in the plateau, this is the natural distribution point of the Augustan aqueduct into Pompeii. The drains are less conspicuous and a practical necessity. Even today, the characteristic sudden downpours of the region can turn the streets into small torrents draining water from much of the city. Each gate on the downslope side of Pompeii was a natural drain because they tend to be on the topographical depressions that formed the easiest access routes into the city. Rather than having the water pass through the gates, authorities eventually built formal drains through the *agger*. At the Porta Stabia, Nocera, and Sarno, engineers would modify basalt street pavers to direct runoff waste into the sewers passing next to the gate to keep the passageways free of wastewater.[110] Consequently, the drains and aqueduct strengthened the idea of control on all things entering and exiting the city. This concept was enshrined in the *sulcus primigenius* ritual, in which gates were the designated interruptions in the protective boundary where all impure earthly things could pass through.[111]

Assigning a date to the construction of the drains and the associated preoccupation with a proper urban decor is a difficult task. Only Maiuri has attempted to date them, suggesting that their construction occurred after the establishment of the Roman colony because the drains at the Porta Stabia and Nola hindered access to the wall-walk and were therefore built in a period of peace when the walls fell into disuse.[112] Part of the problem with Maiuri's assessment is that the pattern of water drainage likely changed over time. For instance, the foundations of Tower IX built in the late second century BCE cut through an earlier culvert that likely drained water from inside the city.[113] Although the current drain at the Porta Stabia is likely Augustan, there is no reason to discount that it replaced a predecessor.[114]

Most gates indicate that drainage concerns were already current in the late second century BCE, if not earlier. The Porta Nocera preserves a drain on its western flank. Due to its relative height, it clearly predates the lowering of the road passing through it – an event that occurred in the Augustan period. The drain remained in use as engineers altered the flagstones in the road to channel the water into the sidewalk farther uphill.[115] The Porta Sarno also preserves an inlet that drained water from the via dell'Abbondanza incorporated in the gate court wall, pointing to a date coinciding with the construction of the vaults or earlier. On the western side of Pompeii, a drain next to the postern beneath the House of Fabius Rufus dates to the addition of the interior *agger* wall in the late third century BCE.[116] Another outlet on the south side of the Porta Marina is part of the outer curtain, indicating a similar if not earlier date. The drain likely functioned as an overflow for a system of cisterns set slightly farther uphill, but damage wrought on the area by Allied bombing in World War II makes this claim unverifiable.[117] The channel ceased to function with the expansion of the Temple of Venus, when it disappeared behind the masonry of the Villa Imperiale in the course of the first century BCE.

The drain passing through the *agger* at the Porta Nola highlights how the formalization of water channels through the gates achieved an almost monumental status (see plate 18). The inlet is an inconspicuous opening to the left of the steps leading up to the wall-walk. It collected rainwater from most of the via di Nola, which, given its length, must have yielded a considerable amount. The channel through the *agger* is composed of concrete, implying that it is roughly contemporaneous with the gate's vault dating to the second century BCE. The outlet on the opposite end is rather grand, sitting on a high base of travertine masonry extending out from the curtain wall. The resulting petite bastion guided rainwater in a small cascade onto the road below, emphasizing the drainage of the via di Nola every time substantial rain hit the city. Water has cut deep into the masonry, indicating that it functioned for a considerable period. The masonry of the drain extension clearly abuts the curtain wall, indicating its later construction. A striking feature is that engineers built the drain extension in travertine masonry in an effort to match the lower courses of the older adjacent wall curtain. Their aim was to maintain the travertine/tuff ornamentation present throughout the fortifications.

Only a single example exists of a drain passing through the curtain wall other than at the gates. It opened beneath the ornamental engaged colonnade that served to mark the edge of the fortifications and the natural cliff edge along the vicolo dei Soprastanti. The drain was slightly less in the public eye than the gates. Yet engineers formalized the water falling from the city into a cascade draining from the streets into a natural channel passing through the garden of the House of Maius Castricius (VII.16.17).[118] In the course of the first century BCE, workers contained the channel into a series of shallow open-air clearing tanks designed to store and filter the water for reuse.[119] The fortification wall above the formalized channel preserves the triangular peephole in the wall that offered a good view of the water works below. From here, viewers could observe the controlled outflow of water from the city.

Naming the fortifications

The rich archaeological evidence at Pompeii preserves a rare instance for intergroup identification with a city's defenses. The so-called *eituns* inscriptions provide a glimpse into how the population interacted with the enceinte and how city gates would come to function as group identifiers within the community. The inscriptions have attracted a variety of interpretations.[120] The most commonly accepted view sees them as directions for Samnite troops sent to aid Pompeii during the Sullan siege. They get their name from the recurrent word *eituns*, which loosely translates, to *milites* in latin (foot soldiers in Oscan).[121] The inscriptions contain some crucial bits of information, mentioning tower numbers and gate names. Two inscriptions found on the facades of houses VI.2.1 and VI.6.3 direct soldiers toward Tower XII and the *veru sarinu* (salt gate) as a sector commanded by Maras Adirius.[122] A third inscription, located on the House of the Faun (VI.12.1), points the way to Towers X and XI as a sector under the command of Titus Fisanius.[123] A fourth on house VII.6.24 vaguely directs toward the sector commanded by Vibius Seximbrius between the Houses of Maras Castricus and Maras Spurius.[124] A fifth on VIII.6.19 points the way toward the Temple of Minerva in the Triangular Forum.[125] A sixth inscription located on the via dell'Abbondanza III.4.2 mentions a tower and/or road named *mefira*, along with a gate named *urublanu* as part of a sector commanded by Lucius Popidius, son of Lucius and Maras Purillius, son of Maras.[126]

The division in sectors under single commanders, as well as the numbering of the towers as I–XII, carried military as well as social significance. In military terms, they indicate a strict organization of the fortifications before and during the siege. Much more significance lies in the arrangement of the towers. The counterclockwise direction of the numbering reflected a symbolic connection with the direction of the rituals meant to purify the city. It also indicates how engineers envisioned the towers as one single project meant to monumentalize and defend the city. For this, they must have had a rational conceptual understanding of the

city and its urban organization. Tower I was a defining landmark in the conceptualization of Pompeii. Although it remains unlocated, the building probably stood
close to the Temple of Minerva at the tip of the Triangular Forum, where it must
have acted as one of the most visually defining buildings in the urban and rural
landscape. That the tower count started here is a reflection of its status in the conceptualization of the city.[127] The Roman historian Appian supplies some insight
into how towers could visually conceptualize a city and military conquest. In his
account of the triumph celebrating Scipio Africanus at the end of the Second Punic
War, a defining moment in Roman history and identity, Appian describes how
single mock towers, rather than complicated renditions, each represented subju-
gated cities.[128] At Pompeii, the towers must have carried similar associations with
the concept of the city. Each building carried its number inscribed on its exterior,
thereby completing the symbolic barrier protecting the city and projecting their
status as structures defining the city.[129]

As a measure of how towers functioned in the public perception, at least one of
them also had a nickname in addition to its official numeration. The *eituns* inscriptions mention *mefira*, loosely translated as "midway," as a nickname associated
with a tower. The modern view connected the nickname to Tower VIII because
of its location halfway between the so-called Porta Capua and the Porta Nola.
After the recovery of Tower IX disproved the existence of the Porta Capua, a new
hypothesis assigns the *mefira* name to Tower VII and the street leading up to it
because of its placement midway between the Porta Sarno and Nola.[130] Regardless
of which building it refers to, the nickname highlights how the towers acted as
familiar landmarks in the urban landscape. Tall landmarks such as towers are a
focus for navigation and orientation through any city. This aspect was amplified in
antiquity when city streets often did not have signs, much less formal names.[131] The
official and the casual nomenclature associated with the towers is a product of their
function. The administrative numeration refers to the civic and military definition
of the building, whereas the popular names reflect their role as landmarks in the
urban matrix.

Further evidence for intergroup identification with the fortifications comes from
the Pompeian gates. The *eituns* inscriptions name two of them: the *veru sarinu* and
the *veru urublanu*. The *veru sarinu* (salt gate), identified as the Porta Ercolano,
received its name because it opened onto nearby salt (*sarinu* in Oscan) flats.[132]
The *veru urublanu*, identified variously as the Porta Nola, Porta Sarno, and Nocera, gave or received its name from a rural settlement it led to or its associated
quarter in the city.[133] Although tracing the dynamics of how and why these gates
received such names remains difficult, an eloquent explanation lies in the political and social organization of Pompeii. Electoral inscriptions hint that Pompeii
was divided into five tribes responding to an independent electoral college. Four
names and the locations of these tribes survive: the *urbulanenses* in Regiones III
and IX, the *campanienses* in Regiones IV and V, the *salinienses* in Regio VI, and
the *forenses* in the area around the Forum.[134] This setup neatly divides the city
into five regions, which may relate to Pompeii's Oscan name *pentapolis*, (city of
five). The names closely relate with the neighboring gates: the Porta Nola as the

veru urublanu (*porta urbulana* in Latin), Porta Vesuvio as *porta campana* (in Latin), Porta Ercolano as *veru sarinu* (*salienses* in Latin), and Porta Marina as the *porta forensis*.[135] The Porta Stabia remains unnamed because the fifth tribe that occupied Regiones I and II is still unknown. Remarkably, most of these Oscan names would carry through into the Latin, which was officially adopted during the colonial period, suggesting a strong element of cultural memory associated with the fortifications.

The gate names indicate a strong intergroup identification with an architectural landmark in urban districts. It seems plausible to assume that the city gates may have had individual identifiers, such as emblems or subtle embellishments high-lighting these relationships. The present evidence does not immediately point to any such explicit connections. The abundance of travertine masonry at the Porta Stabia, as opposed to the tuff present at other gates, and the slightly oblique layout of the Samnite Porta Ercolano may be minor exceptions. Rather, the gates show a remarkable diachronic unity in design, decoration, and layout, as do the towers. This unity points to a single architectural vision underlying the presentation of the fortifications from their first construction through their subsequent upgrades. In this role, the fortifications acted as a unifying architectural force encompassing the social plurality of Pompeii.

The extensive building program of the second century BCE attests to a desire to establish Pompeii as a city. Its defenses played no small part in constituting its image. This development occurred with a local Samnite elite exposed to Roman and Hellenistic influences. Throughout the period, the fortifications were a driv-ing force for the rapid urbanization of Pompeii. They connected with the city by using similar materials and subtle decorative additions that further highlighted the architectural landscape they enclosed. The defenses acted as a crown, creating a constant visual reminder to those inside and outside their boundary of the city within. As the city grew around them, the walls also interacted socially, acquiring nicknames and acting as landmarks for navigation through the urban landscape. Politically, they emphasized the independence of the city and its elite, and allowed a measure of control on the local population. Their design and continuous upgrades responded to the latest military developments, projecting the power and connec-tions of Pompeii with the rest of Italy and the Mediterranean. Each intervention also marked the ambitions of successive generations of Pompeians. Like any other major monument, fortifications carried social, political, and religious meaning. These elements become clearer in the Roman period when new kinds of evidence further highlight the deep connections between fortifications and the city.

Notes

1 Wallace-Hadrill 2008, 73–212 discusses the subjects of Hellenization and Romanization in terms of architecture.
2 See Sakai 2000/2001 for a discussion of the various dates. De Caro assigns it to the late fourth century BCE; De Caro 1985, 106.
3 Livy *ab ur. cond.* IX.38.2–4.
4 De Caro 1991, 26.

5 Guzzo 2007, 62; Scatozza Höricht 2005, 666; Pesando 2008b, 221–246.

6 Guzzo 2007, 62.

7 Pesando 2010b, 241–242.

8 Diod. Sic. *Bib. hist.* 14.18.

9 I take these measurements from Gasparini and Uroz Sàez 2012, 18; Krischen estimates an overall height of about 9 meters; Krischen 1941,11.

10 Cifani 2010, 41.

11 For the division of labor, see Lugli 1968, 68–70; Winter 1971, 61.

12 See Dion. Hal. *Rom. Ant.* 4.44.

13 Cifani 2016, 82–91.

14 Maiuri suggests an unspecified distant source; Maiuri 1929, 131. Pesando and Guidobaldi identify the area of Regio VI as the source; Pesando and Guidobaldi 2006, 167.

15 Miller 1995, 74–84.

16 Lorenzoni et al. 2001, 38.

17 Kastenmeier et al. 2010, 50–56.

18 Richardson 1988, 369.

19 Jackson and Marra 2006, 429.

20 Adam 2007, 100.

21 Pesando and Guidobaldi 2006, 175.

22 Richardson 1988, 371; De Caro 1991, 27; Kastenmeier et al. 2010, 47–56.

23 Kastenmeier et al. 2010, 43–48.

24 Pesando and Guidobaldi 2006, 37; Carafa 2011, 95.

25 These first two houses have been dated to the late fourth century BCE, although Fulford and Wallace-Hadrill question the accuracy of this chronology; see Fulford and Wallace-Hadrill 1999, 37–144; Wallace-Hadrill 2005, 101–108; countered by Pesando 2013, 117–125.

26 Peterse and De Waele 2005, 197–220.

27 Fulford and Wallace-Hadrill 1999, 114; Pesando and Guidobaldi 2006, 119; Peterse 2007, 377. For the House of the Surgeon, see Jones and Robinson 2007, 391; for the House of the Naviglio, see Cassetta and Costantino 2006, 250.

28 Guzzo 2007, 91; Coarelli and Pesando 2011, 51; for the so-called Hoffmann houses, see Hoffmann 1979, 111–115.

29 Fiorelli 1873, 78–86; Nissen 1877, 669–680; Mau 1902, 37–44.

30 Maiuri 1929, 225–227; Maiuri 1973, 8–12; Eschebach and Eschebach 1995, 158; Cassetta and Costantino 2006, 250; Pesando 2010a, 225.

31 See chapter 1 of this volume.

32 Peterse and De Waele 2005, 205. They also identify houses VI.14.39 and VI.14.40 as having almost identical facades.

33 See Coarelli et al. 2006 for the individual entries and archaeological results.

34 Vitru. *De arch.* 1.3.1.

35 De Caro 1992b, 79.

36 Jashemski 1979, 202–281; De Caro 1992b, 77; Geertman 2001, 131–135; Geertman 2007, 82–97; Guzzo 2007, 72; Nappo 2007, 347–372. For Regio VI, see Jones and Robinson 2005a, 272. For Regio I, see Ellis and Devore 2010, 1.

37 Aris. *Pol.* 7,1330A–1331B.; Vitr. *De arch.* I.4.1–7; Castagnoli 1971, 61.

38 Castagnoli 1971, 61.

39 Malnati and Sassatelli 2008, 429–470; Briquel 2008.

40 Nissen 1877, 466–478; Della Corte 1913, 261–308; Carrington 1932, 5–23.

41 Guzzo 2007, 52.

42 Castagnoli 1971, 62.

43 De Caro 1992b, 82; Guzzo 2007, 67. Pesando and Guidobaldi 2006, 20 suggest a purposeful alignment of the via Stabiana with Vesuvius.

44 De Caro 1992b, 83. Recently countered by Felice Senatore, who points out that the axes actually miss the source of the river and that the sanctuary did not exist at the time; see Senatore 2001, 240–245.

45 Guzzo 2007, 67. The continuation of the via di Nola indicates a similar land parcellation.
46 Le Gall 1975, 287–320; Adam 1999, 11.
47 Eschebach and Eschebach 1995, 56–58.
48 Richardson 1988, 41; Holappa and Viitanen 2011, 169–190; Giglio 2016, 34–48.
49 De Caro 1992b, 79.
50 Zevi 1982, 357; Nappo 1997, 94–96.
51 Zanker 2000, 29 describes the process as starting halfway through the fourth century BCE.
52 Only the Porta Marina, perhaps, lay slightly farther back at the tip of the hill. A few recovered blocks on the eastern end of the precinct of the Temple of Venus may represent an earlier version; Arthurs 1986, 29–44.
53 Pesando and Guidobaldi 2006, 7; Guzzo 2007, 87. This may have included refugees coming from areas devastated by Hannibal; see Lepore 1989, 163; Nappo 1997, 120.
54 The donation implies Pompeian support perhaps in the form of troops or supplies or both; see Pesando and Guidobaldi 2006, 8; Guzzo 2007, 96.
55 Wallace-Hadrill 2008, 131–132.
56 De Caro 1991, 23–46; Zanker 1998, 32–60; Pesando 2006b, 227–241; Wallace-Hadrill 2008, 132. Carafa recently dates the colonnade around the Doric Temple to the post-earthquake period; see Carafa 2011, 99.
57 See Coarelli 1998, 187.
58 Curti 2008, 52–54.
59 Hoffmann 1979, 91; Ling 2005, 36–37.
60 Adam 2007, 100.
61 Hartnett 2008, 91–119.
62 Hoffmann uses the thin layer of plaster on the facade on the House of the Faun as evidence that all tuff facades carried a similar veneer; see Hoffmann 1990, 491. On colonial plaster, see Maiuri 1958, 210.
63 Pocetti ascribes their preservation to a plaster veneer that fell off after excavation. See Pocetti 1988, 321. Similarly, Mau mentions the discovery of one on the north side of VIII.5.19–20 after the plaster peeled off; see Mau 1902, 240. See also Cooley and Cooley 2004, 19. Sgobbo states that excavators recovered the first examples on the naked tuff facades; see Sgobbo 1942, 32.
64 For the House of Pansa: the right external pier preserves an *eituns* inscription, whereas graffiti and an electoral dipinto of P. Cipio next to doorway 19 date to the years before the eruption; see Fiorelli 1875, 104; Pesando and Guidobaldi 2006, 175; Pagano and Prisciandaro 2006, 2, 111. For the House of the Little Fountain, see Varone and Stefani 2009, 320.
65 This circumstance may even reflect an element of facadism for the structures involved.
66 Eschebach and Eschebach 1995, 70–71.
67 Zanker 1998, 32–60.
68 The best preserved examples are the House of the Faun (VI.12.2), House of Pansa, House of Sallust (VI.2.4), VI.1.16–18, House of the Large Fountain (VI.8.22), House of the Ancient Hunt (VII.4.48), VII.4.32, most of the north side of VII.12 (via degli Augustali), and the southern facade of V.1 including the House of the Young Bull (V.1.7.) and the Domus Cornelia (VIII.4.23). The shops lining the west side of the via dei Teatri (VIII.5.31–35) dialogued with the *propylon* of the Triangular Forum. Almost the entire western stretch of the via dell'Abbondanza is uniform, with the exception of VIII.4 and VII.14, which show signs of postearthquake repairs. Illustrative examples are the House of the Lime (VIII.5.28), the southern side of Insula VII.13, and VIII.3.10 where tuff pylons with carved pilasters and a doorway architrave prefigure the decorative scheme of the Eumachia building and the *Comitium*. Finally, Insula VII.6.20–28, including the House of the Peristyle.
69 Wallace-Hadrill 2008, 133.
70 Zanker 1998, 33.
71 Richardson 1988, 370.

72 Moormann 2011, 69–85.
73 Laidlaw 1985, 15–19.
74 Clark 1831, 69; Gell and Gandy 1833, 96.
75 Laidlaw 1985, 307–309.
76 Laidlaw 1985, 303; Garrison 1992, 9–21.
77 Bruno 1969, 305–308.
78 Mols 2005, 245.
79 Laidlaw 1985, 307–309.
80 Onians 1988, 8–18, 27–38.
81 Vitr. *De. arch.* I.2.5.
82 Gabba 1972, 108–110; Jouffroy 1986, 25. On the concept of walls as a symbol of independence see, Aristotle *Pol.* 7 1330b; Livy, *Ab ur. con.* XXIV.37.2–9.
83 CIL I² 1722.
84 Thomas 2007, 108–113.
85 CIL IX 2230.
86 Ramanius 2012, 114.
87 See Van der Graaff forthcoming.
88 Conway 1897, 45; Vetter 1953, Ve 22.
89 The inscription first went to Paris; see Mazois 1824a, 53; Overbeck and Mau 1884, 46; Maiuri 1929, 213. It is now in the British Museum; see Nissen 1877, 511; Conway 1897, 45.
90 Vetter 1953, Ve 13. Guzzo points out that the inscription might instead refer to the Triangular Forum; Guzzo 2007, 98.
91 McNicoll, 2007, 48 and 60.
92 See chapter 7 of this volume.
93 Lancaster 2005, 3–18.
94 On the representational aspects of arches at gates, see Böttcher-Ebers 2016, 277–287.
95 Conway 1897, 39.
96 Mau 1902, 184. The inscription led to the erroneous identification of the current Temple of Asclepius behind the Small Theater as dedicated to Zeus Melichius, which scholars now identify with the sacred area of the Fondo Iozzino; see Sgobbo 1942, 20; Pesando and Guidobaldi 2006, 67.
97 CIL X 1064.
98 Mau 1902, 243; Cooley and Cooley 2004, 127 translate the inscription as follows: Lucius Avianius Flaccus Pontianus, son of Lucius, of the Menenian tribe and Quintus Spedius Firmus, son of Quintus, of the Menenian tribe, duumvirs with judicial power, paved the road at their own expense from the milestone to the station of the carriage drivers, where it is in Pompeii's territory.
99 See CIL IV 1193 and CIL IV 1194. Also, PAH 1 Pars Prima, 155–156, April 14 and 28 1764. Romanelli mentions many inscriptions painted in red on the Porta Nola; Romanelli 1817, 274–275. See also Mazois 1824a, 29 and pl. XI for some of the remaining election notices. See also Breton 1855, 240. De Jorio describes a notice advertising games offered by Rufus, which included two gladiatorial fights, and an animal hunt with the addition of a velum to provide shade; see CIL IV 1187; see also CIL IV 1204; CIL IV 1220–1222. De Jorio 1828, 47. For painted inscriptions and graffiti on the Porta Marina, see Fiorelli 1875, 315; CIL IV 652–655, 1751–1757.
100 Chiaramonte Treré 1986, 30.
101 For the inscription theory, see Pocetti 1988, 315. For the tower theory, see Chiaramonte Treré 1986, 30. For the tower number hypothesis, see Pesando and Guidobaldi 2006, 33.
102 On the towers, see Pocetti 1988, 314; Thomas 2007, 109.
103 Pinder 2011, 72–74.
104 On the taxing and policing functions of gates, see Palmer 1980, 217–233; Van Tilburg 2007, 85–126; Van Tilburg 2008, 133–147.
105 Ain. Tact. v.1.

106 Mau 1902, fig. 111. Excavators recovered the body of a soldier in the niche next to the Porta Ercolano. Scholars widely believed that the eruption killed him while standing guard at the gate. The niche, however, is an honorary tomb; see Clark 1831, 73; Adams 1873, 50; Overbeck and Mau 1884, 44.
107 Palmer 1980, 217–233; Van Tilburg 2007, 85–126; Van Tilburg 2008, 133–147.
108 Müth 2016, 159–172.
109 For drainage shafts, see Lugli 1968, 73–75.
110 Poehler 2012, 101–102.
111 Briquel 2008, 124–125.
112 Maiuri 1929, 211.
113 Etani 2010, 208.
114 Ellis and Devore 2007, 124–125.
115 On the paving stones, see Poehler 2012, 95–120.
116 Grimaldi 2014, 24.
117 Eschebach and Eschebach 1995, 78.
118 Aoyagi and Pappalardo 2006, 491–493.
119 Grimaldi 2014, 38.
120 For a summary, see Sakai 1992, 1–13.
121 Antonini 2004, 279. Others see them as representing commercial signposts; see Conway 1897, 69–71; Degering 1898, 124–146.
122 Vetter 1953, Ve 23 and Ve 24.
123 Vetter 1953, Ve 26.
124 Vetter 1953, Ve 25. The World War II air raid destroyed the dipinto; see García y García 2006, 105.
125 Vetter 1953, Ve 27.
126 Vetter 1953, Ve 28.
127 Pocetti 1988, 324–327. He particularly mentions the *Argei* procession in archaic Rome.
128 Appian *Bell. pun.* 66.
129 Pocetti 1988, 326. For the fragments of the tower number VII, see Chiaramonte Treré 1986, 30.
130 See Chiaramonte Treré 1986, 26. Coarelli reassigned the *mefira* nickname to Tower VII; Coarelli 2000, 106.
131 Ling 1990, 204–214, especially regarding Pompeii. On landmarks and identity in the neighborhoods of Pompeii, see Laurence 2007, 34–45.
132 Ribezzo identified the Porta Sarno with the *veru sarinu*; Ribezzo 1917, 58. For the arguments identifying the *veru sarinu* with the Porta Ercolano, see Sogliano 1918, 161–164.
133 Sogliano 1918, 168; Coarelli 2000, 107. See Pesando and Guidobaldi 2006, 33. Ribezzo argues that the *urbulanenses* came from the town of Urbula, which they abandoned in a time of crisis to settle in a *pagus* outside Pompeii. The road to the *pagus* then gave the name to the gate; Ribezzo 1917, 55–63. Pesando and Guidobaldi suggest that the name *veru urublanu* translates to "gate to the *urbs*," the *urbs* and gate being Pompeii and the Porta Nola; see Pesando and Guidobaldi 2006, 33. For the Porta Sarno, see Della Corte 1921b, 87–88; Spano 1937, 276. For the Porta Nocera, see Sakai 1992, 9.
134 Sogliano 1918, 168. More recently, see Coarelli 2000, 107. See Pesando and Guidobaldi 2006, 33. G. Amodio suggests that the electoral colleges included both *vici* and *pagi* united into *tribus*; see Amodio 1996, 458–468; also Castrén 1975, 80–82.
135 Coarelli elaborates Sgobbo's thesis using the compital altars to divide of Pompeii into five electoral regions; Coarelli 2000, 97. For other Latin names to the electoral colleges, see Eschebach and Eschebach 1995, fig. 39.

5   The fortifications and the Roman colony

During the Social War, Pompeii joined the ranks of the cities revolting against Rome. Sullan forces besieged the city, capturing it in either 89–88 BCE or 82 BCE.[1] The Roman army spared Pompeii the fate of the neighboring rebel town of Stabia, which it completely flattened as punishment for its affiliation with the revolt.[2] Instead, Roman authorities transformed Pompeii into a colony, renaming it *Colonia Cornelia Veneria Pompeianorum* in 80 BCE.[3] The title carried multiple personal meanings to Sulla. *Cornelia* referred to his nephew and cofounder of the colony, P. Cornelius Sulla, and referenced the Sullan *gens* Cornelia. *Veneria* alluded to Venus as the personal protective goddess of the dictator.[4] The new name is an indicator of social change. Although their total number is unknown, the colony would receive an influx of Roman veterans and their families, who would undoubtedly influence Pompeii's social structure. They would share the city with the remaining Samnite population, which would provide a measure of continuity that blurs any neat cultural and social distinction that the date 80 BCE invites to establish. The result would, temporarily at least, broadly divide Pompeii into the pro- and anti-Roman camps, whose differences had led to the Social War.[5]

The history of colonial Pompeii traditionally divides into the late Republic, the early Empire, and the postearthquake period from 62 CE to the eruption of Vesuvius. The first division is political: The strife that ended the Republic must have affected Pompeii. It presumably ended with the advent of empire. The second division is the consequence of a natural event that devastated the city. The changing roles of the fortifications roughly align with these events. The historical division afforded by the foundation of the colony is one of convenience. Many preexisting notions related to the defenses, including civic pride and independence, remained true for the colony. Although similar associations may have existed in the Samnite period, the fortifications, to varying degrees, would incorporate the Roman values of *virtus*, *securitas*, and *romanitas* as social and political changes affected the city. Like any other monument, the meaning of the Pompeian fortifications changed with the developing social reality. The advent of the colony would lead to a gradual process of negotiation and transformation that continued up to the eruption of Vesuvius.

How Pompeii managed its fortifications during the Roman period is an open question. The traditional consensus has treated this period as one of inexorable decline for the fortifications. This was certainly not the case initially, since Pompeii repaired and rebuilt significant sections of the fortifications after it became a colony. Although the scope of this reconstruction is contentious, the primary material used for this project is the same *opus incertum* employed in the towers and gates of the city. After this event, the fortifications indeed lost part of their military value. However, the defenses and especially the gates would remain the object of patronage, changing their function right up to the final days of the city.

After the battle

Powerful reminders of the battle that led to the fall of Pompeii still mark the city. The area of *Insula* VI.1 and the garden of the House of Fabius Rufus (VII.16.21) have yielded a plethora of catapult balls, slingshots, and scorpio bolts. They attest to the violence of the siege since the garden is in a sector of the fortifications that, because of their natural strength, would not seem prone to a sustained attack.[6] Pockmarks in the masonry near the Porta Vesuvio and Ercolano indicate a violent bombardment of ballistae and catapult missiles. The *eituns* inscriptions still point the way for the defending troops, whereas catapult balls, allegedly belonging to the siege, adorned the gardens of Pompeii at the time of their excavation. The preservation of the scars in the walls seems almost deliberate, perhaps part of a collective memory marking a particularly traumatic event.[7] Parallel examples of such a role for the fortifications exist in Athens, where the architect Mnesicles deliberately preserved a section of the Bronze Age wall adjacent to the Propylaea. Workers also incorporated parts of the old Parthenon destroyed in the Persian sack into the foundations of the new Acropolis.[8] The damage wrought upon Pompeii remains difficult to quantify, but the battle debris must have lingered as a marker, positive or negative, of the changes that it induced.

Despite an interim period of uncertain rule between the battle and the installation of the colony, attention quickly turned back to the fortifications.[9] In one of the first construction events after the siege, the *duumviri* Titus Cuspius and Marcus Loreius restored the walls. The inscription celebrating their accomplishment mentions that the work concentrated on the *murum* and *plumam*. It reads:[10]

CVSPIVS · T · F · M · LOREIV(s) M · F
DVOVIR (d) D S MVRVM ET
PLVMAM · FAC · COER · EIDEMQ · PR(o)

The inscription carries a set of problems. Excavators found it in fragments embedded in a threshold and the floors of the House of M. Caesius Blandus (VII.1.40).[11] In its original context – presumably a highly visible part of the restored walls – the inscription would have given a better sense of the extent of the restorations, which are at best unclear. The term *murum* clearly refers to walls, but the translation of the term *plumam* (plumes) as related to fortifications remains uncertain; it may refer to the battlements.[12] The date of the reconstruction is only approximate to the early years of the colony when the two men were in office.[13] Their intervention on the fortifications probably relates to areas damaged during the Sullan siege, but the full extent of restoration, as well as the reason for it, is vague. The most recent hypothesis identifies the revolt of Spartacus around 70 BCE as a major catalyst for the reconstruction. The social unrest also led to the contemporary reinforcement of the fortifications in nearby Cuma across the bay.[14] Nevertheless, more mundane issues such as protection against brigandage, as well as Roman notions correlating city walls with the proper image of a colony, must have played a role.[15]

Part of the problem is that the extent of the damage wrought by the Sullan siege and the subsequent conquest remains unknown. The happenings throughout the siege are mired in obscurity, as are the weapons used. Although the evidence indicates the use of the catapults and ballistae, only rams and mining operations could effectively bring down large sections of the curtain wall. The use of such weapons at Pompeii remains uncertain. The demilitarization of city walls after a siege was a common punishment, whether by full demolition or the removal of battlements. The aim was to prohibit their effective reuse, prevent renewed uprisings, and dismiss any form of civic independence or bargaining power.[16] The historian Appian recounts how the purposeful destruction of citadels and city walls were among the punishments that Sulla inflicted upon communities like Pompeii that had rebelled during the Social War.[17] Such measures are not immediately evident for Pompeii.

Much of the debate concerning the reconstruction of siege damage focuses on the use of *opus incertum* on the curtain sections of the city walls. These are undeniably part of some sort of rebuilding event. However, repairs and reconstructions using similar construction materials, such as tuff or travertine, are difficult to pinpoint, let alone to date. Attributing the employment of one material to all of the restoration events or even identifying all of the work as occurring in a single event is equally problematic. For instance, freshly carved blocks dating to the post-Sullan period compose the large tract of travertine wall next to Tower XI.[18] A tenuous theory suggests that the presence of drainage spouts in the tracts of fortifications built in *opus incertum* is the result of the postsiege reorganization of the battlements and the demolition of a previous roof.[19]

Despite these uncertainties, the majority of the *opus incertum* tracts surviving in the city walls probably represent postsiege repairs related to the battle and its immediate aftermath. My assumption comes primarily from the location of these tracts in the most vulnerable areas of the enceinte, their association with large construction projects such as the amphitheater, as well as masonry analyses and excavations. The remains at Tower IX point to a large post–Social War reconstruction effort, including the full reconstruction of the building and 20 meters of adjacent western curtain wall.[20] On account of the employment of Roman rather than the Oscan foot in the construction process, the tract of *opus incertum* east of the Porta Ercolano represents a postsiege repair.[21] The city side of the same tract saw the construction of a new retaining wall and a pomerial street.[22] The tract south of the amphitheater must relate to the arena's construction, an event that also led to the partial removal of the *agger* and the demolition of the defensive internal wall. Sections outside the Porta Nola, Porta Stabia, and north side of the city probably necessitated extensive repairs. They must have received the brunt of heavy attacks during the Sullan siege because of their vulnerable position. Other sections functioning as terraces immediately west and east of the Temple of Venus must relate to housing construction or the refurbishment of the sanctuary that occurred in the first century BCE.[23]

Reconstructing the walls

Considering the importance of public munificence in Roman society and the symbolic nature of city walls, efforts to integrate or explicitly differentiate the tracts with the rest of the circuit may represent implicit political statements. However, the original appearance of the new *opus incertum* masonry is uncertain because of its rough state of preservation. Although some sort of plaster coating usually accompanied the application of *opus incertum*, much of the surviving masonry is completely bare. The parapet has largely vanished, complicating any further interpretation. The current state of the tracts is the result of their exposure to the elements for about 170 years before the eruption and since their excavation, as well as the accident of survival after their burial.

Two sections of *opus incertum* stretch on either side of the Porta Stabia. No trace of the battlements survive, including any spouts that drained rainwater from the wall-walk. West of the entrance, a tract of *opus incertum* masonry quickly gives way to an entirely demolished area of the defenses looted for construction material in the postearthquake period. To the east, the relic of an earlier internal Sarno travertine pier interrupts an otherwise uniform 60-meter stretch of *opus incertum* masonry. The pier corresponds to a slight reentrant in the curtain and the eastern enclosure wall of the *schola* tomb of Marcus Tullius.[24] The masonry adjacent to the gate is slightly rougher. It may represent a refurbishment or damage of the wall face associated with the tomb construction or a later partial collapse.[25] At its foot, this section of wall displays a well-finished surface that must have accommodated a lost plaster veneer. East of the pier, the masonry includes clear horizontal construction seams that mark the sequential deposition of the concrete during the construction process.

The next large section of *opus incertum* supports the southern side of the amphitheater between Towers IV and V (see Figure 5.1). Maiuri believed that it was pre-Sullan and identified some early Imperial repairs to strengthen the wall against the *agger* behind it.[26] It seems more probable that the concrete section relates to the later construction of the amphitheater, which also led to the lowering of the *agger* and the demolition of the internal wall. Today, modern imitation masonry composes much of the stretch and only the last 10 meters leading up to Tower V are partially original. A couple of images from Maiuri's excavations around the amphitheater in the late 1950s confirm this layout. They show a low wall before its reconstruction, and the ancient masonry is clearly visible next to the tower.[27] The surviving section displays the remains of a plaster base coat that was part of a lost embellishment.

Two large tracts of *opus incertum* survive between the amphitheater and the Porta Vesuvio. The eastern bastion of the Porta Nola is perhaps one of the most imposing stretches of the circuit, towering above anyone passing through the gate.

Figure 5.1 Section of wall supporting the amphitheater. Note the remains of the plaster
         coating.

In the Augustan period, a lowering of the road carried out to ease access into
the city revealed the foundations of the bastion. The result added further height
to the defenses and added an impression of impenetrability to the gate.[28] Only
scant remains of a plaster base coat survive, along with neat masonry surfaced
to accommodate a plaster finish. Farther west, a section of *opus incertum* dating
to the colony is associated with Tower IX. It originally reached some 6 meters in
height, of which only three remain.[29] This section of the defenses does not include
an *agger*. Instead, four counterforts on the city side of the masonry strengthen the
curtain. Upon excavation, the lowest meter of the masonry displayed traces of a
plaster coat, but any other remains had likely disintegrated soon after their first
exposure in the 1800s.[30]

Three large sections of *opus incertum* masonry survive between the Porta Vesuvio and Ercolano. None display traces of a decorative scheme, but they do present
a better preserved parapet. A small section stretching 25 meters east from Tower XI
features seven regularly spaced drainage spouts and a unique decorative architrave
supporting a continuous battlement. A smaller 10-meter patch of masonry also
survives roughly 22 meters from Tower XI, but it is less well preserved. To the
west, an extensive section of masonry starts some 15 meters east of Tower XII
and covers most of the ground toward Porta Ercolano (see previous figure 1.1).
Travertine blocks compose the parapet east of the tower. Farther west, the parapet

transitions into a series of vertical tuff blocks standing on an architrave. Both the travertine and tuff parapets preserved regularly spaced drainage spouts set into the masonry.[31] A vertical seam in the wall facade further marks the transition in the parapet; it may represent successive construction events or the work of differing construction crews.[32] The wall face displays clear horizontal construction seams dating to the Roman colony. They are strikingly similar to those found east of the Porta Stabia. Patches of rubble fill throughout the section point to secondary emergency repairs carried out in the postearthquake period.[33]

Further stretches of masonry survive on the western side of the city. The first creates a terrace for the vicolo dei Soprastanti and is incorporated into the later expansion of the House of Umbricus Scaurus (VII.16.15). Its southern edge displays the same spouts and tuff parapet as the curtain east of the Porta Ercolano. A large patch of fired brickwork at its base points to similar postearthquake repairs. A recent hypothesis posits that a colonist acquired the house immediately after the Sullan conquest. The new owner demolished the previous travertine curtain and substituted it with the current *opus incertum* wall as part of an effort to build rooms on the new lower terrace.[34] Perhaps as an implicit statement of conquest or as a cost-saving measure, he reused much of the travertine masonry as a building material for the refurbishments in the house.[35] South of the Porta Marina *opus incertum* superimposed on a few older travertine courses acts as a terrace for the Temple of Venus. A lonely block of tuff suggests that the masonry

Figure 5.2 Rebuilt section of the wall to the east of the Temple of Venus.

replaced an earlier travertine/tuff curtain. The so-called Villa Imperiale (VIII.1.a) enveloped this section of the fortifications in the late first century BCE, covering them with Third Style frescos and burying the external pomerial street.[36]

East of the temple, another large section of fortifications emerges from beneath later buildings (see figure 5.2). It includes encased drainage spouts and the remains of a tuff parapet similar to the wall east of the Porta Ercolano. The section displays signs of later reuse where someone sheared off and modified the tuff blocks to support floor beams. A terracotta sewer, later cut into the masonry, drained excess water from the gardens above. Smooth white plaster and a surviving patch of embedded amphora pieces, used as a foundation for a veneer, still cover the masonry. It is unclear whether they are part of the original wall decoration or later buildings.[37] Finally, a stretch of *opus incertum*, perhaps related to the construction of the yet unlocated Tower I, partially replaces original ashlars that compose the terrace of the Doric Temple, but no parapet or drainage system survives.

With a few exceptions, the reconstruction effort largely preserved the layout of the battlements and *agger* system of the fortifications. The presence of drainage spouts in the *opus incertum* is a case in point since they probably are reused elements of the tuff curtain. Along with the tuff parapet and the consistent horizontal construction seams, their recurrence indicates that the largest curtain tracts are part of a single construction event. Only the substantially lower level of the spouts below the continuous parapet east of Tower XI stands out. This situation must be the result of the unique battlement requiring a different drainage system and may relate to a different construction event. As far as any sort of plaster embellishment is concerned, the current remains, besides a few scant traces, offer little assistance.

Clues on the original appearance of the fortifications come from the individual descriptions and engravings of the earliest publications on Pompeii. These publications are part of an era that had far less detailed scientific standards, resulting in divergent pictures of the surviving remains. Several authors describe that a plaster veneer covered the *opus incertum* sections, but none mentions whether it was smooth or replicated ashlar masonry.[38] For example, William Gell describes the plaster as deliberately imitating the *opus quadratum* masonry, but it is unclear whether he refers to the curtain, the towers – where it still survives – or both (see Figure 5.3).[39] Thomas Dyer, referencing Gell, is equally vague, describing the *opus incertum* sections as covered in plaster to better fit in with the adjacent masonry.[40] Johannes Overbeck is more specific, describing sporadic remains of plaster covering the *opus incertum* masonry in the middle of the nineteenth century.[41]

Illustrations scattered throughout early publications offer further insight. These are sometimes simple vignettes, but more often prints of engravings derived from field drawings. Their accuracy can be questionable. The prints may have lost some detail due to the draftsmanship or the process of copying the drawings to the plates.[42] The majority of the prints depict the exposed northern part of the fortifications from various angles, and only a handful seem exact. Gell furnishes two drawings: The first depicts the section between Tower XI and the Porta Ercolano, whereas the second is a close-up of Tower XII. Both show the lower sections of

Figure 5.3 Towers X, XI, and XII, etching. (After Gell 1817, pl. XVI.)

the curtain covered in a smooth plaster coating following the horizontal seams present in the masonry. A few patches of *opus reticulatum* masonry appear, which are clearly not present today. Further details are freely adapted from other exposed towers. For instance, the postern for Tower XII, visible in the illustration, was not visible until Maiuri exposed it in 1929.[43] Henry Wilkins, a contemporary of Gell, also renders plaster on the masonry west of Tower XII. He clearly delineates the *opus quadratum* near Porta Ercolano but omits the current drainage spouts.[44] Publishing soon thereafter in 1825, Paolo Fumagalli shows the entirety of the curtain between Towers XI and XII, including the ashlar masonry as completely covered in plaster. However, the rendition is inaccurate: The author describes the *opus quadratum* masonry as bare in the associated text.[45] Le Riche similarly depicts smooth ashlars on the *opus incertum* running west from Tower XII but omits the plaster ashlars on the tower.[46] A vignette published by Breton in 1855 shows the area almost completely devoid of plaster, with the exception of a few remaining patches that seem to imitate ashlar masonry.[47] The Porta Nola and its adjacent bastion in *opus incertum* has been featured frequently in early prints and illustrations (see Figure 5.4). Today, all of the *opus incertum* masonry is completely bare. However, the majority of authors depict the remnants of a smooth plaster coating covering the masonry. None of them indicates any additional colors on the surface or further First Style embellishments such as imitative ashlar blocks.[48] Mazois, who as an architect and royal draftsman under Murat faithfully reproduced the

Figure 5.4 Exterior view of the Porta Nola, etching. (After Mazois 1824a, pl. XXXVII.)

ruins, also details the presence of a secondary coat of plaster, suggesting a refurbishment of the decorative scheme.

A recent hypothesis proposes the presence of a First Style veneer on the entire post-Sullan circuit. This theory relies on a study of the pockmarks created by the Sullan war machine on the wall sections on the north side of the city. Along with a handful of less clear examples, a hole in the fifth block on the exterior western corner of the Porta Vesuvio contains a plaster plug that is the primary evidence for a decorative plaster coating. Russo and Russo, the two authors of the theory, dismiss the missing plaster elsewhere as the result of its complete disintegration after its exposure. If true, the theory solves both military and aesthetic issues: The plaster presented a uniform facade to onlookers and masked divergent, perhaps weaker, *opus incertum* masonry. However, the problems with this theory are many, not least, the complete absence of substantial plaster remains on the *ashlar masonry*, including the early illustrations as well as the notices painted onto the masonry of the passageway walls.[49] The towers and gates provide ample evidence that, under the right conditions, large tracts of the plaster coating may have survived intact. In addition, the plaster plugs that are the foundation of the argument are otherwise lacking in the circuit. Those at the Porta Vesuvio may represent patches preventing the climbing of the fortifications by both individuals and enemy troops. Roman law indicates that such modifications were necessary to keep out thieves or contrabandists trying to avoid the policing function of gates. City walls constituted a special category known as *res sanctae* (holy things), protected by but not consecrated to the gods. Scaling or crossing them illegally was a sacrilege punishable by death.[50] The new status of Pompeii as a colony perhaps led to the law's enforcement.

The visual and archival evidence indicates that some sort of plaster coating originally covered the *opus incertum* used in the curtain wall, however elusive it might be today. The coating was most likely plain and smooth, although the presence of imitative ashlar masonry is possible. However, by the mid-nineteenth century, any traces of it had disintegrated. This is in stark contrast with the gates and towers, where a few remnants of coating cling to the masonry. The reason for its survival on these structures remains unclear. It may relate to the visibility of the towers, which perhaps ensured the application of better quality and therefore more durable plaster. Another hypothesis is that the same visibility ensured the restoration or reapplication of successive coats, which were more durable over time.

The plaster veneer on the tracts of *opus incertum* would have formed a dramatic contrast with the *opus quadratum* of the previous circuit. Perhaps the reasons for its presence are merely practical. *Opus incertum* was a cheap and fast building material. The plaster may represent nothing more than a coat meant to protect the masonry. Nevertheless, it clearly functioned in unison with the First Style decorative embellishments present on the towers and gate vaults in their overall role as a public building. The new sections of curtain wall created a discrepancy in the visual unity of the wall. They readily gave a visual cue of the renovations carried out on the fortifications. Loreius and Cuspius left a clear mark on the structure that announced their euergetism and, perhaps more importantly, the new colonial social order.

After their initial refurbishment and an occasional repair, the curtain sections remained substantially unchanged until the earthquake of 62 CE. The walls slowly changed roles from an effective defensive line to a barrier with a diminished military function. The area of public land associated with the walls would slowly shrink, overtaken by construction and burial grounds. After the earthquake, many areas in front of the curtain became dumps for the architectural debris resulting from the cleanup efforts.[51] Large tracts adjacent to the Porta Stabia, Nocera, and Nola became open-air quarries spoliated for building material in the ensuing reconstruction.[52] Although they lost much of their imposing appearance, the fortifications contributed both psychologically and physically to the rebirth of Pompeii.

They acted as functional markers beyond which the population placed the ruins of the devastated city. The old defenses would provide the necessary material to contribute to its reconstruction. The materials would preserve the memory of the walls in the reconstruction effort and celebrate the old fortifications in the newly rebuilt Porta Ercolano.

The development of the gates

In the period between the installation of the Roman colony and the eruption of Vesuvius, most gates would receive a substantial structural makeover. Some carry the signs of multiple interventions, but it is hard to ascertain when they occurred. This contrast between the curtain walls and the more dynamic gates undoubtedly relates to the high visibility of the entrances in and out of the city. Even as the military role of the curtain walls diminished, they would still create a formidable barrier into the city. The gates would continue to function as the main entrances into the city and as tax stations. These factors ensured that the gates would remain important representational elements in the concept of the city.

Porta Marina

The Porta Marina opens on the western edge of the Pompeian plateau, where it formed a primary link between the coast and the Forum (see Figure 5.5). Most of its structure is composed of a large barrel-vaulted corridor built in *opus quasi reticulatum* and brick quoins set inside the travertine masonry of the first Samnite enceinte. Two vaulted entryways extend in front of it: a smaller passage for pedestrians and a large entrance reserved for pack animals. The smooth flagstones of the road leading up to it imply that carts could not reach the gate because of the steep angle of the road. The exterior southern wing contains a niche that preserved a broken terracotta statue of Minerva. The corridor features a walled-up door on the north side; another still opens on the south side to the modern Antiquarium, which is an unidentified ancient space.[53] On the city side, a large *opus incertum* terrace, built perhaps even as late as the postearthquake period judging by the masonry and supporting a portico heading to the houses above the gate, abruptly cuts off the sidewalk, forcing would-be pedestrians onto the basalt pavers.

A considerable debate has concerned the date and construction sequence of the gate.[54] The current view identifies the barrel vault as a modification built in the early years of the colony, with the smaller outer vaults added soon thereafter. The main barrel vault replaced an elongated version of the typical Pompeian tripartite gate that stretched to the top of the plateau near the current entrance of the Temple of Venus.[55] The current layout of the gate has led many to see it as primarily ornamental, functioning mostly as a police and tax barrier (see Figure 5.6).[56] The barrel vault creates a long cavernous layout that is difficult to defend, whereas the double entrance would facilitate passage through the gate. Nevertheless, the design responded to the circumstances of the terrain, which gave it considerable military capabilities in tune with the wider reconstruction

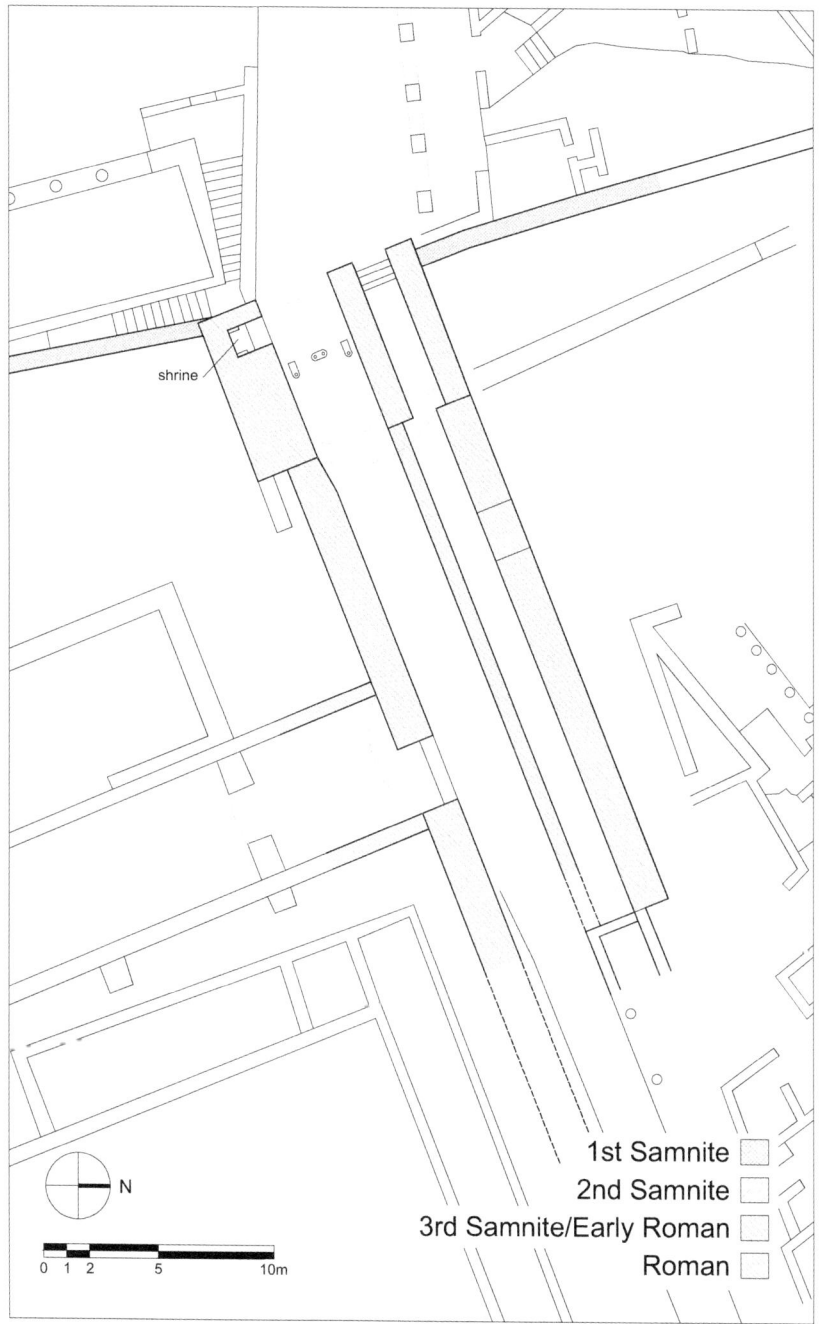

1st Samnite ☐
2nd Samnite ☐
3rd Samnite/Early Roman ☐
Roman ☐

shrine

N

0 1 2   5   10m

Porta Marina

1:300

Figure 5.5 Plan of the Porta Marina. The key highlights the phases. (Drawing by T. Liddell.)

Figure 5.6 Overview of the Porta Marina.

of the civic defenses.[57] The steep road approaching the gate – inaccessible to carts, let alone heavy complicated siege engines – naturally added to its military strength. The elevation transformed the gate into a powerful defensive tower-like structure, where the added height resulted in better views and weapon reach.[58] The Porta Marina, then, displays an adapted design to suit the strategic as well as ornamental needs of the structure.

Two key factors in the urban context explain why the Porta Marina was the object of such an extensive refurbishment in the early colony. Although it was a short street, the gate opened on the via Marina, which links the Forum with a side entrance of the Basilica and the Temples of Venus and Apollo. As if to highlight its importance, builders inserted marble chips between the basalt pavers – a unique circumstance in Pompeii. At about the same time as the refurbishment of the gate, the adjacent Temple of Venus received new ornamentation (see Figure 5.7). This was an overt political statement in the new colony since Venus was the personal protectress of Sulla.[59] Outside the walls, the road that passed through the gate led to the *Navalia*, a harbor or ship storage facility built soon after the installation of the colony. Roman navy ships would use these facilities to dock on land at night and during the winter months. Although the strategic importance of the facility remains unclear (and even the existence of a harbor), it probably played a key role in the region at least until the development of the large naval base across the

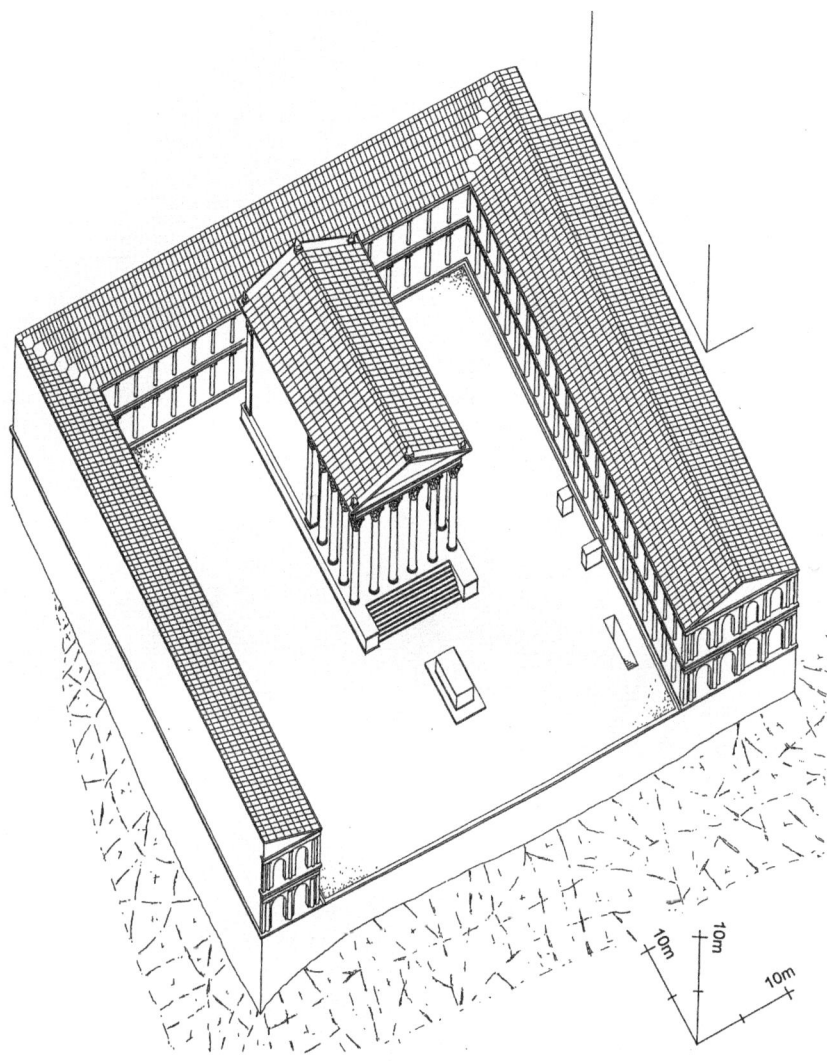

Figure 5.7 The Sullan Temple of Venus. (After Wolf 2004, fig. 5.)

bay at Misenum under Augustus.[60] The proximity to the Temple of Venus and the development of the naval facility led to the refurbishment of the Porta Marina as a marker of the new colonial realities.

During the course of the first century CE, the Suburban Baths, the Villa Imperiale (VIII.1.a), and private houses eventually engulfed the gate, dramatically changing the layout of the area. A terrace of the villa completely buried a stretch of the pomerial street linking the gate to the river harbor farther south.[61] Along with the development of Misenum, this factor led to a loss of importance for the gate. Its proximity to the coast left little space for further urban development, making it the only gate in Pompeii that lacks tombs in front of it. This situation is similar to Rome's where urban development eventually surpassed the gates of the old Servian Wall. The gates would acquire more of a ceremonial role, marking the crossing points through the symbolic boundary of the old defenses. They would come to function as monumental honorary arches, a form that architects would increasingly use to connect urban districts visually and architecturally.[62]

Although its military role might have decreased, the gate clearly maintained a lively and vibrant role as a monumental boundary marker in the city. At the time of excavation, workers found the remains of wooden doors and an iron gate that closed the main and subsidiary passageways, implying that the opening could still be closed when necessary. Authorities also carefully maintained the embellishments. The gate still carried most of its plaster coating in the late nineteenth century. Today it has disappeared after an Allied aircraft bomb pulverized the outer vaults in World War II. Drawings and photographs in early publications and guidebooks indicate that a smooth plaster coat with a slightly raised socle covered the masonry in a layout consistent with the other vaults of the city. At the time of the eruption, the plaster featured many graffiti, including one referencing the prostitute Attica and her fees.[63]

The Augustan gates

In the following Augustan period, many of the gates received overhauls aimed mostly at facilitating passage into the city. At the Porta Stabia the renovations conducted by the *duumviri* L. Avianius Flaccus and Q. Spedius Firmus included the construction of a raised causeway in front of the gate to ease cart traffic from the bustling river port nearby.[64] The new roadway compromised the natural drainage of rainwater through the passageway, eventually leading to the construction of a sewer through the western *agger* that demolished the stairs leading up to the wall-walk.[65] The causeway also bridged the defensive ditch in front of the gate, thereby significantly reducing its military effectiveness. The necessity to build the causeway perhaps relates to the construction of the nearby Small Theater, which would complement the already existing theater to the west.[66] The causeway would formalize access to the city and the adjacent theater district. A few other gates experienced a similar change. Workers lowered the roads through the Porta Nola, Nocera, and Marina and elevated those at the Porta Ercolano and Vesuvio.[67]

The changes to the roadways may seem subtle, but they included important infrastructure works that were radical interventions, such as cutting back scarps, filling in defensive ditches, and repaving the roads. The result was a new emphasis on the gates of Pompeii and an added monumentality. In practical terms, the road works eased the passage into the city as a result of a general reduction of defensive concerns after the establishment of the *pax augusta*.[68] However, the gates would maintain their role as effective filtering points of access into the city and barriers against unwanted elements. At those gates where engineers lowered the roadways, viewers would now experience an added height to the gates and walls as one passed through the *agger*. The tall walls compounded the cavernous effect of the passageways already inherent in the design of forecourt gates. The result enhanced their monumentality and the experience of passing through the limits of the city.

Augustus promoted the construction of many enceintes and gates as part of his program of civic renewal on the Italian peninsula. A parallel development to Pompeii occurred in nearby Cuma, where the median gate that opened onto the via Domitiana – a major regional artery – received a large water collector in the Sullan period to allow for proper drainage, followed by its architectural transformation into a double arch in the Augustan age.[69] In colonies such as Rimini, this program included a range of projects such as sewer networks, roads, and public monuments. The emphasis on defenses as an integral part of this program indicates that these fortifications legitimized urban communities as cities. In cases such as the Arch of Augustus in Rimini, a gate would acquire almost triumphal overtones with the emperor portrayed as the chief benefactor. Pompeii was not isolated from this development. In the early Imperial period, many politically inspired buildings show how the Pompeian elite aligned itself with the new realities of power. The Eumachia building, the Temple to the Genius of Augustus, the Sanctuary of the Public *Lares* on the east side of the Forum, and the private Temple to Fortuna Augusta farther north all celebrated the emperor and his cult. The Temple of Venus, who was the patron deity of the emperor, would receive extensive refurbishment.[70] Although the military function of the fortifications decreased, the modifications on the Pompeian gates were part of the wider urbanization trend and the Augustan message of urban renewal occurring on the peninsula.

Porta Vesuvio

The Porta Vesuvio witnessed extensive changes when it became the designated site for the arrival of an aqueduct. In order to meet proper Roman living standards, the colony needed an aqueduct to bring a constant supply of freshwater to a city that up to this point had relied on cisterns and wells.[71] A recent hypothesis has proposed a Sullan date for a first intervention in the area, but further studies have convincingly refuted the idea.[72] The water *castellum* (distribution tank) on the west flank of the gate is the *terminus* of a branch of the Serino aqueduct built by Augustus to supply the Roman naval base at Misenum around 35 BCE (see Figure 3.4).[73] With aqueducts driven by gravity, the Porta Vesuvio lent itself naturally to the task as the highest point in the city. Its status as a public structure obviated the necessity of costly expropriations. The new *terminus* was as much a monument as it was a redistribution point for aqueduct waters. The rectangular brick structure displays blind arches and pilasters that must have featured some sort of plaster cover. The repetitive arcaded design of the exterior is reminiscent of the arcades that carried Roman bridges and aqueducts through the landscape. A new wide plaza in front of it further stressed the presence of the *castellum* to onlookers. Its location adjacent to the gate only served to highlight the notion of control on all things entering and exiting the city.

The setting created around the *castellum* was largely symbolic. The fortifications – the *agger*, gate, and nearby Tower X in particular – framed the building and further stressed the status of both structures as civic monuments. The works necessary to build the *castellum* would weaken the defensive capabilities of the walls (see Figure 3.4). They included the shortening of the western gate court by some 6.5 meters, the complete demolition of its internal wall, and the construction of a new terrace to hold back the *agger*.[74] The new aqueduct also affected the street network. It closed the access to the pomerial street running at the base on the city side of the defenses. On the exterior, the aqueduct channel approached the city slightly above ground, essentially cutting off the route to the Porta Ercolano. Engineers would reestablish the road by raising the entire area between the two gates with a massive 2-meter-deep earth fill overtopping the aqueduct.[75]

The consequences for the fortifications in this sector were dramatic. The external fill reduced their relative height, buried the tower posterns, and eliminated any outworks in front of the walls (see Figure 1.1). Construction of the channel through the fortifications posed further challenges. The architect had to negotiate the masonry of the curtain and the old buried tower adjacent to the gate. One hypothesis envisions the careful dismantling and reconstruction of the curtain wall to build the channel, but the outer curtain preserves no trace of this event unless workers carefully put back every single block in place. Such a scenario would set a precedent for the purposeful preservation of a public monument. However, an event where workers tunneled through the *agger* seems more probable. Construction of the aqueduct would significantly decrease the defensive capacity of this sector to the point that the walls almost entirely ceased to function militarily.[76] Nevertheless, the decision to locate the *castellum* within the fortified perimeter and to bury the channel implies a desire to keep them out of reach and within the protection of the fortifications.

The aqueduct and its *terminus* was a new civic monument that affected the fortifications that had constituted a symbol of Samnite Pompeii. The exterior earthwork brought in to cover the channel also buried the travertine socle of the curtain wall, thereby cancelling out the juxtaposition with the tuff masonry above it. This hallmark of Samnite Pompeii was now obfuscated in one of the most visible and militarily powerful areas of the fortifications. The gate itself, once a potent symbol of independence, was partially demolished to frame the new Roman waterworks. This purposeful dismantling of an urban artifact is particularly poignant assuming that funding for the aqueduct originated in Rome as part of the Augustan program of civic renewal.[77] Similar connotations extend to a scenario where private local donors, which by this point were part of a well established colonial framework, financed the aqueduct supplying Pompeii. Although the memory of the symbolism of the travertine and tuff juxtaposition had likely faded somewhat by this stage, the partial burial and dismantling of the fortifications transformed the image of the city. The Porta Vesuvio would become a landmark of the Augustan urban renewal. It would maintain this role through the rest of Pompeian history.

The postearthquake interventions on the Porta Ercolano and Porta Vesuvio

After these events, the gates remained largely untouched until the 62 CE earthquake. In the ensuing reconstruction of the city, the Porta Ercolano would receive the most radical reconstruction (see Figure 3.9).[78] Remarkably, this seems to be the first major intervention on the structure after the siege despite the fact that the area was subject to heavy bombardment. Pockmarks still litter the masonry next to the gate, whereas many adjacent houses flattened during the siege were rebuilt.[79] Today, as in antiquity, the scars and the subsequent reconstruction of the enceinte to the east acted as visual reminder of the attack and conquest of the city. The role of the gate, however, remains unclear. Considering the destruction wrought on the area, the gate must have needed some sort of repairs after the battle, which presumably followed its previous Samnite layout.

The current remains of the Porta Ercolano depart radically from its Samnite predecessor (see Figure 5.8.) The gate features a large central roadway open to cart traffic with two smaller flanking pedestrian passageways on either side. Barrel vaults covered the side passages, whereas the central bay was open to the sky. A portcullis on the exterior and a double leaf gate on the interior closed the passageway on either end. The result was a typical open court plan similar to the *cavaedium* gate types adopted in the Augustan period at Fano, Torino, and Spello and in the fourth century CE in Porta Nigra in Trier.[80] Open court gates had a clear military function to trap attackers and allow defenders to bombard them from an elevated position. Further defensive measures typically included adding towers to flank a central core. Such gates also carried monumental overtones through their sheer scale and were often embellished with architectural details such friezes and engaged columns. The Porta Ercolano, as a slight variant of the type, lacks the flanking towers. It never achieved full military functionality since the current remains do not indicate an easy access to the second floor. Instead, the gate is almost exclusively a monumental structure that resembles a triumphal arch.

Excavated in the 1760s and therefore on modern view for several centuries, the Porta Ercolano is the most discussed gate among Pompeianists in terms of its

Figure 5.8 The Porta Ercolano: interior view (left) and exterior view (right).

original appearance. Its monumental layout has invited discussion whether it was primarily a defensive or ornamental structure or both.[81] However, few of these discussions have concerned the First Style embellishments that once covered the gate. The ruined piers preserve only a fraction of the original coating. Illustrations in Mazois and Rossini point out how much decorative detail has disappeared. The stucco ornamentation on the facades once included lost pilasters applied in low relief on the outer edge of each pier. On the interior, the passageways featured a painted black socle, which Mau described as a typical Augustan Third Style variant of the First Style.[82] Drawings published in the volumes of Gell and Overbeck-Mau illustrate two Ionic marble columns framing the exterior of the eastern pedestrian entrance, but their presence at the time was the result of an erroneous reconstruction using pieces from a nearby tomb.[83] In similar fashion to the other gates, the stucco was otherwise a plain white. It would function as an *album* displaying graffiti, city ordinances, and announcements of gladiatorial games. Successive paint coats cleared the clutter, attesting to both the careful maintenance of the gate's appearance as well as its performance as a billboard in the city.[84]

Despite the poor state of conservation, a few authors have attempted rather fanciful reconstructions. Although the author admits a rather speculative approach, William Gell published a reconstruction where he projected the gate as a full-fledged Roman triumphal arch crowned with a great *biga*.[85] Subsequent reconstructions varied the theme only slightly (see plate 19). Niccolini lowered the height of the two pedestrian entrances, whereas Delaunay published a more reserved version without the *biga*.[86] These somewhat overzealous reconstructions clash with the physical remains. For example, Gell reconstructs full double parapets on either side of the gate, but an inn and stable flanked it to the south, whereas a staircase rather than the internal parapet abutted it to the north.

Although the military capabilities of the Porta Ercolano are a matter of debate, its primary function as a monument is evident. In a radical departure from the previous Pompeian gate designs, architects explicitly designed the new gate as a triumphal arch almost as a tribute to the postearthquake reconstruction effort. In the landscape of Pompeian monuments, the choice of rebuilding the Porta Ercolano is no coincidence. It straddled one of the most important routes of the city, connecting it directly with Rome. This same circumstance led to the construction of the well finished adjacent *opus quadratum* tuff masonry of the previous structure. At the time of the gate's reconstruction, opulent villas, funerary structures, and a concentration of elite houses in Regio VI all lined the road, indicating that it retained its status as a prominent thoroughfare in the city.[87] The stucco embellishments and architectural layout of the Porta Ercolano are a testament to how the fortifications and the gate still were active monuments in the urban landscape. The application of the First Style decoration is part of a tradition that on the fortifications stretched back some 180 years before the new gate. If the architectural layout signaled a radical departure from the past, the gate's ornamentation would reference Pompeii's resilience and history.

Similar considerations explain why, at the time of the eruption, construction had also begun at the neighboring Porta Vesuvio after the earthquake had demolished it in 62 CE.[88] The intended layout is difficult to trace. One hypothesis even proposes that the final design of the Vesuvian gate was a carbon copy of the Porta Ercolano. The ruins propose otherwise, where the Porta Vesuvio would have featured an elaborate double arch similar to the Porta Marina. It would have one significant difference: The minor pedestrian passageway would be largely formal, covering a blind sidewalk that dead-ended on the travertine bastion and an earlier *lararium*. The new arch respected this older religious space, indicating a continued concern with the ornamental and religious connotations of the gate. Considering the Porta Marina also had a shrine, then the similarity between the gates is hard to dismiss despite the 160 years that separate their construction. It implies a continued concern of the unified conceptual presentation of the gates despite the fact that, after the earthquake, parts of the enceinte had fallen into disuse. The area between the Porta Ercolano and Vesuvio was symbolic because it acted as a large open-air dump for much of the architectural debris associated with the postearthquake cleanup.[89] If finished, the two gates would have bookended the space only to reinforce their status as symbols of the reconstruction effort.

Although the lack of epigraphic evidence keeps the arguments uncertain, it is likely that imperial coffers largely financed the reconstruction effort of Pompeii, including the gates.[90] It is unclear whether Nero, Vespasian, or even Titus supplied funds for the gates. Both Nero and Vespasian had plausible reason to do so since they each had further links to the city. Nero's wife Poppaea, whose *gens* came from the Pompeii, personally connected the emperor to the city. He even dedicated a golden lamp to Venus Pompeiana as a symbol of his sponsorship in reconstructing her temple.[91] Vespasian's involvement in the postearthquake period included the addition of a temple in his honor on the east side of the Forum. A recent reevaluation points to his heavy participation in the reconstruction of the city, to the extent that most of the public buildings were rebuilt by the time of the eruption.[92] He also ordered the restitution of all illegally occupied public land to the city, including its walls.[93] The donation of the gates eminently expresses the completion of this process.

The reconstruction of the Porta Ercolano in such a monumental fashion was a key element of imperial patronage. In the course of the first century CE, the practice of building fortifications and, more specifically, gates would become closely associated with the figure of the emperor whether by direct imperial intervention or through local figures seeking approval. The commemorative Arch of Augustus that functioned as a gate in the enceinte of Rimini is an early example of the process. The monument celebrated the city's new colonial status as Colonia Augusta Ariminensis and the accompanying renewal of its infrastructure.[94] The arch also commemorated the completion of the Augustan renovations of the via Flaminia that connected Rimini with Rome. The arch was one of a pair voted by the Senate in 27 BCE. Its twin, now destroyed, marked the same event on the Milvian Bridge just outside of Rome.[95] The resulting structures created an interregional connection with Rome. Porta Ercolano, through its position on the via Consolare facing Rome and its commemoration of the reconstruction effort, must have had a similar role. City gates continued to function in their role as tax barriers, which by extension

carried strong associations with political and legal authority. The proper maintenance of the embellishments and the continuous interventions on the Pompeian
gates were a reflection of that authority. If the Porta Marina was associated with
the new colony and if the refurbished Temple of Venus and the Augustan gates
were part of a broader overhaul of the city, the postearthquake emphasis on the
gates projected Pompeii's reconstruction and reasserted imperial authority. The
message was clear: No other buildings could symbolize the rebirth of the city more
effectively than the renewal of the markers on its boundaries.

Change and continuity in the towers

The trends evident in the curtain wall and city gates are also present in the towers.
Many carry the marks of repairs presumably carried out in the early colony. In the
absence of excavation data, much of this assumption is based on the application of
specific construction techniques – admittedly a somewhat problematic approach.
Tower II on the south side of the city displays radical alterations to its original
structure. Today, only the main lower chamber remains which, judging from the
clear seams in the masonry, includes some modern restorations in the upper portion. Enough of the ancient masonry survives to confirm that the tower featured
brickwork laid in regularly spaced toothed quoins, in a technique adopted and
perfected in the first century BCE.[96] Unless the tower was built later than previously
assumed, the masonry indicates that the building was heavily restored after the
siege. The unusual floating postern on its western flank – sitting some 2 meters
above the level of 79 CE – points to a drastic lowering of the surrounding terrain in
antiquity. Equally unusual are the rectangular openings in the facade and back wall
of the main chamber. Assuming that the modern reconstruction is largely correct,
the colonial restorations of the tower radically changed its function. The circumstances of Tower II are peculiar enough to merit a separate discussion in chapter 6.

Tower III to the east displays quoins composed of *opus vittatum mixtum* on the
surviving corners of the main chamber and a patch of brickwork on its western
wall (see plate 20). Maiuri suggests that the quoins represent early Imperial
repairs and describes Towers III, IV, and V as *mozzate*, or chopped off, during the earthquake and never rebuilt. However, following Fröhlich's assessment
of the *opus vittatum mixtum* technique as typical of the postearthquake period,
then much of this repair masonry likely dates to after 62 CE.[97] A small patch of
surviving plaster on the eastern quoin indicates that this repair work included the
application of a new coat of plaster on the building. It is unclear what prompted
these late interventions on the tower. Perhaps they were meant to merely shore
up the building. They may also represent an effort to renew the urban decorum
connected with the reconstruction of the nearby amphitheater or a resumption of
the burial activities in the necropolis below.[98]

Heading toward the amphitheater, modern masonry almost entirely constitutes
the remains of Tower IV. Nearby Tower V to the east presents extensive toothed
brickwork in its quoins and southern flank. The construction technique is similar
to that of Tower II, implying that its refurbishment occurred in the early colony.

The ground floor inside the tower has no pavement. When or what event caused its removal is unclear. Modern masonry composes much of the upper structure. If restored correctly, the tower displays a floating rear entrance and side doors that are the result of a lowering of the *agger* associated with the construction of the amphitheater in 70 BCE. A sliver of surviving *opus vittatum mixtum* masonry, set below a modern restoration on the first corner of the stairway, hints at a postearthquake intervention. Considering that the entire upper arcade of the neighboring amphitheater displays similar brickwork, it seems likely that both structures received contemporary postearthquake repairs.[99]

Adjacent Tower VI does not indicate any major refurbishments in its surviving masonry (see Figure 5.9). Nevertheless, in a similar fashion to that of Tower V, its rear and northern entrances are floating high above the ground level of 79 CE. Once again, construction of the amphitheater must have lowered the *agger* enough to impede their direct access, leaving the southern doorway as a direct entrance to the building. The main chamber inside the tower is also missing a ground level pavement. Curiously, the door that opened at the rear of the chamber sits some 1.5 meters higher than the floor level of the corridor it gave access to, making it accessible only by stairs. These steps are otherwise missing, due perhaps to their original construction in perishable materials. Of the three towers surrounding the amphitheater, Towers V and VI display clear signs of measures that reduced their

Figure 5.9 Tower VI as it stands next to steps leading up to the Amphitheater. Note the high floating entrance because of the lowering of the *agger*.

military effectiveness, whereas the modern masonry of Tower IV precludes further assessment. Instead of a military role, the ancient restoration effort indicates that the towers acquired a more symbolic and representational function.

To the north, beyond the Porta Sarno, only the main chamber of Tower VII survives. It is unique in comparison to the others in the circuit. The quoins on each side of the building display toothed *opus vittatum* masonry composed of small tuff blocks (see plate 21). This construction technique as applied to the toothed quoins is rare at Pompeii. Its earliest use is otherwise attested to only in the large lower arcades of the amphitheater – its application to Tower VII remains unnoticed.[100] Despite the different material – in this case, tuff stone rather than fired brick – the concept behind the technique is similar to the brickwork quoins of Towers II and V. The masonry on Tower VII may therefore represent any of three scenarios: it may be part of the same reconstruction event, a variation of the construction technique, or the tower being roughly contemporary with the construction of the amphitheater in the early colony. Given the evidence elsewhere in the circuit, it seems that this masonry coincides with the reconstruction effort of Towers II and V.

Farther west, Tower VIII is in an equally ruined state, rising no farther than its lower main chamber and the staircase leading down from the first floor. It stands out because the ruins still preserve most of their plaster coating. The chamber shows signs of alteration and demilitarization. Later masonry partially fills in the arrow slits in an effort to make them smaller. Extra plaster applied on the exterior masked the alterations. On the interior, white plaster frames indicate that the windows also functioned in their previous unaltered state. Workers also walled up the postern and covered the exterior with plaster to hide the modification (see plate 22). Earthquake debris dumped in front of the postern supply a *terminus ante quem* of 62 CE for its closure. The earthquake subsequently damaged the building. It effectively ceased to function when a shabby concrete wall closed off the staircase to the main chamber from the city side. A few fragments of the tower, recovered in volcanic debris, indicate that parts of the ruined building still stood at the time of the eruption.[101] The peculiar circumstance of the walled-up postern has led Johannowsky to suggest that defensive strategies called for its express camouflaging and the necessity to keep it closed. He imagined that troops would only smash through it in the case of a sortie.[102] This scenario seems unlikely – the effort to break through the masonry would negate any element of surprise. Furthermore, the arch jambs still carry traces of an earlier plaster coating, confirming that the doorway was previously open. Although it is unclear when the transformation occurred, the closing of the postern and modifications of the arrow slits indicate the tower's changing strategic role in antiquity. The windows seem adjusted for smaller weapons, indicating that defenders no longer possessed the organization and weapons or needed to worry about withstanding well organized sieges. The careful application of a new plaster coat to mask these alterations marks a clear effort to preserve the representational role of the tower.

Nearby, the ruins of Tower IX differ significantly compared with the other towers. Its remains are part of a post-Sullan reconstruction of the building that included a stretch of the adjacent curtain wall. The ruins display the traces of a barrel vault that once covered a large chamber. In the back, a door surmounted by a small window gave access to street level on the city side of the walls. This design is unique because it implies that the fortifications did not have an *agger* in this area of the circuit. The lack of stairs and further windows implies that the new building only had a ground floor and changed its function to a warehouse. Nevertheless, the rebuilt structure featured the same First Style embellishments as the other towers, suggesting that it retained some sort of military or public function in unison with the fortifications.[103] Fragments of a previous tower complete with First Style ornamentation have emerged as they had fallen during the Sullan siege. They provide direct evidence of the continuous use of the First Style on the towers, as well as proof for the (re)construction of the towers and sections of the enceinte before and after the siege.

The three towers on the north side of Pompeii do not display repairs in the masonry. The deep fill that buried the exterior of the ground chamber and their posterns was the most radical change that reduced their military capabilities. The postern and the stairs heading down to the main chamber of Tower XI was walled up in antiquity in a similar fashion to Tower VIII, but it remains unclear when or why this happened.[104] The earthquake demolished large portions of the two western towers (XI and XII). Tower XII is missing its entire exterior facade, and crude walls seal the passageways. Tower X seems to have fared better. The arrow slits opening on its flanks were closed at some point in antiquity at an unspecified date. However, unlike the two towers farther west, much of the building was probably standing at the time of the eruption. Large parts of the structure lay collapsed in volcanic debris on the exterior of the fortifications at the time of excavation.[105]

The changing role of the fortifications

After an initial refurbishment, the city walls of early colony would gradually weaken militarily, and emphasis would shift to the gates, where it would remain even after looters started to dismantle the fortifications to rebuild the city after the earthquake. With each transformation, the defenses changed their meaning and symbolism with the shifting social and urban environment. Initially, security must have been an immediate concern that motivated Cuspius and Loreius to rebuild the walls in the early colony. With the recent memory of the Social War, the revolt of Spartacus, and the Catiline incident, the Italian peninsula was unstable, and brigandage was rampant. Cicero, in his orations against Catiline and his failed revolt, provides some contemporary textual evidence on Roman thinking regarding city walls and notions of citizenship. In 62 BCE, after Cicero uncovered his coup attempt, Catiline fled from Rome with his supporters and pitched camp outside the town of Fiesole. In his speech, Cicero labels Catiline and his followers as bandits who, from the Roman point of view, included lowlifes such as gladiators, prostitutes, gamblers, and others. Cicero refers to the walls of Rome as the protective barrier of the civilized city and its institutions against the dangers posed by Catiline and his followers camping in the uncivilized countryside.[106] City walls were the conceptual barriers protecting civilized institutions defining the state against the unknown "other."

Ideological concerns must have influenced the decision to rebuild the Pompeian defenses. In addition to the same Hellenistic notions, popular in the previous Samnite period, which included city walls in the image of the ideal city, fortifications were intrinsic to Roman colonial identity. Early colonies symbolically shared their plan with Rome, also known as *Roma Quadrata* (square Rome).[107] In these foundations, colonists would march to the designated destination laid out by surveyors and carry out the elaborate *sulcus primigenius* ritual as a symbolic connection to the legendary inception of Rome. In myth, Romulus used this Etruscan ritual to establish the city limit using a plough pulled by an ox and a bull.[108] It is here that city walls acquired a religious character as the liminal boundary between the sacred space of the city and the profane exterior, thereby creating a connection between the gods and the new city. In the earliest foundations, colonies often functioned as bridgeheads in dangerous enemy territory and were fortified because of their exposed position.[109] Their strict Hippodamian layout reflected the plans of army camps. The walls projected an inward sense of security to citizens and an outward message of dominance upon local populations.[110]

Polybius best describes how such orthogonal layouts functioned psychologically when he compares standard Greek vs. Roman army camps. Greek commanders chose easily defensible places in the landscape, adapting the encampment to the local conditions. By contrast, Roman generals chose less defensible sites and consistently adapted the landscape to repetitive Hippodamian plans. This approach carried significant advantages. Individual units always billeted in the same buildings. Upon entering the gates, each soldier returned to his established quarters irrespective of the camp's physical location. The designs strengthened "facility and familiarity," thereby adding a sense of community and confidence to camps in distant territories.[111] Upon first foundation, the rigid rectangular Hippodamian plans of early *colonia* must have functioned in a similar way. Civic populations consisted primarily of poor landless Latins or Roman citizens and of army veterans who received a plot of farmland in return for their years of military service or willingness to move from Rome. The concepts of security and dominance associated with the walls resonated strongly specifically with those who had served in the army.[112] The walls of the colony symbolically embodied those messages irrespective of whether they carried any further ornamentation.

The city walls would acquire symbolic meaning as the crowning element of a controlled landscape. As a colony became established, the surrounding landscape would change with it. Much of this change began with the construction of roads and the implementation of centuriation where surveyors would mark out precise plots of land for distribution to the colonists. At the heart of this practice was the intersection of the main *decumanus* and *cardo* – the axial roads dominating colonial urban layouts – which ideally would be an open space accommodating the Forum and Capitolium.[113] From this charged location surveyors measured out the city (*urbs*), as well as the division of the countryside (*ager*). The two roads extended into the regional highway network connecting the heart of the colony to the outside world and, by extension, to Rome. Such regional roads integrated the urban image into the surrounding landscape, forming new avenues of commerce and venues of ostentatious display. Burial monuments, centuriated land, villas, and sanctuaries lined approach routes to the city and established dramatic backdrops to anyone approaching the *urbs*.[114] It is unclear whether such a redistribution of land occurred at Pompeii, but luxury villas and ostentatious tombs appeared rather suddenly after the foundation of the colony to form a new colonial landscape.

The villas of the aristocracy that dotted the Italian countryside serve to illustrate how city walls were an integral part of Roman ideals. Already Scipio Africanus, the great general who had defeated Hannibal, upon his retirement would go to his villa fortified with a wall and towers. Perhaps he ordered it fortified out of military habit, but Scipio set a precedent for later adaptations.[115] Tangible remains come from a group of three agricultural villas built in the first century BCE near Cosa. The two best preserved structures, known as the Villa delle Colonne and the Villa Settefinestre, each displays the remains of precinct walls featuring round miniature towers that explicitly mimic those found on real city walls. The precinct walls marked the edge of the property where they fulfilled a clear protective role against brigands and other dangers such as wild animals. The model towers had no military function. Instead, they were a nod toward ideological notions associated with city walls and the model landscape developed throughout the Italian peninsula.[116]

It is from this perspective that city gates and walls projected the concepts of *securitas* and *romanitas* – ideas that translate loosely to "security" and "Roman-ness." Although these ideas generally apply to Roman city walls, they were particularly significant as part of the colony's position in hostile territory. Major city gates would thus herald the entrance back into civilization and the protection of Rome, whereas the fortifications provided an indispensable sense of security and order to both newcomers and the local population. The presence of any towers, especially if their spacing followed a prescribed distance or rhythm, as Vitruvius recommends, served to underscore visually the concepts in the landscape. Possessing fortifications signified both a colony's allegiance to Rome, but also its status as an independent community. The defenses would thus project the *romanitas* of the town they protected. Although it is hard to gauge how local populations would have viewed the establishment of new colonies, there must have been some hostility. The defenses in their strategic and symbolic role would have also acted as engines of mutual assimilation.[117] Certainly, Pompeii was not an *ex novo* foundation, but many of these concepts must have accompanied its refoundation and status as a colony. These factors led to a paradigmatic necessity to rebuild the fortifications in a way that had little bearing on their intrinsic military value.

Loreius and Cuspius operated in an equally paradigmatic system in its association between politics and euergetism. Their actions are part of both the effort to establish a proper colony and their own political career. The decision to finance particular public buildings, including their style, types, and placement in the city, implicitly reflected a patron's ideology and what he considered beneficial for his own career and the community.[118] Such actions were imperative if an individual hoped to pursue a further political career either at home or in Rome.[119] A parallel example of how this process also involved fortifications comes from the nearby town of Telesia, which, like Pompeii, was a Samnite town that became a colony of Sullan veterans after the Social War. Surviving inscriptions attest that six separate pairs of *duumviri* built towers or sections of fortifications in its early years. At Pompeii, Quintus Valgus, who was a member of the Sullan elite, adopted a similar course of action. In order to promote his political career outside of Pompeii, he engaged in the construction of fortifications in nearby towns. Originally from Hirpinia, he served as *patronus municipii* (city patron) in the town of Aeclanum and as a *duumvir quinquennalis* for an unidentified settlement near Abellinum. In both settlements, he paid for the construction of the city walls as part of his political ambition and Sullan patronage.[120] Given the recent conflict on the Italian peninsula, it is hard to imagine how settlements would be able to build fortifications without the knowledge of the Roman dictator. These examples highlight the importance of fortifications for colonial identities in the post–Social War period.[121] These powerful notions undoubtedly motivated Loreius and Cuspius to rebuild the Pompeian city walls.

Political, social, and military factors led to the further transformation of the defenses with the advent of empire. At Pompeii, the period saw a progressive reduction of the defenses with easier access routes into the city and increased construction on and outside of the fortified line. In immediate practical terms, an explanation for the degraded military capabilities of the Pompeian walls perhaps lies in the absence of a regular body of troops to maintain them. The army had stationed three cohorts at Pompeii during Caesar's Civil War who had offered Cicero control of the city as he passed through on his way to join Pompey in Greece in 49 BCE.[122] The presence of army units explains some minor modifications in the defenses, such as those present on Tower VIII. Presumably, the establishment of the *pax augusta* led to their withdrawal and redeployment elsewhere as the focus on defending the empire shifted to faraway frontiers. A military presence reappeared in Pompeii only after the riots of 59 CE between the Nucerians (Nocerans) and Pompeians that prompted the Senate to send the Praetorian Guard to restore order. The new deployment explains some of the minor repairs carried out on the towers and wall curtain around the amphitheater in the last years of the city. Troops reinforced the defenses in the area where the riots occurred, and large crowds could gather to threaten the social order.

The reduced military importance of the defenses has induced some to see their outright abandonment. This approach fails to take into consideration the continued social interaction with the fortifications and their role in safekeeping the community. Despite their faltering, the fortifications of the early Imperial period still formed a formidable barrier that remained largely complete without any glaring gaps in the perimeter. The emphasis on the gates in the Augustan and postearthquake period is a result of their persistent role as formalized filtering points for the continuous movement of goods and people and the population.[123] The Porta Marina even continued to function as a boundary marker despite the urbanization process occurring around it. A parallel situation developed in Rome where the expansion of the city eventually engulfed the line of the Servian Wall built in the fourth century BCE.[124] A few gates of the old wall continued to operate as formal points of passage into the city, with some receiving an overhaul in the Augustan period.[125] Eventually, gates and customs points would become synonymous. In the early Imperial period, Augustus designated new customs points in Rome for areas now well outside of the old Servian Wall. The architects of the later Aurelian Wall would include many of these existing toll points into their circuit rather than designating new gates to the city. A further redefinition of the custom points occurred in Rome during the censorship of the emperor Vespasian and his son Titus in 73–74 CE. This act was part of a wider effort where the emperor would dispatch Suedius Clemens to Pompeii and other places to return illegally occupied public land to the state.[126] Suedius would mark this event with a series of *cippi* installed on the public land outside the Pompeian gates for everyone to view.[127] Despite the abandonment of the curtain walls at this stage, the placement of the *cippi* serves as a reminder of how the gates functioned as formal points of passage reflecting authority.

The city walls built during the Augustan period in Italy fulfilled more of a practical symbolic role rather than a defensive one. They functioned primarily to keep out brigands, wild animals, and thieves.[128] For instance, wild wolves spotted within the perimeter of the walls were one of the worst omens that could befall a Roman city. Their presence was an insult to the city's protective gods and symbolized death, war, and the collective threat of a dangerous species.[129] Such unpredictable forces were among the chief reasons that communities first chose to defend themselves.[130] Bandits were similarly undesirable elements living beyond the civic confines.[131] Uncontrollable rogue elements that fell outside the social order, such as Spartacus in the Vesuvian area, were particularly dangerous to the city. Political adversaries such as Catiline, who were members of the upper-class circle of peers, posed a challenge to urban social order. Cicero condemned and exposed Catiline as a traitor conspiring against the state, essentially framing him as a dangerous outsider. In his speeches, the orator referred to city walls as forming the last line of defense protecting the civilized *urbs*, whereas the countryside where the "traitor" dwelled harbored the improbabilities and dangers of disorder.[132] Certainly, the charged climate and multitude of political adversaries in the final decades of the Republic emphasized the role of city walls as the guarantors of order and stability. Without the immediate danger of organized armies attacking the city after the advent of the empire, their makeup changed to reflect this role. The fortifications of Pompeii were still formidable enough to challenge any such threat.

Notes

1 On the ambiguity of the dates, see Appian *Bel. civ.* I.39 and I.50; Vell. Pat. *Hist.* II.16; Paul. Oros. *Hist ad. pag.* V 18; Lo Cascio 1991, 123. As a testament to the importance of the battle, Sulla himself may have led the siege; see Amery and Curran JR 2002, 17.
2 On Stabia, see Pliny *Nat. hist.* 3.70. For a discussion on the extent of the destruction, see Castrén 1975, 50.
3 Lo Cascio 1991, 122.
4 Descoeudres 2007, 16.
5 Guzzo 2007, 118–119.

6 Burns 2003, 1–9; Grimaldi 2014, 25–26.
7 Van Buren 1925, 110–111; Van Buren 1932, 14–17; Jashemski and Meyer 2002, 7; García y García 2006, 15; Pesando 2011, 10–11.
8 Rhodes 1998, 36.
9 Lo Cascio 1991, 123.
10 CIL X 937. Loosely translated the inscription states how the Duoviri Loreius and Cuspius oversaw and completed the construction of the walls.
11 Fiorelli 1862, 96; Fiorelli 1873, 89.
12 Nissen 1877, 511.
13 Castrén 1975, 161, 184.
14 D'Agostino 2013, 218.
15 For the association with the revolt of Spartacus, see Castrén 1975, 88; Zevi 1996, 129; also Pesando and Guidobaldi 2006, 30; Guzzo 2007, 118.
16 Rowlands 1972, 448; Lawrence 1979, 115; Kern, 323–351; McK. Camp II 2000, 48.
17 App. *Bel. civ.* I. 96.
18 Hori 2010, 288–289.
19 Russo and Russo 2005, 71–75. Hori is more reserved, seeing their presence as generally typical of all *opus incertum* tracts; Hori 2010, 286.
20 Sakai and Iorio 2005, 329; Etani 2010, 308.
21 This conclusion is tenuous because it assumes that only Roman builders used this measurement; see Hori 2010, 288–289.
22 Excavations north of Insula 2 in Regio VI have identified and dated these remains; Garzia 2008, 1–3; D'Auria 2008, 103–106.
23 Curti 2007, 72; Hori 2010, 288.
24 Mau 1902, 422; Pesando and Guidobaldi 2006, 259.
25 The recent collapse at the Porta Nola demonstrates how *opus incertum* masonry tends to shear off in sections ("Pompei, ancora un crollo – Corriere della Sera" 2011, www.corriere.it/cronache/11_ottobre_22/pompei-crollo_9626fc92-fc8e-11e0-92e3-d0ce15270601.shtml?refresh_ce-cp).
26 Maiuri 1939, 233.
27 Maiuri 1960, figs. 19 and 20.
28 Guzzo 2007, 115.
29 Etani 2010, 94.
30 Etani and Sakai 1995, 60.
31 Maiuri 1943, 281–286. Nineteen spouts survive west of the tower.
32 Perhaps one crew focused on reconstructing the tower and the other on the curtain section.
33 Hori 2010, 289.
34 Pesando 2006, 148–150.
35 Cassetta and Costantino 2008, 200–202.
36 Pappalardo 2005, 331; Guzzo 2007, 118.
37 Similarly, a door cut into the travertine masonry offered access from the House of Championnet II (VIII.2.3).
38 Bonucci 1830, 81; Niccolini and Niccolini 1862, 8; Adams 1873, 49; Mackenzie and Pisa 1910, 20; Sogliano 1937, 280.
39 Gell 1832, 90.
40 Dyer 1867, 58.
41 Overbeck 1854, 39.
42 See Gell 1832, 2, 161–164.
43 Gell 1832, pl. XVI and XVII.
44 Wilkins et al. 1819, pl. IX.
45 Fumagalli 1828, 36.
46 Le Riche 1827, pl. 35.
47 Breton 1855, 179.

48 Rossini 1831, pl. LXXV; Mazois 1824a, pl. XXXVI, figs. 1 and 2; Gell 1832, pl. XV. Gell describes a piece of cornice on the ground in front of the north bastion as part of a decorative feature that had collapsed from the parapet, but no further evidence exists to place it there. Such a cornice may have belonged to a nearby funerary monument. See also Breton 1855, 240.

49 Russo and Russo 2005, 71–75. The only plaster surviving on the *opus quadratum* is on the western bastion of the Porta Vesuvio, where it remains stuck behind the later arch jamb. The niche of the Porta Stabia also still carries plaster. In both cases, the embellishments are limited to the shrines rather than covering the entire masonry.

50 Just. *inst.* I.2,1,10; Gai. 2, 8–9. For further discussion on the matter, see Seston 1966, 1489–1498; Smith and Tassi Scandone 2013, 469–471.

51 Maiuri 1943, 279–280; De Vos 1977, 29–47; Chiaramonte Treré 1986, 57–59. Similarly, excavators recovered the *schola* tombs outside the Porta Stabia, which had been covered in rubbish and debris deposited after the 62 CE earthquake. See Anonymous 1889, 281; Richardson 1988, 254–255.

52 Maiuri 1939, 232.

53 Perhaps they were a series of water tanks and later horrea, but its original function remains unknown; Eschebach and Eschebach 1995, 77.

54 See Brands 1988, 184; Wallat 1993, 364. Ippel goes as far as identifying a colonial phase where the Porta Marina had three openings to resemble a triumphal arch like the Porta Ercolano. No authors have since elaborated on this observation, and it is absent in the surviving masonry; see Ippel 1925, 181–182.

55 Guzzo 2007, 116–117.

56 See Mau 1879, 233–234.

57 Palmer 1980, 217–218.

58 Parallels exist in Perugia and Ferentino. See also Guzzo 2007, 117.

59 Jacobelli and Pensabene 1995–1996, 45.

60 The arguments for and against the presence of a harbor in this area of the city are many. I tend to accept that the stone rings in the wall had some sort of naval association. I therefore follow Curti 2005, 59. He counters Descoeudres, who identified the quay wall as a fortification; Descoeudres 1998, 216.

61 Guzzo 2007, 116.

62 Scagliarini Corlàita 1979, 29–72; MacDonald 1988, 99–110; Favro 2006a, 193.

63 Until the bombing the plaster coat had hidden much of the masonry, leading to a debate on the date of the gate; Breton 1870, 232. Also for the closing mechanism Fiorelli 1875, 215–216; Von Rohden 1880, fig. 24; Molesworth 1904, 27; Fischetti and Conforti 1907, 26; Van Buren 1925, 106–107; Brands 1988, 184; García y García 2006, 167–171 and figs. 408 and 409.

64 CIL X 1064; Maiuri 1929, 198. The date is somewhat uncertain. Castrén sees them in office in the early Julio-Claudian period. Castrén 1975, 141, 223. The new causeway may coincide with the closure of the pomerial street leading to the Porta Marina. A Claudian *sestertius* found beneath the paving stones at the Porta Stabia suggests a slightly later date; see Spano 1911, 377. Otherwise, see Van der Graaff forthcoming.

65 Recent investigations indicate a late date based on wear patterns on the nearby fountain; see Ellis and Devore 2007, 124–125. However, this discounts the possibility that wooden boards covered the drain.

66 Construction on the Small Theater may have begun before the siege; see Wallace-Hadrill 2008, 131.

67 At the Porta Marina and Nocera, the works also led to the blockage of the sewers. For the Porta Vesuvio, see Seiler et al. 2005, 252.

68 Maiuri 1929, 219; Guzzo 2007, 115.

69 D'Agostino 2013, 219.

70 Pesando and Guidobaldi 2006, 23; Guzzo 2007, 170.

71 Guzzo 2007, 136.
72 Ohlig 2004, 93–102; Keenan-Jones 2015, 191–215.
73 See Maiuri 1943, 277–279; Eschebach and Eschebach 1995, 84; Keenan-Jones 2010, 250–252.
74 Seiler et al. 2005, 224.
75 Ohlig 2001, 272–273.
76 Maiuri 1943, 277–279; Ohlig 2004, 93–102.
77 If we follow a Sullan date for the first aqueduct, the new complex also carried an implicit colonial statement. For the Sullan date, see Ohlig 2004, 104; otherwise, see Keenan-Jones 2015.
78 Here I follow Fröhlich 1995, 153–159.
79 For siege damage to the houses, see Burns 2003, 1–9; Jones and Robinson 2005b, 259.
80 Frigerio 1935, 93; Brands 1988, 190.
81 Brands 1988, 188–190. The debate has focused on the closing mechanisms of the thresholds. The exterior passageways lack a closing mechanism, whereas the main arch displays grooves for a portcullis. The presence of plaster in the grooves has induced some authors to believe that there was no portcullis because it would have ruined the coating; see Niccolini and Niccolini 1862, 9; Overbeck and Mau 1884, 55. Monnier sees the absent portcullis as a ploy to draw in attackers; see Monnier 1865, 101; Monnier 1870, 96. In any event, the gate closed on the interior and could still function to keep unwanted elements out.
82 Mazois 1824a, pl. XI, figs. II and III; Rossini 1831, pl. XII; Mau 1882, 58; Overbeck and Mau 1884, 55.
83 Gell 1832, pl. XIII. See also Overbeck and Mau 1884, fig. 14.
84 Mazois 1824a, 29; Bonucci 1830, 79; Breton 1855, 240; Dyer 1867, 58; Fiorelli 1875, 76–77; Delaunay 1877, 84; Niccolini and Niccolini 1896, 14.
85 Gell 1832, pl. XIX, reused in Overbeck and Mau 1884, 42; Coarelli 2002, 32.
86 Niccolini and Niccolini 1896, pl. XIV; Delaunay 1877, 81. Delaunay is not very accurate: He describes the side entrances on one of the towers as being the Porta Stabia; Delaunay 1877, 12. See also Fischetti and Conforti 1907, 80.
87 Senatore 1999, 95. The Porta Ercolano necropolis seems to have a higher incidence of elite tombs, perhaps because it served Regio VI where we find a concentration of opulent houses. On the importance of key roads, see Zanker 2000, 25–41.
88 Fröhlich 1995, 151.
89 Maiuri 1943, 278–280.
90 Ling 2007, 125.
91 De Caro 1998, 239–244; Zevi 2003, 856–864.
92 Pesando 2009, 378–385.
93 See chapter 6 in this volume.
94 For further discussion, see chapter 8 of this volume. Also, Ortalli 1995, 469–529.
95 Cassius Dio *Hist. Rom.* 53.22–2. On this role for arches, see Scagliarini Corlàita 1979, 29–72.
96 Adam 2007, 105–107.
97 Maiuri believed that the quoins represent an early Imperial reconstruction based on the employment of *opus vittatum mixtum* at the Porta Ercolano; Maiuri 1929, 232; countered by Fröhlich 1995, 153–159.
98 On the burial activities, see Guzzo 2007, 151–153.
99 On the amphitheater repairs, see Pesando and Guidobaldi 2006, 72.
100 Pesando and Guidobaldi 2006, 72; Adam 2007, 108. Both sources ignore Tower VII, which may have formed a precursor.
101 Chiaramonte Treré 1986, 28–31.
102 Johannowsky 1994, 131.
103 Etani and Sakai 1999, 125; Etani 2010, 307–309 and pl. 20.

104 Maiuri 1943, 292. Perhaps engineers closed the opening to facilitate the deposition of the earth fill.
105 Maiuri 1943, 286, 292.
106 Cic. *Catil*. 1.13.32–33, 2.1.1–13, 2.9.20.
107 Gros 1996, 33.
108 Salmon 1970, 24; Rykwert 1988, 135.
109 Cornell 1995, 126.
110 Salmon 1970, 15, 17.
111 Polybius *Hist*. 6. 41–42.
112 Zanker 2000, 25–29.
113 Salmon 1970, 20–28; Rykwert 1988, 137.
114 Zanker 2000, 25–33.
115 The location of the villa is unknown. Seneca describes it two centuries later; Seneca *Epist. mor. ad lucilium* LXXXVI.4.
116 Calastri 2004, 186; Poggesi and Pallecchi 2012, 166.
117 Février 1969, 286; Zanker 2000, 25–41; Pinder 2011, 72.
118 Laurence 2007, 36.
119 Castrén 1975, 87–92.
120 CIL I² 1722; ILLRP 598. Castrén 1975, 89; Adamo Muscettola 1992, 80; Zevi 1995, 9–10.
121 For continued trends during the empire, see Thomas 2007, 110–112.
122 Cic. *Fam*. 7.3.1; Cic *Att*. 10.16.4; Castrén 1975, 91.
123 Pinder 2011, 73.
124 Favro 2006a, 191.
125 E.g., the Arch of Dolabella; CIL VI 1384.
126 See Palmer 1980, 217–218 for the walls of Rome as custom points.
127 See the next chapter of this volume.
128 Rosada 1990, 364–368; Rosada 1992, 171–183.
129 Trinquier 2004, 85–118.
130 Rowlands 1972, 453.
131 For a definition, see Grünewald 2004, 161–165.
132 Cic. *Catil*. 1.13. For an analysis, see Habinek 1998, 71–74.

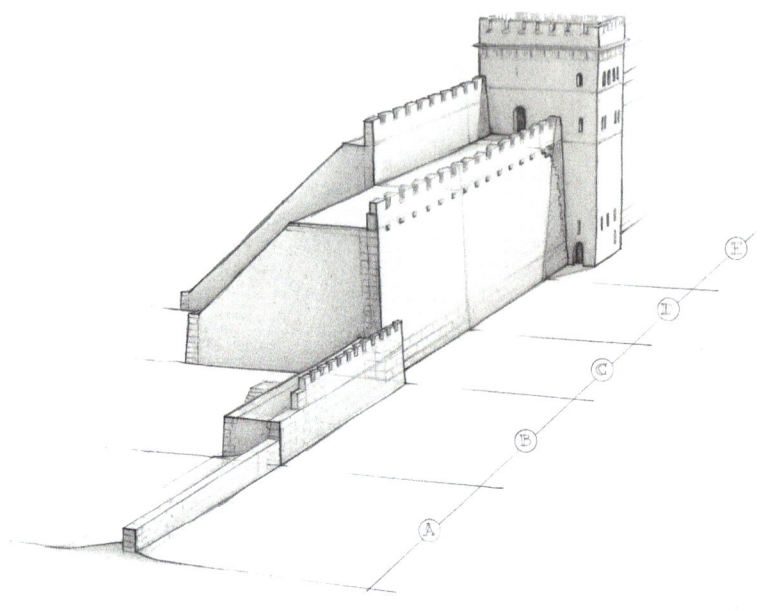

Plate 1 The phases of the Pompeian fortifications marked A–E. The main circuits are (A) Pappamonte, (B) Orthostat, (C) Samnite. Letters D and E mark upgrades on the Samnite fortifications. (Drawing by L. Kukler.)

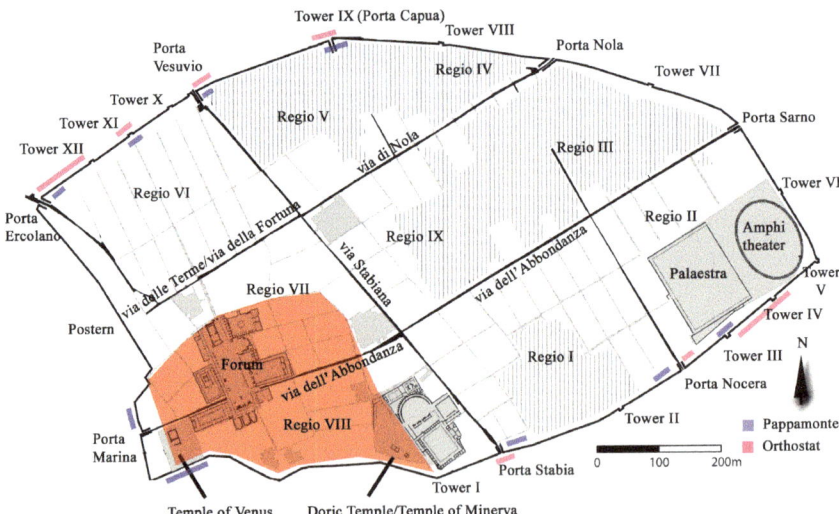

Plate 2 Plan of Pompeii highlighting the fortifications as well as the *Altstadt* (in red). The colored sections mark the approximate locations where excavations have recovered the Pappamonte and Orthostat fortifications. (Base plan courtesy of S.J.R Ellis, modified by author)

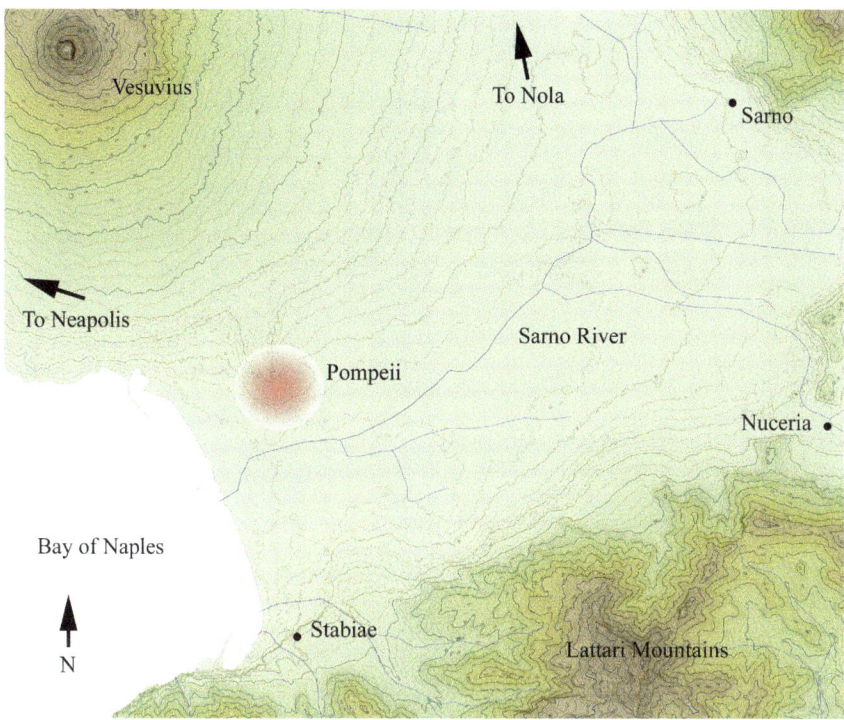

Plate 3 Digital elevation model of the Sarno valley with Pompeii marked in red. (Courtesy Giovanni Di Maio.)

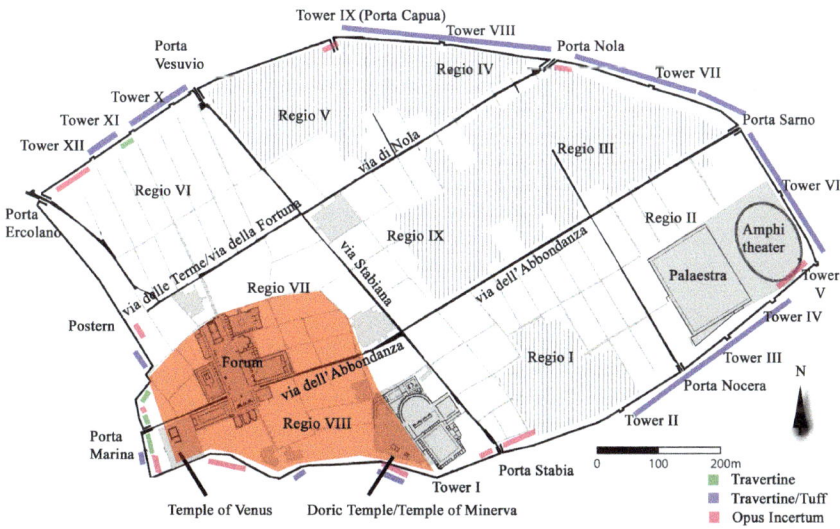

Plate 4 Plan showing construction techniques present in the Pompeian fortifications. The *Altstadt* is marked in red. (Base plan courtesy of S.J.R Ellis, modified by author)

Plate 5 Section of the fortifications beneath the House of Fabius Rufus. The lines indicate the construction seams.

Plate 6 Section of fortifications west of the Porta Nocera. The red lines mark the exposed foundations and the remaining course of tuff masonry.

Plate 7 Section of masonry west of the Porta Nola showing the juxtaposition of the travertine and tuff masonry marked with a red line. Note how the tuff masonry was robbed out at a later date.

Plate 8 Wall curtain between Towers XI and XII highlighting (by the red line) the differentiation between travertine and tuff.

Plate 9 Reconstruction of the Porta Nola during the first Samnite phase. (Modified by author after Krischen 1941, pl. IV.)

Plate 10 Reconstruction of the Porta Nola. The inset shows its current state. (Drawing by L. Kukler.)

Plate 11 The Porta Nocera and its exposed foundations. The red lines indicate the concrete and the exposed tuff foundation blocks.

Plate 12 The cross section of the Porta Vesuvio. Note the juxtaposition of the masonry and the *castellum* in the rear as well as Tower X (right).

Plate 13 Reconstruction showing the upgraded wall to include the towers. (Drawing by L. Kukler.)

Plate 14 Repairs in the ashlar masonry west of the Porta Nola.

Plate 15 Reconstruction drawing of the Porta Nola in its final phase. (Drawing by L. Kukler.)

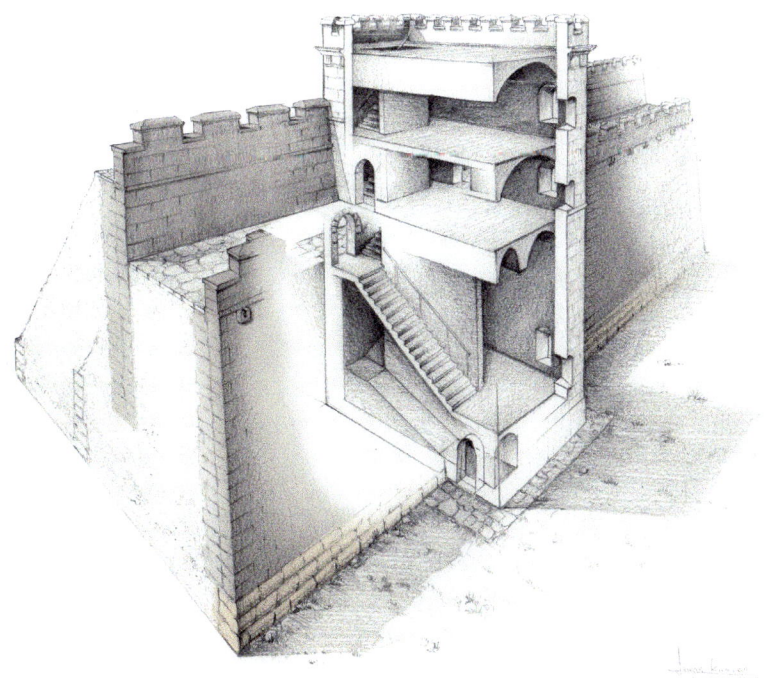

Plate 16 Cross section of Tower X on the north side of the fortifications. (Drawing by L. Kukler.)

Plate 17 Tower VI near the Amphitheater. The arrow points to the entrance from the left. The red line (on the left) distinguishes the travertine and tuff masonry.

Plate 18 Drain of the Porta Nola. The arrow marks the outlet. The red line marks the travertine/tuff distinction; the inset shows the drain inlet.

Plate 19 Reconstruction of the Porta Ercolano. (After Niccolini and Niccolini 1896, pl. XIV.)

Plate 20 Tower III. The red circle denotes the brick patchwork. The arrow indicates the quoin repairs and the line of exposed foundations.

Plate 21 The ruins of Tower VII. The circle marks the *opus vittatum* tuff quoin. The red line marks the transition between travertine and tuff.

Plate 22 Tower VIII with the walled-up postern and modified arrow slits highlighted. The inset shows the interior with the newly finished windows (left) and the walled-up postern in the rear.

Plate 23 Frieze of the Trojan cycle in the House of the Lararium of Achilles.

Plate 24 Apollo and Poseidon oversee construction of the walls of Troy, House of P. Vedius Siricus.

Plate 25 Early Imperial *emblema* from the floor mosaic in the House of the Centenary. (By concession of the Ministero dei Beni e delle Attività Culturali del Turismo – Museo Archeologico Nazionale di Napoli.)

Plate 26 Pompeii, House I.3.23, Riot at the Amphitheater. (Photo M. Larvey, by concession of the Ministero dei Beni e delle Attività Culturali del Turismo – Museo Archeologico Nazionale di Napoli.)

Plate 27 The Polychrome Tumulus at Cerveteri. (Photo by N.A. Van der Graaff.)

Plate 28 The curtain wall adjacent to the Porta Romana at Segni.

Plate 29 The entry corridor into the town of Trebula Balliensis.

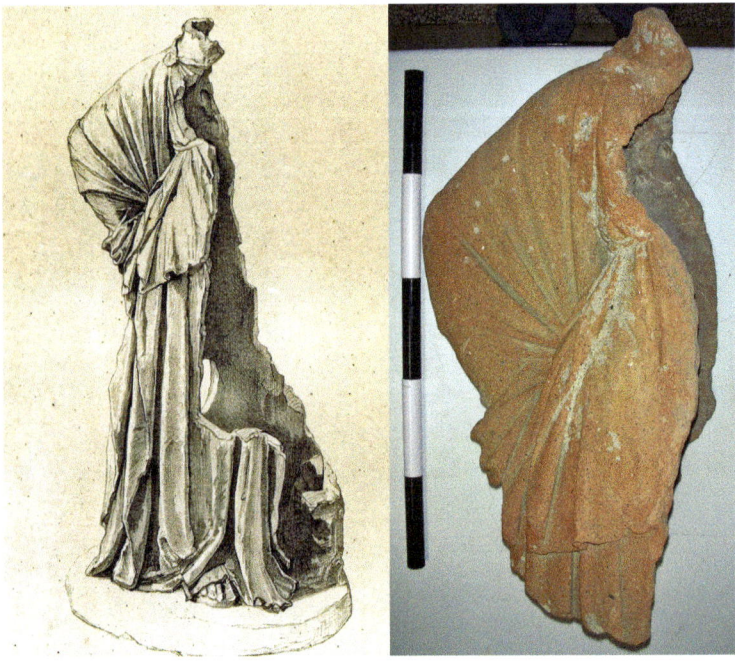

Plate 30 The terra-cotta statue of Minerva from the Porta Marina: the reconstruction (left) (after Von Rohden 1881, pl. 31) and its largest fragment as recovered in 2010 (right).

Plate 31 Fresco of a procession from the Shop of the Carpenters. Note Minerva on the top left. (Photo M. Larvey, by concession of the Ministero dei Beni e delle Attività Culturali del Turismo – Museo Archeologico Nazionale di Napoli.)

Plate 32 The remains of the Porta Venere in Spello. The inset shows a reconstructive drawing. (After Frigerio 1935, fig. 118. Photo by N.A. Van der Graaff.)

6   The fortifications and Roman Pompeii

After the establishment of the colony, Pompeii would flourish for approximately 160 years until the eruption of Vesuvius in 79 CE. Much of the ruins on view today are a product of this period. Their presence limits any approach to the elusive urban matrix of the older Samnite city. However, the rich remains of the colony allow for a different perspective on the fortifications. Mosaics, (funerary) architecture, and frescoes, as well as a clear picture of Pompeian society, help to contextualize the social role of the walls. Admittedly, the evidence tends to belong to a very particular group of wealthy freedmen and citizens who could afford to decorate their houses and commission tombs. Nevertheless, it is precisely within this group who benefitted most of the system that the connections between the city walls and wider notions of Roman citizenship and ideals are most evident. Practical, political, and ceremonial factors had shaped the design of the fortifications of the Samnite city. The defenses of the Roman colony would purvey notions of *romanitas*, *securitas*, and *virtus*. Some of these ideals must have been part of the fortifications previously, but they come into focus after the installation of the colony. The extent to which these notions were associated with the walls changed over time, achieving various degrees of complexity that hinged upon the changing urban, social, and political developments.

Adjusting the civic image

The establishment of the colony came with an influx of Roman veterans that induced a dramatic shift in the demographic and social make up of Pompeii. Estimates put their numbers anywhere between an entire legion, that is, 4,000–5,000 men, and about half that number, all bringing with them family and dependents.[1] In terms of social and urban organization, the event must have included a measure of property reallocation toward the new settlers that typically accompanied the establishment of a colony. Where the new settlers lived and how they interacted with the local population are matters of contention. Some elite houses remained the property of the old aristocracy, but proscriptions did occur and many assets changed hands. A number of colonists settled somewhere northwest of the city in the newly founded *Pagus Augustus Felix Suburbanus*.[2] The most affluent settled in three main areas: the large agricultural villas north of the city, a series of houses lining the via

Marina/via dell'Abbondanza axis near the Forum, and the houses along the southern and western ridge of the city. These concentrations reflect the new order: The reallocation of the villas represents an economic takeover, the dwellings near the Forum put the new elite at Pompeii's political heart, and the houses along the ridge would become mansions exploiting the magnificent views.[3] The new owners would renovate the interiors of these houses or build them from scratch.

The development of these dwellings and their interior decorations provide a measure for the association between the fortifications and the new social order. As an indicator of this process, a number of the distinctive tuff facades of the old elite Samnite houses disappeared beneath a thin coat of white plaster. The older tuff facades would thus fall within a wider scheme. The elaborate carved blocks, cornices, and capitals were now almost indistinguishable from the imitation plaster otherwise applied to facades built in *opus incertum*. New houses would increasingly adopt this new construction material – their preponderance is probably because they were easier and cheaper to build. The most elaborate facades with ornamental First Style schemes included a low socle, surmounting ashlar zone, and door pilasters featuring tuff capitals applied to reflect the social prominence of the owner.[4] Less prominent facades would feature smooth plaster and a brightly painted socle in an arrangement similar to the vaults of the city gates. The mechanisms of the adaptation of tuff facades is by no means clear or ubiquitous; houses such as that of Pansa would remain without a plaster veneer. In some cases, such as the House of the Faun, the plaster masked earthquake repairs. In others such as at the entrance VIII.5.19–20, the plaster would serve to cover an otherwise intact wall face.[5]

The retention or covering over of tuff facades may be a reflection of an owner's identity, status, or even political allegiance. It could relate to the concept of *facadism* where the owner deliberately chooses to preserve the facade as an urban and social artifact.[6] Such political associations are evident in the Augustan period when the First Style as applied on the exterior of buildings saw a resurgence as a reflection of the renewed emphasis on social *mores*. This connection relates to the proposed origin of the style as an imitation of Athenian civic architecture and the Augustan exaltation of the Classical Greek aesthetic.[7] It is also an effort to replicate the effect of white marble, which became a newly abundant status marker surpassing tuff after the opening of the Carrara marble quarries under Augustus. These quarries allowed the Emperor to boast that he transformed Rome into a city of marble and went hand in hand with the development of the new imperial ideology.[8] If the application of the First Style in the Augustan period responded to the fashion and politics of the time, it must have carried some sort of similar ideological weight in the previous Sullan period as well. House facades and public buildings such as the fortifications displaying the First Style presented Pompeii as a proper Hellenistic city.

Together with the houses, the edifices erected and modified in the early colony indicate an effort to claim the city as Roman. Pompeii would experience a construction boom as the local elite attempted to achieve a complete civic image as part of their own political aims. New public buildings included the Amphitheater, the so-called Odeion, and baths immediately north of the Forum. Public temples also received renovations: The Temple of Mefitis Fisica would transform into that of Venus Fisica Pompeiana, the Temple of Jupiter in the Forum would change to accommodate the Capitoline triad, and the Temple of Apollo would receive a new altar and staircase.[9] Each of these buildings would carry its own ornamentation, but the public temples would resonate most with the city walls. As opposed to the brightly painted private sanctuaries, these shrines included white First Style decoration on their exterior to highlight their role as houses of the public gods. The Temple of Minerva (Doric Temple) and the Temple of Venus received new terracing walls in *opus incertum* that were part of the fortifications. The masonry in such a prominent location undoubtedly carried a plaster embellishment and renewed the connection among the walls, the deities protecting the city, and the sponsors of the reconstruction effort. The collective import of these buildings and refurbishments, which would supplement already existing Roman spaces such as the Basilica and the Forum, transformed Pompeii into a late Republican colony.[10] The First Style exterior embellishments reflected its status and new social reality.

The significance of an entertainment district: the fortifications, the Amphitheater, and the Palaestra

The Amphitheater, built on top of the fortifications in the southeastern corner of the city, is the most overt symbol of the Roman colony. Dating roughly to 70 BCE, the dedicatory inscription announces that the *duoviri quinquennales* C. Quinticius Valgus and Marcus Porcius built the structure and donated it in perpetuity to the colonists of the city.[11] Its placement created a direct architectural dialogue with the enceinte for the remainder of Pompeii's history (see plate 2). The traditional view sees the location of the amphitheater as a matter of expediency: There was a general lack of urbanization in that area, and the *agger* could support over half the structure.[12] However, the process to build the amphitheater was more complicated than has been proposed to date. The area was in private hands with agricultural houses occupying the district. The *duoviri* needed to conduct costly expropriations, demolitions, and perhaps even receive special dispensation from the city council to occupy the city walls.[13] Rather than a matter of simple convenience, the patrons must have had an implicit desire to build the amphitheater in this part of the city. The location on the city walls, a symbol of previous independence, may very well have acted as part of a political statement of conquest and subjugation.[14]

Much more than a simple entertainment structure, the Pompeian amphitheater carried a symbolism tied to the status of Pompeii as a colony, the Roman army, and its veterans. Katherine Welch has convincingly argued that the amphitheater was part of a wider entertainment complex with military overtones that included an early version of the neighboring Palaestra.[15] Although an early version of the Palaestra remains uncertain, the notion of a complex designed to entertain veterans is plausible. A building this large carried intrinsic meanings that included a strong relationship between the army and the Roman colonies of the late Republic. By the first century BCE, soldiers trained in gladiatorial techniques and watched spectacles in amphitheaters built in army camps to introduce them to the brutality of battle.[16] The Pompeian amphitheater even included architectural references to army camps that are lost to the modern viewer. Visual similarities existed between the ramps that offered access to the parapet flanking camp gates and the stairways that are iconic of the arena.[17] The appearance of the form at Pompeii with a contingent of veterans is a clear result of these associations and the new colonial reality of the city.

On a regional scale, the amphitheater would act as a landmark that amplified its colonial message by drawing large crowds from in and around the city. The fortifications – although weakened militarily to function as support for the building – would form an architectural backdrop carrying implicit messages to those attending the games. For visitors arriving from the Sarno plain, the towers and walls set the stage for the amphitheater and city behind them. The spectators funneled through the Porta Sarno and the Porta Nocera, which symbolically marked the entrance into the colony and the new entertainment district. The large influx of crowds is probably the primary factor that led to the lowering of the Porta Nocera since the road going through it connected directly to bridges over the Sarno River toward the neighboring towns of Stabia and Nocera. The works related to this lowering included cutting back the scarp in front of the walls, thereby exposing their foundations. Although the result effectively weakened the fortifications from a military perspective, the lowering added to their height and created a further impression of strength and impenetrability.[18]

Once a viewer was inside Pompeii, Towers IV, V, and VI framed the arena acting as an architectural crown as visitors approached the building. As spectators climbed the stairs and reached the upper platform of the amphitheater and their seats, they had spacious views visually punctuated with towers that defined the edge of the city and projected its image onto the countryside. Inside the building, spectators of the early colony sat according to military rank – an arrangement that would transition to separation by class in the Augustan age. The result was a microcosm of Roman social order.[19] Gladiatorial games, as representatives of social control, were venues of encased ritualized violence reflecting the brutality of life, empire, and Romanization.[20] At Pompeii, the fortifications surrounded the gladiatorial spectacles – the fights among animals and criminals that represented the very irrational forces the walls were meant to keep out. The walls would thus express the Roman character, or *romanitas*, of the city beyond their immediate associations with the proper image of the colony.

Anxiety about social order and crowd control must have led to the reconstruction of the towers around the amphitheater. The revolt of Spartacus, quelled only after a long and bloody conflict that directly involved Campania and the environs of Pompeii, had exposed the dangers related to the gathering of gladiators. The riot of 59 CE between the Nucerians and Pompeians at the amphitheater is a reminder of the dangers of large crowds. Social tensions were rife in the years after the establishment of the colony. Cicero in his pro-Sulla speech clearly identifies a degree of mutual distrust between the local inhabitants and the new arrivals.[21] Sulla's nephew Publius Cornelius Sulla, one of the founders of the colony, stood accused of siding with the Catiline conspiracy. In his defense, Cicero praised

Sulla as an exemplary citizen, pointing to his role as arbiter quelling dissension and tensions between the two populations in the colony, which simmered along the pro-Roman and pro-Italian divide of the Social War.[22] The construction of the amphitheater led to the lowering of the *agger*. This event would isolate the main entrances of the towers, which engineers rebuilt contemporaneously or soon thereafter as part of the general refurbishing of the walls. Given Pompeii's previous allegiance, the intensity of the conflict, and its lingering effects, authorities probably rebuilt and isolated the towers as a security measure to prevent their hostile takeover during an uprising or riot. The towers would fulfill a policing role, providing a powerful reminder of civic order when visibly manned during gladiatorial contests.

Construction of the Palaestra radically changed the district with the advent of the empire. The building is almost of equal size to the amphitheater, occupying some six regular *insulae* (city blocks) to the east. A precinct wall defined its western edge, whereas three porticoes to the east framed a large garden with a central pool. Although the benefactor of the building remains unknown, the Palaestra would become a defining element of the district. Its role in expressing the realities of the new empire is apparent. The Palaestra probably functioned as a *porticus* similar to those built by the new emperor in Rome, open for citizens and, in this case, spectators of the games to enjoy.[23] The building also worked as a *campus*, or training ground, for a paramilitary youth organization known as *iuventus*, formalized under Augustus. This organization specialized in training young men to become ideal citizen soldiers who would demonstrate their skills in annual open drills known as the *iuvenilia* that took place inside the adjacent amphitheater.[24] The two buildings would operate almost as an extension of each other, in the process acquiring a mutual symbolic relationship connected with the military.

The spatial proximity between the Amphitheater, the Palaestra, and the adjacent city walls led to a natural architectural dialogue that architects would exploit.[25] The Palaestra's western precinct wall displays diminutive merlons mirroring those of the city walls nearby (see Figure 6.1). The imitative fortification was purely symbolic, imparting the image of a fortress and acting as backdrop to the basic military training expected of the youth enrolled in the *iuventus*. The merlons symbolized the concept of *virtus*, or excellence and bravery of the individual, to those training within its grounds.[26] *Virtus* was a well established concept and an essential quality in the character of the Roman aristocrat because it promoted combat skills, tactical insight, and courage to command units in battle.[27] It was a central Augustan ideal that endorsed the proper physical fitness and moral character of Roman citizens. *Virtus* was the strongest belief connected to military life and – along with the notion of *pietas* – fundamental to the Augustan moral values accompanying the program of civic renewal and construction of city walls on the Italian peninsula.[28] The associations among the nearby amphitheater, the fortifications, and the embellishments on the Palaestra produced a wider symbolism in the area connected with the realities of power. Together, they projected the Augustan renewed emphasis on social *mores* accompanying the image of the proper colony.

Figure 6.1 Surviving merlons on the exterior of the Palaestra wall.

After the earthquake of 62 CE, the towers and large sections of the fortifications around the amphitheater collapsed and were spoliated for use in the reconstruction effort. Although the towers display some repair work, the powerful messages that the fortifications had projected on the area lay in ruins. In the Sullan period, the construction of the amphitheater district gave it a symbolic role that followed the ideals of the new colony. With the construction of Palaestra in the following Augustan age, the area became an encompassing reflection of Roman order to city and countryside inhabitants alike. Throughout these periods, the city walls remained a constant presence translating their protective roles as reflections of changing civic institutions and ideals. From the symbol of Samnite independence to the Roman appropriation and projection of the new social order, the meaning of the fortifications transformed along with the changing architectural and social landscape.

Claiming public land

In the period between the installation of the colony and the eruption of Vesuvius, the public terrain associated with the line of the fortifications would witness a progressive occupation. The establishment of a military zone around the fortifications was a matter of strategic importance. Surviving Hellenistic inscriptions carved on the fortifications of Nisyros, Paros, Pylos, and Ephesus, for instance, specify leaving an open space, ranging between 5 and 40 feet, in front of the fortifications.[29] Its purpose was for defenders to see and engage the enemy, and to prevent attackers from using any obstacles as cover to their advantage. The effect of letting the space fill up with buildings over time could be disastrous. In his description of the siege of Cremona in 69 CE, Roman historian Tacitus vividly describes how attacking troops of Marcus Antonius Primus used nearby buildings, which overtopped the walls, to dislodge the defending Vitellian forces. The attackers used their superior position to pelt the defenders with raw materials including beams, roof tiles, and flaming missiles. Cremona would fall soon thereafter.[30]

Recovered inscriptions confirm that much of the Pompeian circuit included a similar military zone. Some are no more than small *cippi* inscribed with L.P.P. (*locus publicus pompeianorum*) discovered at the suburban baths.[31] These examples supplemented those of Suedius Clemens placed in front of the Porta Ercolano, Vesuvio, Nocera, and Marina.[32] Each *cippus* had a place at roughly the same distance from the gates, implying that the military zone began roughly at the base of the *agger* and moved out to some 30 meters in front of it. They all displayed the same inscription:

> EX · AUCTORITATE
> IMP CAESERIS
> VESPASIANI AUG
> LOCA · PUBLICA · A · PRIVATIS
> POSSESSA · T · SUEDIUS · CLEMENS
> TRIBUNUS CAUSIS COGNITIS · ET
> MENSURIS · FACTIS · REI · PUBLICAE
> POMPEIANORUM · RESTITUIT

Mau translates it as follows:

> By virtue of authority conferred upon him by the Emperor Vespasian Caesar Augustus, Titus Suedius Clemens, tribune, having investigated the facts and taken measurements, restored to the city of Pompeii plots of ground belonging to it which were in the possession of private individuals.[33]

The *cippi* announced the work of the Tribune Titus Suedius Clemens, whom Vespasian had sent to Pompeii to restore illegally occupied public lands to the city. Nissen and Della Corte identified this stretch of land as the *pomerium*, or sacred boundary, thinking that the tribune acted solely on the public space associated with the fortifications. Such a narrow approach ignored that Suedius's mandate encompassed the whole of Pompeii as well as its hinterland as part of a wider effort to reclaim public land that also involved Rome, Cuma, and Cannae.[34] The Vespasianic intervention indicates that the occupation of the land was sometimes illegal. Rather than indicating a sacred boundary, the *cippi* at Pompeii marked the extent of public land associated with the town behind the walls. They are emblematic of how the fortifications functioned as symbolic markers of civic and public boundaries.

The dispensation of the public land associated with the Pompeian fortifications must have carried with it some sort of political association or other mechanism lost to us. The advent of the colony ensured that those stretches of public domain associated with the fortifications would fall under the jurisdiction of the *ordo decurionum* (town council). This was probably merely a matter of continuity from the previous Samnite council. The *schola* tombs located in front of the gates are examples of how the town council erected honorific funerary monuments for model citizens on highly visible public land. Coordinated transfers of land associated with the defenses must have occurred on the southwestern side of Pompeii, where the houses are too massive and well organized to represent haphazard illegal construction activity.[35] Much of the land was valuable real estate offering spacious views of the bay and a cool summer breeze. The process of housing expansion began in the second century BCE, but it still allowed for a measure of defense since many houses had wide-open terraces able to accommodate defensive troops. With the advent of the colony, the houses eventually spilled over to incorporate the defenses, rendering them useless militarily.[36] The exact phasing of this process remains elusive, but it must have occurred on an individual property basis continuing well into the Julio-Claudian period and beyond.[37] Identifying the old line of the walls became increasingly difficult, and owners deliberately covered up the masonry of the city walls with wall paintings.[38] The House of Umbricius Scaurus (VII.16.15), straddling the walls slightly north of the Porta Marina, included a new luxury wing built to resemble a defensive tower adjacent to the fortifications. Although the patron of the extension remains unknown, the fact that this was an elite house implies that this architectural nod toward the fortifications carried some sort of symbolic association with citizenship and civic ideals.[39]

On the north of the city, new development also started chipping away at the fortifications. In Regio VI, the construction of the Augustan *castellum* led to the closure of the inner pomerial street at the base of the *agger*.[40] This event changed the traffic pattern of the neighborhood, essentially turning most of the north–south roads of the Regio into private dead-end streets. Without an easy access route, the military effectiveness of this section of the enceinte would decrease accordingly. As opposed to the expansion of housing on the south side of the city, only the House of the Vestals (VI.1.7) eventually incorporated a large part of the *agger* between the Porta Ercolano and Tower XII.[41] The architectural effect upon the city walls was less dramatic, leaving much of the exterior curtain and towers intact. As the urban expansion continued, the military function of city walls faded away. With it went the symbol and identity of Samnite Pompeii.[42]

Fortifications and the tomb: dialogues in social order

Despite their slow occupation and military degradation, the fortifications would become potent symbols of citizenship and social identity as the markers separating the worlds of the dead and the living. In antiquity, most cemeteries found a place in close proximity to fortifications as part of a common prohibition where dead must be buried outside the city walls. At Pompeii, the practice was already current in the Samnite period when burials were modest without ostentatious markers.[43] The arrival of the colonists heralded some dramatic changes: The burial practice shifted from inhumation to cremation, and monumental tombs appeared along the main roads outside the walls.[44] The funerary landscape would become a deeply competitive venue for the promotion of oneself and the *gens*. Tomb owners would vie for the most prominent locations close to the city walls where those transiting to and from the city would see them. The *schola* tombs built starting in the Augustan period were the most coveted because the city council would award them solely to prominent citizens in the space directly in front of the gates.[45] On occasion, the council also reserved prominent spots near gates such the Porta Nola to the members of the Praetorian Guard who died at Pompeii after their deployment in 59 CE.[46] By contrast, the lower classes, who could not afford lofty monuments, found a resting place in the so-called *sepulture dei poveri* (poor man tombs) located in the remote areas of the city walls west of the Porta Nola, near the Porta Sarno and close to Tower VII. A few names scratched into the ashlars of the curtain wall preserve some of their identities.

Inevitably, the proximity between the new monumental tombs and the fortifications led to an architectural dialogue between the two that would highlight the relationship between the image of the city and the ideal citizen. As a reflection of the deeply personal nature of the monuments, tomb forms are notoriously difficult to differentiate chronologically. Those at Pompeii seem to follow a few distinct trends.[47] The so-called *aedicula* tombs found outside the Porta Nocera were popular in the early colony (see Figure 6.2). They feature a high base supporting an *aedicula* or *cella* usually displaying statues of the deceased.[48] New monumental types emerged in the Augustan period in conjunction with the rise of wealthy freedmen who replaced some of the established elite. Increasingly, elaborate forms appeared, including the *schola*, circular superstructure, chamber, and altar-type tombs. The tombs of the freedmen expressly mimicked civic honorific architecture in order to connect the deceased with the image of the city.[49] In an ornamental arrangement adopted from the earlier *aedicula* tomb types, the dialogue between funerary monuments and city walls is apparent in the use of the First Style applied on the exterior of tombs. Patrons adapted its layout by eliminating the lower orthostat-ashlar sequence otherwise characteristic of the First Style in favor of a variant displaying continuous stretches of imitative ashlars.[50] Their aim was to evoke notions of citizenship by staging a direct allusion toward the masonry of the city walls and the embellishments on the towers and gates.

Philon of Byzantium provides some textual reference elucidating a clear relationship between model citizenship, identity, military honor, funerary monuments, and fortifications. In his manual on siege warfare (*Poliorcetica*), he writes:

> It is also necessary that (individual) tombs of honored men and common burial-places (of soldiers) be built as towers, in order that the city become

Figure 6.2 The *aedicula* tomb 23OS of the Vesonii.

secure and that both those (men) on account of their deserts and those (others) who died for their country may be buried honorably in their very fatherland. [51]

The passage describes what Philon believes is the proper way to build monumental tombs for soldiers and model citizens at or near the fortifications. The distance of time has made this passage vague, and it is unclear whether he means that cities should use towers as funerary structures. Only one known tower built in the much later Hadrianic period at Attaleia (Antalya) may have served such a purpose.[52] Perhaps Philon's concerns are somewhat practical: He simply may be warning that, due to their size and proximity to the fortifications, architects should incorporate monumental tombs into the defenses to deny an advantage to attackers. Yet Philon implies an acute awareness of the symbolic relationship between model citizenry, military life, and city walls.

At Pompeii, the architectural dialogue between the shape of the *aedicula* tombs and the nearby towers may shed more light on Philon's passage. Most known examples of the *aedicula* tomb type lie on the western side of the Porta Nocera necropolis at the foot of Tower II. The architectural resonance between the two is striking. The structures employ an almost identical construction technique of the brick tooth quoined *opus incertum* pioneered in the early years of the colony.[53] In a similar format to the tombs, the tower stood on a high isolated podium, as did its neighbor standing within view to the east (see Figure 3.14). The large opening on the facade of Tower II may reflect its later transformation into an *aedicula* similar to the tombs or, for that matter, some of the other funerary structures featuring high podia nearby. Excavation notebooks report the discovery of an epitaph in the main chamber of the tower in the 1950s. The report describes its recovery in a layer of backfill deposited inside the building sometime after its first recovery in the 1800s.[54] Although this is a secondary context, the epitaph presumably came from the near vicinity – presumably even from the building itself. The notebook tells us its dimensions, 42 × 28 × 4.5 centimeters, and the literal transcription reads:

P. TIN. TIRIVS. P. T. ADIVTOR. ET
TIN. TIRIAE. ESTAE
ILIAE. SVAE. V. ANNVII
ET. SIBI ET. PONTIAE. HE
DYMAE. LENI XV. SO
RI. SVAE. ET SVIS

Unfortunately, no other information exists regarding the epitaph, and the transcription in the excavation diary seems to contain a few errors. An aggressive translation may read as follows:

Publius Tintirius Audiutor, son of Publius [made this] for
his daughter Tintiria . . . esta [beginning of name apparently lost] (she lived
    7 years)
and for himself
and Pontia Hedyma, his gentle wife
and their [family?][55]

The Tintiria *gens* mentioned in the epitaph is not well attested in Pompeian epigraphy with the exception of N. Tintirius Rufus, who served as an aedile in the year 2 BCE. This was a comparatively low office among those available in the governance of the city, but it attests that, for a time at least, the Tintirii were part of the Pompeian elite competing for prominence in the funerary landscape.[56] Considering the inherent height and dominant position of the tower, its reuse as a tomb is possible in a period when individuals increasingly appropriated the terrain belt associated with the fortifications. The competition for space in the Porta Nocera necropolis rapidly increased as it expanded first toward the Porta Stabia and later toward the amphitheater in the Julio-Claudian period.[57] The tower carried the distinct advantages of prominent display and architectural drama essential in tomb construction. Its transition into private hands may also explain its derelict state if the expropriations conducted by Suedius Clemens led to the partial demolition of the building. A nearby *schola* tomb in front of the Porta Nocera displays the signs of a similar fate, perhaps as the result of similar expropriations or even as an effort to erase the memory of the dead.[58] Admittedly, the evidence for a complete transformation of the tower into a tomb is somewhat circumstantial, but it does help explain the curious remains.

The correlation between fortifications and funerary architecture continued more tangibly in the precinct walls that fenced off tombs starting in the first century BCE.[59] Outside Pompeii, a series of unique late-Republican miniature funerary reliefs, found mostly in Campania and central Italy, further elucidates the connection[60] (see Figure 6.3). The reliefs separately depict the typical monumental *cavaedium* (court) gates and towers similar to those of Torino and Spello that appeared in

Figure 6.3 Relief depicting a city gate in the Avellino Museum. (By concession of the Museo Archeologico Irpino.)

Italy during the Augustan program of civic renewal.[61] The reliefs stem from an Etruscan figurative tradition where gates often represented the separation between the dead and the living and were sometimes represented as city gates.[62] Although most reliefs are now without context, they probably appeared in tomb enclosures to mimic the city walls that were often located nearby. In a similar fashion to the role of city walls, the precincts symbolically separated the dead from the living and signaled the social status of the tomb owners.[63]

At Pompeii the enclosure walls – in addition to their practical role of keeping unwanted things out – functioned in a similar manner. Among the surviving examples, many have doorways offering access to the tombs, whereas others featured jogs to secure a ladder, or included steps in the concrete to climb over the wall.[64] The most prominent enclosures, located primarily outside of the Porta Ercolano, featured further decorations, including scenes of gladiatorial combat, which seem particularly apt given their origin as commemorative funerary display.[65] Another common theme is the presence of small merlons, similar to those of the Palaestra precinct wall, marking the enclosures.[66] They alluded to the real fortifications nearby, acting as small fortresses to protect the tomb and to separate the deceased from the living. On occasion, the merlons featured small detailed ornamentations related to the funerary realm. Those on the Cenotaph of C. Calventius Quietus, a member of the *augustales*, included depictions of Oedipus

Figure 6.4 The Cenotaph of C. Calventius Quietus. The inset shows the lost merlon images.
(After Mazois 1824a, pl. XXVI.)

and the sphinx, scenes of offering, and the goddess Victoria carrying a Celtic horn that was a motif borrowed from triumphal military art (see Figure 6.4).[67] The presence of the merlons on this tomb is particularly significant because the monument specifically commemorated a man who belonged to the *augustales* a college of priests dedicated to the cult of emperor. The *augustales* were wealthy *libertini* (freedmen) who, because of their status, could not hold elected office. Yet they strongly identified with citizenship and the city as part of the system that had given them wealth and freedom. As is the case with the Palaestra, the merlons and the precinct walls referenced the ideal of *virtus* and commemorated the deceased as an ideal citizen in a manner that permeated both funerary and military architecture. The fortifications continued to form a powerful reference point for civic identification even in death.

City walls in the domestic sphere: reflections of urban and social ideals

In addition to the correlations with civic architecture, Pompeii preserves evidence for the social significance of the city walls through representations in the domestic sphere. It supplies a glimpse into how select individuals – in this case, mostly members of the urban elite – conceptualized the fortifications in the wider context of the city. In Roman art, simple representations of city walls first appear on a fresco in the early third century BCE Tomb of Fabius or Fannius on the Esquiline hill in Rome, where at least one scene shows a simple monolithic wall with merlons representing a city.[68] In typical Roman fashion, the scene is historical, displaying some sort of meeting – whether it be a parlay or a surrender – occurring between leaders outside a city. Although the identity of the settlement is unclear, the simple depiction of a wall is enough to evoke and define the concept of the city.

Such representations acquire political significance in the Augustan period, where they connect with the real reconstruction efforts that the emperor sponsored in Rome and the Italian peninsula. The sculpted frieze on the Basilica Aemilia, restored in the early empire, once depicted the legendary foundation of Rome (see Figure 6.5). Of the few surviving scenes, one shows a goddess, perhaps Pales, overlooking the construction of city walls, presumably those of Rome itself. Her left hand rests on a boundary stone in reference to the *Parilia* – an annual festival celebrating the foundation of the city.[69] To the right, men build the city walls including part of a gate with a flanking tower. The scene is clearly schematic, and the artist renders proportionally large figures to enhance the relief's legibility for the viewer below. Almost identical contemporary scenes survive from a fresco recovered from the Tomb of the Statilii in Rome. The pictorial sequence depicts the mythological beginning of Rome where two scenes depict the foundation of Lavinium and Alba Longa.[70] In each image, a seated goddess personifying the city looks upon workers toiling to build her wall.[71] Despite the different media and the well attested rituals associated with urban foundations, including that of Rome itself, the artists in both cases chose to depict the foundational act with the construction of the defenses rather than any other symbolic moment.[72] They are a

Figure 6.5 Relief from the Basilica Aemilia. (By concession of the Ministero dei Beni e
delle Attività Culturali del Turismo – Parco Archeologico del Colosseo.)

vivid representation of the relationship between patronage, city walls, and divine
protection that at Pompeii found architectural expression in the fortifications' acting as terraces for its temples. The two images are so similar that the cycle on the
basilica must have served as a model copied in the tomb as part of an ideological
framework. The imperial coffer helped fund the restoration of the basilica, whereas
the Statilii were among the most loyal supporters of Augustus. The frieze and
fresco cycles are part of an official visual language related to Augustan ideology
and a redefinition of *romanitas* as an individual's duty to the collective state. The
foundation scenes in particular connected to the image of Augustus as Romulus
and founder of the new Rome after the civil wars.[73]

The later Trajanic fresco recovered in a cryptoporticus beneath the Baths of
Trajan in Rome displays a Roman city complete with characteristic monuments,
such as temples, theaters and city walls that were part of the urban ideal.[74] A
roughly contemporary depiction known as the Avezzano relief shows a city and its
walls dominating the developed landscape. Although workers recovered the relief
in a secondary context during the drainage works of Lake Fucino in the 1800s, its
material and composition imply that it had a commemorative function on a public
or private funerary monument.[75] Such civic representations fall within a wider
trend of establishing an ideal urban setting associated with the establishment of
Roman order that began in Republican Italy. It included the construction of a new
landscape with the installation of colonies, city walls, roads, sanctuaries, tombs,
and monuments that reflected a Roman ideal.[76] The Avezzano relief depicts the
culmination of the process in the Imperial period. The prominence of city walls
underscores their association with the image of the ideal city.

In the Bay of Naples, artists use fortifications to conceptualize the city in a
trend that runs parallel to the personified Tychai representing cities present in the
eastern Mediterranean. Fortifications appear in harbor motifs, where they serve as

metaphors for refuge and security as a reflection of the human-made harbors that served the villas developing in the area.[77] In other examples, they directly reference cities. The fresco of the Fall of Icarus in the House of the Sacerdos Amandus (I.7.7) at Pompeii still decorates the wall of a *triclinium*, where it served to foster erudite conversations during elaborate meals (see Figure 6.6). The compressed narrative depicts the moment the son of Daedalus plunges to his death after nearing the sun and the recovery of his body.[78] The artist depicted Crete, or Knossos, in the background as a heavily fortified city showing its walls surrounding a mass of unidentifiable buildings. The fortifications are robust ramparts built in regular ashlar isodomic masonry, presumably inspired by the *opus quadratum* of the Pompeian enceinte. Rather than the amorphous collection of buildings they enclose, it is the gates and towers, which prominently face the viewer as the most obvious markers of the city.

Figure 6.6 Fresco with the Fall of Icarus, House of the Sacerdos Amandus.

The House of the Lararium of Achilles (I.6.4) contains a more schematic example in the stucco frieze depicting the Homeric cycle in a *sacellum* dedicated to the familial cult. Three separate scenes depict the departure or arrival from Troy: Hector in the act of leaving to confront Achilles, the dragging of his body around the city, or Priam recovering his remains.[79] In each instance, the artist – clearly constrained by the limited space of the frieze – chose to render Troy as a simple large city gate flanked by two large towers (see plate 23). The gate vault is overly large to accommodate the figures, but the flanking towers create a familiar shape reminiscent of the *cavaedium* tower-flanked gates built in the Augustan period throughout the Roman west. Architects would incorporate their core design into the later Porta Ercolano. In both the fresco and the frieze, the visual and imaginative concept of the city translated to the fortifications more than any other architectural element.

A more complex example found in an exedra (10) of the House of P. Vedius Siricus (VII.1.25) carries strong ideological connotations. Today, the fresco has faded, but nineteenth-century drawings show us the original composition where a seated Poseidon and a standing Apollo in the foreground look upon a group of workers completing the fortifications of Troy in the background (see plate 24).[80] The workers use a crane to lay the wall courses, while bulls are transporting the stone. There is no need to doubt the artist's competence at depicting ashlar masonry, but the city walls located just up the street from the house perhaps formed an inspiration.[81] The scene tells the story of the two divinities whom, after they plotted against him, Zeus banished from Olympus. They would set to work for King Laodemon in building impregnable fortifications for Troy.[82] The king's cruel treatment would lead Apollo to send a pestilence and Poseidon to send a snake to devastate early Troy, but the walls would survive to serve in the later battle against the Greeks. Two scenes on the adjacent walls accompanied the fresco: Hephaistos presenting the armor of Achilles to Thetis and the scene of a drunken Hercules with Omphale. Put together, these scenes prefigure the Trojanic cycle and the Aeneid. Hercules and Omphale had a son named Tyrrhenus who, in one version of the myth, led the Etruscan people from Lydia to Italy.[83] In the adjacent triclinium (8), another scene displayed a wounded Aeneas who was the ancestor of the Roman people. This collection of episodes must be part of a broader ideological thread of citizenship associated with the owner. The house probably belonged to the Vedii, who had just commissioned artists to repaint the house when the eruption halted their efforts in 79 CE.[84] They were a prominent family: P. Vedius Siricus was a *duumvir* in 60 CE and a candidate for *quiniennalis* in 75 CE, whereas P. Vedius Nummianus ran for aedile in the final years of Pompeii.[85] Such a strong political background implies that the scene showing the construction of Troy's walls – and by extension notions associating fortifications with ideal cities – fits into the owner's values of citizenship and Roman myth.

Images of fortifications also gained popularity as framing bands in black-and-white mosaics during the first century BCE. They originate from the imitative carpet fringes bordering Hellenistic mosaics.[86] Such elaborate carpets were status symbols. The translation of their borders into representations of city walls reflects similar connotations of elite ostentation and citizenship. Some of these mosaics must have functioned as mere decorative motifs or as functional dividers separating architectural spaces.[87] On other occasions, they carried ideological weight. The earliest examples of fortified bands dating to the late third and mid-second centuries BCE come from elite houses in the Roman colonies of Atri and Suasa in Italy.[88] They must have referred to the same notions of *securitas* and *romanitas* associated with the fortifications of Roman colonies. In the late Republic and early Principate, the bands were popular in Italy and the Roman west as emblems of *romanitas*, reflecting the real fortifications built in new colonies and the Augustan program of civic renewal.[89] The mosaics were also metaphors of protection and safety, functioning as apotropaic devices particularly when they framed elaborate labyrinths.[90] Each band included a varying mix of these meanings depending on their immediate contexts and commissioners.

In the region of Campania, Pompeii alone holds seven known examples of city wall mosaics in the villas of Diomedes and P. Fannius Synistor and in the Houses of the Menander (I.10.4), M. Caesius Blandus (VII.1.40), the Centenary (IX.8.6), and two in the House of the Wild Boar (VIII.3.8).[91] The depictions range from simple tower or city gate motifs in the House of M. Caesius Blandus to the elaborate wall systems surrounding labyrinths at the Villa of Diomedes. They were popular with the introduction of the Second Style, placed in panels and thresholds or as bands functioning as metaphorical fortifications protecting spaces.[92] The similarity and extensive tradition of the type, ranging from the third century BCE to the Flavian period, suggests that many motifs were decorative and widely available in pattern books.[93] The House of the Menander, for example, preserves a fragmentary Second Style mosaic panel depicting a fortification lining the *impluvium* of the *atriolum*. Despite the damage, one still can discern the corner of a city gate and a wall rendered in *opus quadratum* capped with an oversized T-shaped merlon. It leads to a tower and another stretch of curtain wall, suggesting that what survives is a fragment of a patterned decorative motif.[94] The examples contained in the villas of Diomedes and P. Fannius Synistor are either completely lost or in a fragmentary state. They functioned in Second Style ensembles in luxury dwellings associated with the arrival of the colonists.[95]

Pompeii also preserves examples that are specific commissions diverging from the typical representations. They exemplify how their patrons engaged with the fortifications and recognized their symbolic roles. The House of M. Caesius Blandus (VII.1.40) preserves a Second Style example in the threshold between the *fauces* and the *atrium* (see Figure 6.7). It is a straightforward symmetrical rendition centered on a city gate flanked with curtain walls and a tower on either side. The gate is vaulted and closed with a two-leafed door that is fortified with a jagged crown and three exaggerated T-shaped merlons. The curtain sections are orthostat walls topped with smaller merlons. The towers each feature two large vaulted windows with surmounting crenellations in a schematic rendition reminiscent of the frieze on the Basilica Aemilia in Rome. Shields above each curtain further stress the military character of the panel. The threshold is set at the back of a wider mosaic spanning the *fauces* that includes framing dolphins, a sea dragon, a trident and central rudder with a perched bird. The combination of the two motifs is atypical for mosaics displaying fortifications. It must reflect an explicit commission connecting personally to the owner of the house.

Figure 6.7 Second Style mosaic in the *fauces* of the House of M. Caesius Blandus.

The mosaic carried both functional and symbolic associations. In functional terms, the vertical composition of the mosaic acted as a guide leading spectators into the house, whereas the threshold, emphasized by the closed gate, invited spectators to stop and view the *atrium* and the *tablinum* beyond.[96] The various elements carry further meaning: The rudder and trident, as references to the sea and navigation, are apt for a port town like Pompeii. The trident was also a symbol of Neptune, whereas the rudder was an attribute of Venus Pompeiana as one of the principal protective deities of the city. The bird species perched on the trident is somewhat vague, but it may represent a woodpecker as an attribute that alluded to Mars in keeping with the military character of the city walls and the shields.[97] Della Corte argued that Neptune was a god particularly popular in Rome rather than in Campania at the time, an assertion that in turn led him to identify the owner's pro-Roman affiliation.[98] However, these clues do not present such clear associations beyond the owner's strong identification with Pompeii and its maritime heritage.

The unique combination of motifs has sparked a broader debate on the house, its owner, and the commission of the mosaic. Originally known as the House of Mars and Venus, the house received its name because of the Fourth Style medallions representing the two divinities painted throughout the dwelling. It has since received a new name after the graffiti scratched in the peristyle, several of which mention M. Caesius Blandus and one of which refers to him as a centurion of the Praetorian Guard.[99] As a military leader, the martial character of the decorations would then be reflective of his rank.[100] The *fauces* mosaic would have referenced Blandus's deployment overseas, including a victory over an unknown city acknowledged by the fortified gate depicted in the threshold.[101] This reasoning does not take into account the fact that the *fauces* mosaic is in the Second Style. The owner must have commissioned it soon after the installation of the colony. Although a Caesius Blandus may have owned the house at the time, the mosaic cannot reflect the exploits of the Blandus who was member of the Praetorians deployed at Pompeii over 100 years later after the riot at the Amphitheater in 59 CE.[102] A recent hypothesis ties the house to either Loreius or Cuspius, one of the two men responsible for restoring the city walls of Pompeii.[103] Excavators originally found the six fragments of the inscription that commemorated this event reused in a threshold and the floor of the house as part of a postearthquake reconstruction effort.[104] The theory proposes that the fragments were commemorative because one of the two *duumvirs* was an ancestor of the house's last owner. The *fauces* mosaic would fit neatly in this context as the commission of a *duumvir* seeking to elucidate his euergetism. As convincing as this idea might be, one would imagine that the owner would display such an ancestral artifact in a more prominent position rather than scatter it throughout the floors. A more cautious approach is to examine the motivations of the Caesii who were a distinguished colonial family. L. Caesius was a pro-Sullan man who, enriched by the proscriptions, financed the Forum Baths during his tenure as a *duovir iure dicundo*.[105] In the assumption of a continuity of ownership, L. Caesius or one of his immediate descendants commissioned the mosaic as a reflection of the new colonial reality and a strong identification with Pompeii's fortifications as signifiers of *romanitas*. Such an expression is equally compelling if the Loreii, Cuspii, or Caesii were the owners of the house. All three were strong pro-Sullan *gentes* supplying *duumvirs* early in the colony.

At the time of excavation, the House of the Wild Boar (VIII.3.8) preserved two large mosaics with city walls dating to the early Imperial period. One of them that featured fortifications framing hippocamps and dolphins seems to have disappeared.[106] The surviving mosaic is composed of a patterned intersection of perpendicular I- shaped forms spanning the entire *atrium*. A continuous band depicting city walls frames it.[107] The rendered curtain wall features oversized regular ashlar masonry and large T-shaped merlons similar to the mosaic in the House of C. Blandus, suggesting that they are part of a repertoire. Towers appear at regular intervals and on each corner of the space. Compared to the curtain walls, they display smaller ashlar masonry, perhaps as an allusion to a similar contrast between the stucco embellishments on the towers and the bare ashlars present in the Pompeian circuit. In Roman art, such a careful distinction of masonry in visual representations began in the late Republic with the advent of the First Style on the exterior of actual buildings.[108] In order to stress their functional importance as part of the entry/exit axis of the house, the artist subtly changed the aspect of the towers placed in front of the *tablinum* and *fauces*. As opposed to the majority of towers, which feature two stories with two windows on the first floor and three on the second, the *tablinum* tower displays only a second floor and no ashlar masonry. The *fauces* display a more pronounced distinction. The tower features a large vaulted opening creating a metaphorical gateway into the *atrium*. This detail added an apotropaic role as well as a functional address, inviting viewers to enter the *atrium* after stopping to admire a mosaic showing two dogs attacking a wild boar in the adjacent *fauces*.[109]

Figure 6.8 Augustan floor mosaic in the *atrium* of the House of the Wild Boar.

The elaborate mosaic is a specific commission that carried further meaning. Matteo Della Corte used it, together with electoral slogans painted on the facade, to link the dwelling to L. Coelius Caldus and the Coelia *gens*. The Coelii were a distinguished family that included a quaestor among their ranks and traced their ancestry to C. Coelius Caldus, a general who in 99 BCE participated as a praetor in the siege of Clunia in Hispania Citerior. Della Corte associated the mosaic with this particular siege: The wild boar in the *fauces* was a symbol of Hispania Citerior, whereas the fortified enclosure denoted the battle.[110] These are tenuous associations. The dwelling was a so-called *ordo decurionum* house lining the via Marina/ via dell'Abbondanza axis. These houses belonged to prominent early colonists – mostly high-ranking officers of the Roman army – who acquired them for their visibility and proximity to the Forum as the political heart of the city.[111] Although the commissioner of the mosaics remains unknown, he belonged to an influential Pompeian family. Their early Imperial date coincides with the period when city wall mosaics reached the apex of their popularity in the Augustan colonial establishments of the Roman west as part of an expression of *romanitas*.[112] In addition to their functional address and the apotropaic symbolism, the mosaics prominently displayed in the *atrium* of the house dialogued with the Pompeian fortifications to express the Roman identity and political affiliation of the owner.

Another mosaic belonging to the Augustan period adorned a *cubiculum* opening onto the main peristyle of the House of the Centenary (IX.8.6).[113] Its main black-and-white geometric pattern consists of swastikas set in large octagons, each separated by squares. The central *emblema* depicts a gorgon head set between two panels: one depicting a nautical scene and the other an image of a fortified circuit (see plate 25). The nautical panel displays a merchant ship and small support vessel moored onto a lighthouse. On the opposing side the fortified circuit faces the back of the room as does the gorgon, implying that they carried some sort of apotropaic value. The circuit includes ten towers and a curtain wall set out in a semioctagonal shape with a square protrusion reminiscent of the wider geometric pattern of the mosaic. An unidentified structure, perhaps a crude version of the tower gate found in the mosaic of the House of the Wild Boar (VIII.3.8), surmounts the walls to the right. It points toward a palm tree and two smaller subsidiary buildings, possibly a *sacellum*, outside the circuit. The fortified circuit and its inclusion in the *emblema* is a motif without known parallels. It must be uniquely associated with the owner. Considering that the house preserves a broad marine decorative theme, the mosaic may have functioned as a reference for the merchant profession of the owner or as a port metaphor for the safety and security of the home.[114] The *cubiculum* mosaic might even depict a city particularly dear to the owner, perhaps Alexandria as referenced through the lighthouse and palm tree.[115] If it indeed represents the Egyptian city, the walls and the lighthouse are a remarkable shorthand for the complex metropolis.

Pompeii also preserves simpler city wall renditions that are typical of the Second Style. They usually depict a series of towers joined with low curtain walls that patrons placed in thresholds or around *impluvia*. The House of Tryptolemus (VII.7.5) contains a prime example of the type where a polychrome double band

Figure 6.9 The Second Style mosaic band acting as a threshold to exedra *u* in the House
       of Tryptolemus.

acted as an extended threshold to a prominent *tablinum*[116] (see Figure 6.9). The
space still contains an *emblema* in *opus scutulatum* that is a rarity for this period,
indicating the house belonged to a prominent figure, perhaps even a founder of
colonial Pompeii.[117] The threshold motif is reversible, having been done in such
a manner that a fortified circuit always faces a viewer crossing into and out of
the space. The merlons of each tower form the crenellations of the opposite low
curtain walls and vice versa, resulting in a double line of fortifications that clearly
announced the limits of the room on each side. A slightly simpler band runs around
the *impluvium* of the House of Cornelius Rufus (VIII.4.15). Now almost destroyed,
a few old photographs capture its original layout. It featured black-and-white
towers alternating to form two lines: one facing the *impluvium* and the other the
atrium.[118] On the southern side, the band jutted out to envelop the wellhead and
a large marble table with back-to-back griffins as leg supports. A herm recovered

in the *atrium* attests that the house belonged to Cornelius Rufus, a member of the Cornelii, who were the descendants of Sullan clients including P. Cornelius Sulla, one of the founders of the colony.[119] In practical terms, the apotropaic value of these two examples as protective bands/thresholds is evident. Their regular pattern also implies that they were a common decorative theme. Nonetheless, considering their placement in public areas related to patron–client rituals and the prominence of the families, the mosaics must have carried associations of *romanitas* and referenced the newly refurbished walls of the city.

The preponderance of city wall mosaics surviving at Pompeii date to the late Republican and early Imperial periods of the colony. They undoubtedly were part of decorative motifs available to artists, and their significance sometimes did not go beyond that. Nevertheless, their boundary defining and protective roles translate naturally from the real enceintes defending the community. The period of their highest popularity coincides with the refurbishment of the fortifications in the early colony first and the gates in the later Augustan period. The connection between the mosaics and the actual fortifications occurred on multiple levels. Those in the Pompeian pavements worked as apotropaic devices and distinct emblems of the new colony. The mosaics would also provide a powerful reminder of the relationship between the city walls, the Pompeian elite, and the political messages of the new empire. The ideals associated with city walls, their status as boundary markers, and their apotropaic and religious values, connected in the domestic and political sphere to help foster the identity of the new city.

In addition to depictions of city walls in mythological or abstract settings, Pompeii stands out because it preserves a few renditions of its actual fortifications. They are a glimpse of contemporary social and artistic perceptions of the circuit. The fresco depicting the riot between Pompeian and Nucerian (Noceran) factions offers an idea of the amphitheater quarter in the years leading up to the earthquake of 62 CE (see plate 26). It is a vivid depiction of Tacitus's narration of the events of 59 CE, which left many spectators dead, and ensured an imperial ban on games in the city for a period of ten years.[120] These events are a testament to the continued importance of the district in the history of the Pompeii. The commissioner of the fresco was likely involved in the happenings of that day and chose to display the event in his peristyle for selected guests to admire.[121] It is a historical document responding to specific demands of the commissioner and thus supplies a rare glimpse of the fortifications in the Pompeian imagination.

The fresco displays the arena quarter from a bird's eye perspective, allowing the best view of the action occurring in and around the amphitheater. It survives because excavators detached it from a wall in peristyle *n* in the House of Actius Anicetus (I.3.23) with the result that parts of it are cropped. The artist stressed the recognizable aspects of the arena, including its prominent stair ramps that are the most conspicuous features of its facade. Above the arena, the artist deliberately represented only part of its covering *velum* awning to allow a full view of the action inside.[122] As the focus of the riot, the amphitheater stands out as the largest structure. By contrast, the Palaestra to the right – which in reality is equal in size – is much smaller. Farther back, the city walls frame the riot. As a measure of their importance, they stand at roughly the same scale as the amphitheater. In an effort to suggest their length, the painter has added an exaggerated curvature to the fortifications. Two small openings on either side of the amphitheater might represent schematic depictions of the Porta Nocera and Sarno. Rather than showing the back of the towers, the artist shows them from the front in their most recognizable view. Their appearance, with large open doors and four windows, is quite different from the real multistoried buildings. Rather than detailing the First Style embellishments on the buildings, the artist renders plain towers in an effort to differentiate them from the curtain wall and highlight their role as architectural reference points in the district. In an empathic choice, the artist conspicuously omits the *agger* in order to isolate visually the arena and the fortifications. Although at this point the exterior curtain was a medley of construction techniques, the painter accentuated its ashlar masonry as the most marked aspect. The result is a conceptualized rendition of the fortifications that emphasizes their exterior appearance. Such an external viewpoint stresses a unified structure, indicating the architectural and symbolic role of the fortifications as a monument defining the city and the district.

A glance at how individuals would have perceived the fortifications comes from a relief that once belonged to a *lararium* in the House of Caecilius Jucundus (V.1.26) (see Figure 6.10). It shows the effects of the 62 CE earthquake on the area of the Porta Vesuvio with the gate toppling over to one side. The *castellum* stands intact to its left, whereas the fortifications extend to the right. In the foreground, panicked animals – mules or bulls – dart to the right, and an altar stands next to a tree that may indicate an unknown *sacellum* outside the walls. Once again, the *agger* is missing, and the sculptor choses to depict the ashlars that compose the external curtain wall despite the fact that the facade of the *castellum* places the viewer on the internal (city side) of the fortifications. The relief is somewhat schematic; the gate's long passageway walls that projected into the city are missing. Nevertheless, it offers a unique glimpse of the gate in antiquity displaying a decorative pediment above the vault, as well as the two open doors that would have shut the passageway. The relief is a clear historical document and as such differs significantly from more generalized civic images. Although found out of context, it is associated with a second depiction that portrays the effect of the earthquake on the Temple of Jupiter and an adjacent arch in the Forum. Both reliefs have drawn much discussion concerning their placement, ownership, depicted buildings, and the motive for their commission, which are identified as commemorative, votive, or apotropaic.[123]

Figure 6.10 Relief of the 62 CE earthquake from the House of Caecilius Jucundus.

The very conceptualization of the disaster with the Temple of Jupiter and the Porta Vesuvio highlights their importance to the owner of the house as monuments defining the city. A series of wax tablets recovered inside the building connect the dwelling to L. Caecilius Jucundus – a wealthy banker and son of L. Caecilius Felix, a freedman of the Caecilii who traced their roots to Roman colonists who settled at Pompeii.[124] As freedmen, both father and son had little in terms of direct ancestry to highlight their social status as Roman citizens. The Caecilii commissioned an ornamental program aimed at creating an ancestor house to dignify their *gens* and make them exemplary citizens.[125] The *lararium* and its reliefs, prominently displayed in the public *atrium* of the house, were a clear part of Jucundus's effort to portray himself as a Roman. Jucundus was a wealthy freedman, a social class that often exhibited a strong identification and loyalty to the status quo that had enabled their social mobility.[126] The placement of the reliefs in a household shrine equates this status quo and therefore its protection to the Temple of Jupiter and the city walls as the symbolic conceptualization of the city. Considering that the gate lies within view of the entrance to the house and that the Porta Campana – the ancient name for the Porta Vesuvio – may have lent its name to the *campanienses* electoral college; its presence on the *lararium* could be the result of the Caecilii identifying with their neighborhood.[127] The choice of associating the disaster with the collapse of the Porta Vesuvio and of making it the object of daily prayers in the private *lararium* of a freedman speaks volumes about the role of fortifications in the concepts of the city and citizenship in the final years of Pompeii.

Notes

1　Lo Cascio 1991; 123–125; Lo Cascio 1996, 118–119; Zevi 1996, 130–132; Savino 1998, 440–444.
2　Andreau 1980, 183–184. For the question of land reallocation, see Savino 1998, 448–455; Zanker 1998, 62. For the *pagus*, see Adamo Muscettola 1992, 76. The epithet "Felix" testifies to its foundation under Sulla, whereas Augustus points to a second phase under the first emperor.
3　Lo Cascio 1991, 122; Zevi 1991, 71–72; Zevi 1996, 130; Savino 1998, 460 base their conclusions on the adoption of the Second Style in many of these houses. Pesando warns against reading too much into this phenomenon; Pesando 2006, 75, 91.
4　Mazois 1824b, 41; Maiuri 1958, 203–211. Notable examples are the House of the Figured Capitals (VII.4.57), House of Julius Polybius (IX.13.1), House of Caecilius Jucundus (V.1.26), House of the Vettii (VI.15.1), House of the Centaur (VI.9.5), House of Messius Ampliatus (II.2.4), House of Cuspius Pansa (I.7.1), House of Meleager (VI.9.2), House of the Dioscuri (VI.9.6–9), House of the Small Fountain (VI.8.24), House of the Suettii (VII.2.51), and House of the Mosaic Doves (VIII.7.34) and House of the Menander (I.10.4), which includes a yellow socle similar to the gates of the city. On the via dell'Abbondanza, shop facades have a similar layout; see the Thermopolium of Asellina (IX.11.2–4) and the Tavern of the Four Divinities (IX.7.1).
5　See chapter 3 of this volume, notes 68–69.
6　On facadism, see Richards 1994, 7. Such elements may have played a part in the deliberate retention of the tuff facades.
7　Mols 2005, 245.
8　Zanker 1988, 105 and 117.

9 See Jacobelli and Pensabene 1995–1996, 43–45; Zevi 1996, 128; Coarelli 1998, 185–190.
10 Wallace-Hadrill 2008, 79–81. The author outlines how allied cities willfully adopted Roman architectural forms. For the buildings, see Zanker 1998, 61–78; Zanker 2000, 126–140. On the temples, Ling 2007, 122; Moormann 2011, 69–85.
11 CIL X 852. Castrén 1975, 89; also Adamo Muscettola 1992, 80; Zevi 1995, 9–10.
12 Zanker 1998, 69.
13 See Jashemski 1979, 202–284; Guzzo 2007, 137.
14 Tybout 2007, 408.
15 Welch 2009, 88–100.
16 With the Marian reforms, the state paid gladiator trainers to instruct recruits; see Goldsworthy 2003, 136–140.
17 Welch 2009, 88–100.
18 Conticello de' Spagnolis 1994, 19–25. The earliest tombs outside the gate, dating to the second half of the first century BCE, are a *terminus ante quem* for the lowering of the gate; D'Ambrosio and De Caro 1983, 29.
19 Zanker 1988, 147–156; Zanker 1998, 69–72.
20 Futrell 1997, 4–9.
21 See Savino 1998, 455–460; Guzzo 2007, 120.
22 Guzzo 2007, 118–119.
23 Ling 2007, 123.
24 Welch 2009, 95–100.
25 Maiuri 1939, 168–169.
26 Zanker 1998, 114–116; Guzzo 2007, 170.
27 Goldsworthy 2003, 25.
28 These twin notions were central to Augustan ideals and the construction of city walls throughout the peninsula; see Gros 1992, 211–224.
29 Maier 1961, 92; Curuni 2012, 12.
30 See Tacitus *Hist.* III.30; Lawrence 1979, 39.
31 See Jacobelli 2006, 67–68.
32 See CIL X 1018 for the Porta Ercolano example; for the Porta Vesuvio see Spano 1910, 399; Porta Marina see Jacobelli 2001, 44–48; Porta Nocera see Sertà 2001/2002, 228.
33 Translation after Mau 1902, 408.
34 Starting with Magaldi 1939, 21–60; Jacobelli 2001, 44 and 49. For Rome in particular, see Castagnoli 1981, 261–275.
35 Jacobelli 2001, 57–58.
36 Noack and Lehmann-Hartleben 1936, 5–15; Tybout 2007, 407–420; Pappalardo et al. 2008, 294.
37 Tybout 2007, 407–409.
38 See Noack and Lehmann-Hartleben 1936, 5–9 and the individual entries.
39 Pesando 2006, 91, 148–150. The building's appearance remains unknown. The *umbricii* who owned this house at the time of the eruption had built their wealth on the trade of *garum*, a famed local fish sauce; see Castrén 1975, 232, note 424.
40 Garzia 2008, 2; see Ohlig 2004, 87–106 for the closing of the street. Keenan-Jones argues for a single Augustan phase to the aqueduct; Keenan-Jones 2015, 191–215.
41 Jones and Robinson 2007, 389–394.
42 Tybout 2007, 408.
43 Della Corte 1916, 287–305; De Caro 1979a, 179–190.
44 Cormack 2007, 586.
45 To date, only the Porta Marina and Porta Sarno lack *schola* tombs. On the type, see Borrelli 1937; Kockel 1983, 18.
46 CIL X 8349 to 8361. See Senatore 1999, 96–100 and 103–110; Minervini in PAH 2 pars quarta, 593–597, April 10–October 17· 1854; Mazois 1824a, pl. XII, fig.5; De Caro 1979b, 85–95; D'Ambrosio and De Caro 1983, 25.

47 Von Hesberg 2005, 59.
48 In particular, the *aedicula* tombs 9OS, 13OS, the tombs of Marcus Octavius and Vertia Philumina, of 23OS of Publius Vesonius Phileros, and of Vesonia and Marcus Orfellius Faustus; also 27OS of Aulus Campius Antiocus, and 29OS of Lucius Caesius and Annedia.
49 Cormack 2007, 593.
50 Mols 2005, 245–246 in particular points to tombs 3S, 13 OS, 9 OS, 29 OS.
51 Philon *Polior.* I. 86, as translated by Lawrence 1979, 89; see also the associated annotation.
52 Idem.
53 Wallat 1993, 363–375.
54 Excavation diary entry June 4, 1952. The excavation reports are vague, describing the recovery of the epitaph in what the excavators designate Tower 1. Assuming continuity in excavation nomenclature, this associates the report with Tower II since Maiuri later describes the tower as 1; see Maiuri 1959, 82–83.
55 I wish to thank Professor A. Riggsby for his invaluable translation of the text.
56 Castrén 1975, 229.
57 D'Ambrosio and De Caro 1983, 24–25; Emmerson 2010, 77–86.
58 This is the only *schola* tomb at the Porta Nocera; its owner is unknown; Varone 1988, 143–154; also Guzzo 2007, 177.
59 Von Hesberg 2005, 59.
60 Von Hesberg 1992, 63–65.
61 See Richmond 1933, 149–174; Kähler 1942, 1–108.
62 Camporeale 2008, 22.
63 Rebecchi 1978, 153–166; Von Hesberg 1992, 63.
64 Typical examples are the tomb of M. Obellius Firmus outside of Porta Nola, tomb ES7 at the Porta Nocera, and South Tomb G at the Fondo Pacifico.
65 E.g., the Tomb of N. Festius Ampliatus near the Porta Ercolano and the Tomb of Caius Vestorius Priscus at the Porta Vesuvio; see Kockel 1983, 75, 84.
66 The best examples are tomb ES5 at the Porta Nocera, the tombs of Marcus Veius Marcellus and Caius Vestorius Priscus outside the Porta Vesuvio, of Naevoleia Tyche (süd 22), of Numerius Istacidius Helenus (süd 21), of C. Calventius Quietus (süd 20) Caius Fabius Secundus (süd 18), and of Aulus Umbricius Scaurus (süd 17).
67 Many have disappeared; Mazois recorded four examples; Mazois 1824a, 45 and also Overbeck 1854, 283, fig. 209. See also the Tomb of Caius Fabius Secundus (süd 18) as described in Kockel 1983, 85–90 and 92–93.
68 For an extensive survey, see Frederiksen et al. 2016, 173–196.
69 For this interpretation, see Albertson 1990, 807; on the *Parilia* and *sulcus primigenius*, Ov. *Fas.* IV 712–860.
70 Borda 1959, 4; Sanzi di Mino 1983, 163–164; Cappelli 1998, 51–58; Moormann 2001, 101.
71 Brilliant 1984, 30.
72 See chapter 2 of this volume.
73 Augustus must have given consent for the frieze; see Kampen 1991, 452–458. For the redefinition of *romanitas* and its association with the basilica and the tomb, see Holliday 2005, 89–129.
74 La Rocca 1984, 31–53; Moormann 2001, 99–100. On the baths of Trajan, see La Rocca 2000, 57–71; Favro 2006b, 20–38.
75 Segenni 2001, 25–27; Facenna et al. 2001, 38.
76 Zanker 2000, 31.
77 Bergmann 1991, 49–50; Clarke 1996, 93; Van der Graaff 2015, 559–564.
78 Von Blanckenhagen 1968,124; Sampaolo 1990, 594–596.
79 Blanc 1997, 37–38.
80 See Fiorelli 1862, pl. 1; Reinach 1922, 28 pl. 4; Bragantini 1996a, 295.

81 Fiorelli mentioned how the fortifications in the fresco referenced those of Pompeii; Fiorelli 1873, 79.
82 Hom. *Ill*. xxi. 443; Apollo. *Biblio*. 2.5.9.
83 Dion. Hali. *Rom. Ant*. 1.28.1.
84 Niccolini and Niccolini 1854, pl. 1.
85 CIL IV 805, Bragantini 1996a, 228. For the Vedii, see Castrén 1975, 234.
86 Becatti 1971, 297–298; Morricone 1973, 507.
87 Clarke 1979, 10.
88 Azzena 1987, 51–55; De Maria 1996, 414. Pensa 1998, 705.
89 Lavagne 1987, 135–143. Pensa 1998, 704–705.
90 Iorio 2008, 289; Kern 1981, 25–26. Kern connects the representations of labyrinths with the *lusus troiae* festival, where equestrians performed labyrinth-like dances also carried out at the foundations of cities. The dances were meant to expel spirits from the area inside the city walls. Evil spirits could allegedly fly only in straight lines. The combination of walls and the labyrinth formed an effective countermeasure.
91 Ling et al. 2005, 290. One of the mosaics in the House of the Wild Boar seems to have disappeared; see Niccolini and Niccolini 1862, pl. 5.
92 Ling et al. 2005, 57–59.
93 Becatti 1971, 297–299; Lavagne 1987, 138–139.
94 Ling wrongly identifies it as a gate Ling et al. 2005, 58; countered by Iorio 2008, 290.
95 An early nineteenth-century drawing from the Villa of Diomedes shows a fortification wall enveloping a labyrinth, see L. Barré 1841, vol. 5, fig. 6.4; Kern 1981, 25–26, 115. The Villa of P. Fannius Synistor had a band mosaic enclosing the small peristyle. Today a fragment survives dated variously to the early Second Style and the Imperial period; see Barnabei 1901, 17; Ling et al. 2005, 57; Iorio 2008, 291. For the owner, see Sauron 1998–1999, 1–28.
96 Clarke 1979, 9.
97 Della Corte 1965, 186.
98 Bernabei 2007, 35–36.
99 CIL IV 1717, 1719, 1733; Los 1995, 167.
100 Fiorelli 1875, 165. Della Corte 1965, 167 identified the figures in the medallions as Caesius Blandus and his wife as a reference to Mars and Venus. He also associated the shields and lightning bolts in the threshold mosaic between the *tablinum* and the *atrium* as the same martial theme in the *fauces*.
101 Della Corte 1965, 186–188.
102 For the date of the pavements, see Pernice 1938, 53–54; Bragantini 1996b, 380. For the deployment of Blandus at Pompeii, see Los 1995, 167.
103 Pesando and Guidobaldi 2006, 212–213.
104 Fiorelli 1862, 96; Fiorelli 1873, 89.
105 Castrén 1975, 146.
106 See Niccolini and Niccolini 1862, pl. 5.
107 Bragantini 1998, 365–366.
108 Lavagne 1987, 135.
109 Clarke 1979, 10.
110 Della Corte 1965, 228–229. On the Coelii, see Castrén 1975, 155–156.
111 Bragantini 1998, 365–366; Pesando 2006a, 149; Pesando 2006c, 91–95; Pesando and Guidobaldi 2006, 231.
112 Lavagne 1987, 137.
113 Sampaolo 1999, 978–980; Coralini 2001, 51.
114 Coralini 2001, 48, 54–56. The house allegedly belonged to A. Rustius Verus or Ti. Claudius Verus in the final years of Pompeii; Della Corte 1965, 133. The commissioner of the mosaic remains anonymous.
115 Pesando and Guidobaldi 2006, 241.
116 Bragantini 1997, 232–233.

117 See Pesando 2006a, 103–105. Others precariously identify L. Calpurnius Diogenes or the Cisonii as the owner; see Della Corte 1965, 220.
118 Bragantini 1998, 518–520.
119 Fiorelli 1875, 340–341; Castrén 1975, 157. The mosaic is undated due to the loss of the associated decorative scheme.
120 Tacitus *Ann.* XIV.17.
121 Clarke 2003, 154–156.
122 On the *velum*, see Graefe 1979, 66–70.
123 On the original context, I follow Maiuri 1942, 17–21 who argues for commemorative; Ryberg 1955, 170–171 makes a case for votive; whereas Spano 1959, 16–19 sees an apotropaic function; see Huet 2007, 142–150 for a summary of the various theories.
124 Castrén 1975, 120, 145.
125 Hackworth Petersen 2006, 181–182.
126 Ain. Tact. i.6, i.7, and v.1; Castrén 1975, 124.
127 See Laurence 2007, 34–45 for neighborhood identity in Pompeii.

7 Contemporary fortifications in Italy and the Mediterranean

Sic deinde, quicumque alius transiliet moenia mea. "So shall it be henceforth with everyone who leaps over my walls."[1] Livy imagines that Romulus uttered these words after his terrible fratricide of Remus at the foundation of Rome. His brother had dared to jump over the line of the future walls of Rome in contempt. Living at least seven centuries after this mythical event, Livy cannot possibly have known this sentence with any true accuracy. Instead, the words speak to the symbolic importance of city walls to the very concept of Rome at the time when Livy was witnessing the birth of empire. Civil strife and war, into which Rome was born in the conflict between brothers, was ending.[2] For Romans, their walls were powerful metaphors of an us-vs.- them mentality, where state and social order were defined by the symbolic limit of the city. The roots of these ideals must come from the centuries-old tradition of city wall construction on the Italian peninsula and in the Greek world. These same traditions and ideals helped shape the Pompeian circuit and the genesis of the settlement.

The construction of fortifications in Republican Italy was a widespread endeavor requiring substantial investments and resources. Despite their prominence at many archaeological sites, the issue of design transmission in fortifications remains understudied. Part of the problem is that any study has to rely on the uncertain date of many circuits. For Italy, only one encompassing survey exists based largely on inscriptions installed in Latin or pro-Roman cities, colonial foundations, and excavations. It registers ninety-four urban circuits securely dated – whether built or refurbished – to the Republic. Of these, forty-two date to the fourth and third centuries, seventeen to the second century, and thirty-five to the first century BCE.[3] Even so, city walls are notoriously difficult to date, and potentially hundreds of small and medium-sized dwellings in the Italian countryside remain unrecorded. Most enceintes are in a terrible state of preservation, after centuries of abandonment or reuse in later contexts, complicating any reading they may give of their development. Pompeii by far preserves the most complete circuit and therefore is a natural departure point for any attempt at comparisons.

Exterior factors must have influenced the layout and ornamentation of the circuit at Pompeii. However, any systematic evaluation between the Pompeian circuit and others in the ancient Mediterranean ended with Maiuri's work, and no single approach has accounted for the ornamentation present in the fortifications. Certainly, the application of any form of ornamentation on city walls is not a strictly Roman phenomenon. Numerous precedents exist in the Greek world, most notably in the care taken to finish the masonry in elaborate circuits.[4] However, the use of polychrome masonry as applied in Pompeii seems more specific to the Italian peninsula, particularly in areas with an abundance of different construction materials such as travertine and tuff. In some instances, patrons would import a foreign material, such as tuff in otherwise predominantly limestone regions, in order to apply it in the most visible areas of the circuit. Such an explicit differentiation of materials for ornamental value seems to originate in the funerary monuments of Etruria and is manifest in Republican circuits, only to disappear with the advent of concrete and the Empire when enceinte construction moved to the periphery of the Roman world. The evidence suggests a pattern of regional usage and construction that, to borrow the ideas of John Onians, establishes an architectural and therefore ornamental value to structures that otherwise seem primarily militaristic.[5]

External influences and military tactics affecting design

Given the defensive nature of enceintes, military matters were primary factors influencing their appearance. The military concerns affecting the design of the Pompeian fortifications have fueled the debate on the external influences shaping the city's identity. The traditional approach views the *agger* system of the first and third circuits as an Italic design with Etruscan precedents, as opposed to the *emplecton* technique of the second circuit that would have tied Pompeii to the Greek cities of Cuma and Megara Hyblaea. A similar assumption has viewed the position of Pompeii on a high plateau near the sea as part of its Etruscan origin because it topographically resembles the location of large settlements such as Veii and Cerveteri farther north.[6] Such assumptions underestimate local agency in the decision to adopt a particular type of fortification or location for a settlement. As Strabo noted, and in a similar fashion to the ports of Gravisca and Pyrgi which served Tarquinia and Cerveteri, Pompeii was the coastal *emporium* to the inland cities of Nola and Nocera. The Sarno River was a natural highway to the interior.[7] This line of reasoning links the design of the Pompeian defenses, at least in their later stage, with Samnite cities. Certainly, close political ties with inland settlements along the Sarno River valley were inevitable. Nevertheless, Pompeii's position and defenses also offered a measure of independence. Rather than ascribing distinctly separate Greek, Roman, Samnite, or even Phoenician influences, the Pompeian defenses built at the end of the fourth century BCE display a medley of techniques – *agger* for the curtain and *emplecton* at the gates – aimed at giving the city the most advanced defenses.

The choice to build and subsequently upgrade a circuit in a particular manner inextricably links to the military developments and tactics of the time. This is the case for the *agger* system used as the basis of the third main circuit in Pompeii. In the second half of the fifth century BCE, armies in Greece and southern Italy developed new siege techniques. The weaponry included early versions of rams, mantelets, catapults, and bolt throwers designed to demolish fortifications.

New tactics also evolved to include sapper tunnels, earth mounds, and siege towers designed to breach or surmount walls. The defenses became more elaborate, responding with larger towers and stronger masonry. Engineers designed outworks with ditches and palisades intended to keep catapults out of range and to allow sortie parties to harass enemy positions. By 350 BCE, armies had fully implemented the use of siege machines, leading to elaborate enceintes and outworks, especially in towns lying in vulnerable open plains where offensive weapons were most effective.[8] These developments required increasingly organized armies capable of building and supporting the units. In the fourth century BCE, authorities in the Hellenistic world introduced the payroll, leading to professional year-round armies capable of mounting longer campaigns, sieges, and permanently occupying territory.[9] In Italy, the Roman army introduced maniples, small flexible units designed to replace the heavy phalanxes that had proved ineffective against the mobile Celtic and Sabine formations it had recently encountered. The versatility of maniples enabled targeted attacks on cities and strongholds, resulting in the obsolescence of the existing defenses.[10] In response to these developments, many Italian cities built anew or revised their defensive systems, including Pompeii.[11] The *agger* system in particular, with its heavy earth embankment, carried some advantages: It easily absorbed the vibrations of artillery hits, and its weight threatened tunneling efforts with cave-ins.[12] If a section of wall did collapse, the embankment remained a formidable obstacle that attackers had to climb to enter the city.

Questions regarding who designed the original *agger* at Pompeii and its subsequent upgrade are more difficult to answer. The mechanisms of transmission for military architecture have not received much attention. Common concepts of siege warfare compound the problem. For instance, the L-shaped merlons of the battlements at Pompeii find contemporary parallels at Chalcis and the attic fortress of Phyle in the Greek world and much later at Wiesbaden, Wimpen, Altenstadt, Trier, and Avenches in the Roman west.[13] A common consensus sees an adoption of Greek architectural elements in the fortifications of Latin colonies starting in the fourth century BCE. These include the arched gate (Paestum, Falerii Novi), the enclosed gate court, the portcullis, and artillery towers (Cosa). Within this transmission and absorption of tactics and design are also the adoption of urban ideals from the east, where a settlement needed an enceinte to achieve urban status.[14] The introduction of Greek forms would occur in an already strong Italian tradition developed in the large enceintes built in the sixth century and earlier in regions such as Etruria.

The Servian Wall in Rome, which is roughly contemporary with the first Samnite enceinte at Pompeii, exemplifies the complexity of the issue. Much of the surviving remains were built using tuff *opus quadratum* in the first quarter of the fourth century BCE. Engineers exploited the natural topography and built an *agger* in the weak areas to maximize the strength of the defenses. This was a familiar layout stemming from a tradition with precedents in the large urban centers of neighboring Etruria. The Servian Wall received a revetment of tuff blocks that Gösta Säflund, based on the block sizes and the similar use of Greek letters as quarry marks, associated with Greek architects and construction techniques coming from Syracuse. He believed that the evidence indicated an exchange of military technology since Rome and Syracuse were allies united against Carthage. Syracuse and its King Dionysius I – a pioneer in the use of siege machines in the Hellenistic world – would have learned the art of defense from their encounters with Carthaginian armies who, in turn, had been exposed to siege warfare in the Near East.[15] At least part of this theory seems plausible since construction seams in the Servian wall are spaced roughly a *plethorn* measure apart, suggesting a similar organization of building crews that Diodorus Siculus mentions for the Epiploae fortifications in Syracuse.[16] Another approach traces the builders of the Servian Wall to the Greek settlements of Paestum and Naples. These two cities first employed the technique, commonly found in Rome, of laying header and stretcher blocks in single courses.[17] These lines of reasoning are difficult to trace with any measure of certainty. Säflund's Phoenician connection is tenuous, whereas the construction technique of the Servian Wall finds ample parallel in Etruscan cities such as Veii and Vulci.[18] Both the Greek and Phoenician theories fail to take into account the enceintes of Etruria and Latium, where the *agger* system was developed and employed locally already in the eighth century BCE.[19]

A remarkable cross-cultural recurrence of construction symbols points to a uniform set of norms or guidelines in the construction process of the time (see Figure 7.1). Both the Servian and the Pompeian circuits display almost identical construction marks preserved in their masonry. Those at Pompeii seem to concentrate on the interior wall, whereas workers chiseled away those on the exterior, or they remain buried inside the earth embankment. Such marks are common on blocks that workers intended to leave out of view. Craftsmen would otherwise remove them when dressing the stone. The marks preserved on the inner wall of the Pompeian circuit are visible because the construction of the later towers led to the lowering of the *agger*, exposing them to view. Similar marks come from a variety of contexts stretching from the Archaic to the late Republican periods. In addition to Rome and Pompeii, examples are present on the defenses of Etruscan Tarquinia and Vulci, Greek Syracuse and Neapolis, and even in Punic Eryx, Palermo, and Carthage.[20] Their presence at Pompeii indicates that whoever built its defenses did so according to a well-established tradition using the most advanced techniques available.

The interpretation and position of the building marks at Pompeii have been matters of debate. One theory even goes as far as giving them an apotropaic role for protection of the community.[21] Until recently, the common consensus saw the marks exclusively as letters of the Greek alphabet or as derivatives thereof, leading to an assumed Greek origin of the masons. The realty is far more practical. Although some letters are indeed of the Greek alphabet, they are more commonly shorthand for construction tools associated with groups of masons. Such an approach demystifies the alleged Greek origin of the builders, allowing for the introduction of local masons into the building process. Many symbols concentrate in different areas of the circuit: A trident symbol recurs in the blocks near the Porta Nocera, whereas asterisks predominate near Tower XI.[22] Such concentrations imply a long drawn-out construction process, or it signals a scenario where

Figure 7.1 Surviving construction marks on the stretch of interior fortifications adjacent
to the Porta Nocera.

individual construction crews worked in separate sectors. Whether these teams
worked simultaneously, in sequence, or in entirely separate events remains an
open question. Nevertheless, each building crew would immortalize their work by
marking their section of wall. Today the marks bring to mind a powerful reminder
on the construction of the walls. Once visible, they must have evoked a similar
memory in antiquity.

The closest parallel to the Pompeian circuit

Of the known circuits in ancient Italy, it is the neighboring city of Nocera – ancient
Nuceria Alfaterna – that displays the closest parallel to the Samnite defensive system of
Pompeii. Located higher in the Sarno River valley, Nocera and Pompeii share a close
history (see Plate 3). Nocera led the Nucerian League, a loose federation of cities
brought together under the umbrella of the Samnite *alfaterni* tribes that included
the cities of Pompeii, Herculaneum, Nola, and Stabia. Just how the league's political
structure worked and the role of its *meddices* – the magistrates representing settlements – and their leader, the *meddix tuticus*, is unclear.[23] Its power seems consigned
to no more than a weak federation of strongly independent *meddices*. However, its
existence may have influenced defensive considerations.[24] The politics of a league

of cities potentially swayed the design and construction of fortifications. Loose alliances, such as the Etruscan League of twelve cities known as the *dodecapoli*, ensured a degree of mutual security against external threats. The agreements reached could involve the siting of settlements, warning systems, the construction of common fortification works, and mutual nonaggression pacts. A central political body could also manage the resources necessary to coordinate the simultaneous defense of several settlements or to conduct offensive operations.[25]

The fortifications and layouts of the two cities indicate a common origin in their defensive considerations. The circuits employed similar construction techniques used in two successive upgrades, and in both cases the materials were ashlar tuff blocks and *opus incertum*. Similarities in design lie in the first upgrade of the Pompeian defenses carried out at the turn of the third century BCE. Ancient Nocera had a rectangular plan with natural scarps defending its northern, western, and eastern sides. The most vulnerable section on the south side received the heaviest fortifications in the years before the Second Punic War. In a similar fashion to the upgraded Pompeian circuit, this section featured a double trace wall built with regular tuff ashlars with an *agger* in the back.[26]

The striking aspect of Pompeii and Nocera is that their circuits transfer the concept of double trace terraced Samnite hill forts to a freestanding *agger* in vulnerable areas of open ground. Double trace designs were common in ancient Samnium and are Italic in origin. Hilltop settlements such as Teano, Trebula Balliensis, Monte Santa Croce, and Saipins (Samnite Saepinum), dating to the end of the fourth century BCE, display a tiered system where lower terraces built in front of the main walls created defensive bastions.[27] Samnite fortifications adopted the most effective defensive designs by exploiting the local landscape. Most strongholds were perched on inaccessible hilltops where such tiered layers of defense made them almost impregnable.

Despite this Samnite tradition, the consensus so far has imperfectly ascribed a lopsided Hellenistic influence to these types of defensive enclosures. Philon's writings are the basis of the arguments to identify a Greek influence in the Pompeian fortifications, with parallels in the defenses of Athens and nearby Neapolis built in the fourth century BCE.[28] Philon recommends the double trace system of fortification in vulnerable areas with gentle terrain, as exemplified in Pompeii's northern flank.[29] Also known as a *proteichisma*, such a structure was a forward independent defensive element parallel to the main wall. The comparison with Pompeii refers to a slightly more elaborate version of a common *proteichisma* in Athens and Neapolis that was no more than a low terrace behind a ditch.[30] The Athenian, Neapolitan, and Pompeian versions carried the same strategic advantage with two lines of defense and extensive cover behind the wall. They also had a handicap, offering an enfilade below the lower wall if the enemy managed to capture the wall-walk. The closest parallels of a double trace system outside Italy are in the circuits of Side and Perge, which included superimposed parapets and in Selinus where two curtains defended its north gate. A critical difference between these examples and Philon's writings is that none has an *agger* attached to the double trace system, as is present in Pompeii and Nocera. Otherwise, very few contemporary examples of a double trace system set in an *agger* survive. The fourth century Servian wall in Rome may have featured a similar setup,[31] whereas the much later walls of Constantinople are a prominent successor of the Pompeian layout. In fact, the only "Greek" aspect of the Pompeian fortifications at this stage is at the gates where we find the *emplecton* construction technique used to strengthen the passageway walls. Instead, the Pompeiian fortifications adopt both Greek and Italic influences, blending them in an almost unique design adapted to local circumstances and strategic necessity.[32]

Nocera and Pompeii share further affinities in the towers that both cities built in the following upgrade (see Figure 7.2). During the Second Punic War, the defenses at Nocera had proved formidable enough that Hannibal resorted to starve the city rather than risk a direct attack. He subsequently destroyed it in 216 BCE.[33] After the war, Nocera rebuilt and strengthened its defenses with the addition of towers in *opus incertum*. The circuit preserves the remains of two distinct types built in successive phases. Their design is part of a purported evolutionary process as predecessors of the Pompeian examples.[34] This dating is tenuous, resting on typological differences, which, as at Pompeii, may respond to the local terrain. A distinct possibility exists that both cities built the towers within the space of just a few years. The first type found at Nocera, dating to the early second century BCE, features a coarser *opus incertum* and a solid base. The second is more intricate and almost identical to the towers in Pompeii. The buildings had three floors with two doors opening on each side of the wall-walk and a rear access from the *agger*. The ground floor featured a vaulted chamber with an exit postern. The elevation differed slightly with five and four windows on the second and third floors, as opposed to the four of Pompeii. The embellishments on the towers indicate strong ties between the cities. In their last phase, each tower at Nocera had a gabled roof with decorative *architectural terracottas*. Workers fashioned them from the same molds as those that once ornamented the Basilica at Pompeii, which was a public court building intimately tied to the administration of the city.[35] The towers in both circuits had almost duplicate First Style embellishments with the imitative ashlars and the Doric frieze having identical dimensions.[36]

The similarities between the towers at Pompeii and Nocera point to a common origin for their design and decoration. They are the product of a long tradition in military architecture that finds its origin in the Greek east and Italy. Tracing the development of their design is complicated since very few examples are complete enough to determine an origin. A few surviving examples, dating as early as the late fourth century BCE in the Greek enceintes of Aigosthena, Oiniadai, Messene, and Perge, display gabled roofs as predecessors of the Pompeian and Nucerian designs.[37] The presence of a roof on towers is more a matter of practicality to shelter delicate machines such as catapults and ballistae from the elements and enemy fire. The introduction of gabled roofs followed similar practical reasons. As opposed to earlier pyramidal roofs, they presented a stronger unified facade to the enemy and facilitated repairs. On the Italian peninsula, it is the circuit of Paestum that preserves towers that are the predecessors to the Pompeian examples. The earliest towers on the north side of the Paestan circuit date to the late fourth century BCE.[38] They stand on a tall solid masonry base with similar side openings on the wall-walk and a surmounting floor. The chamber on the ground floor of the

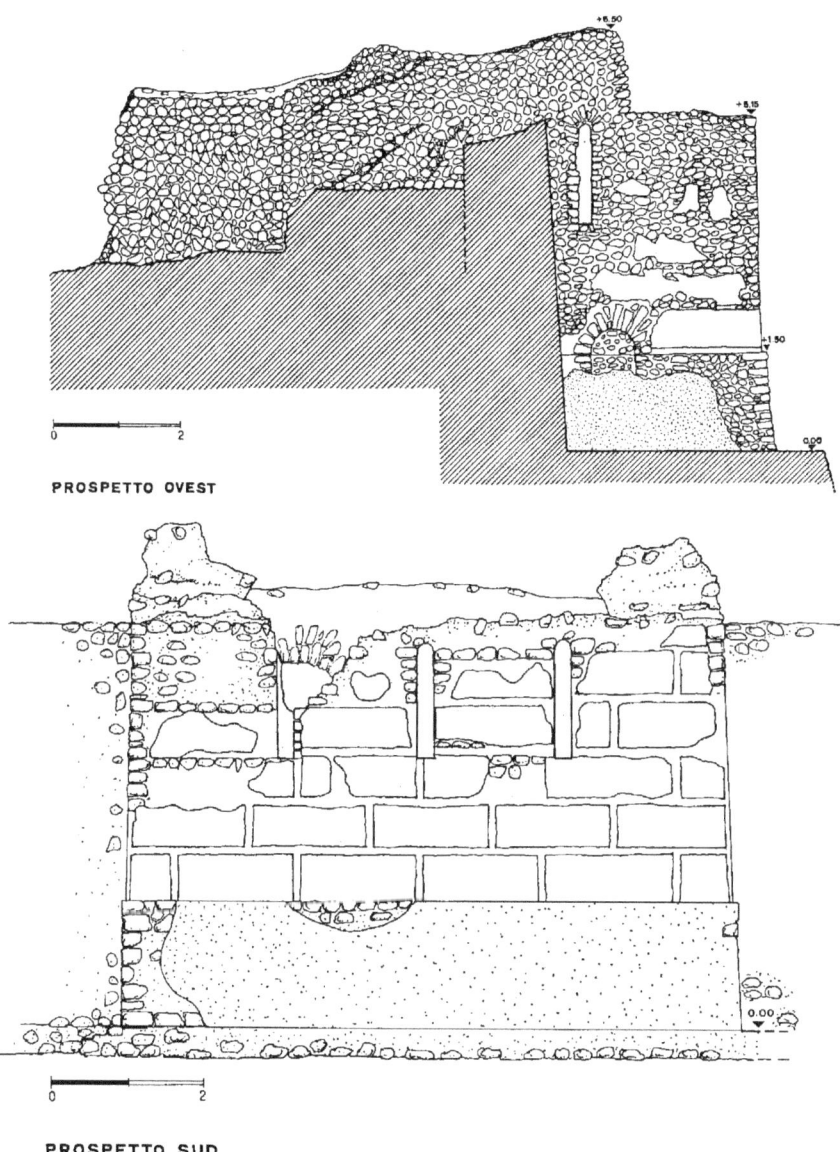

PROSPETTO OVEST

PROSPETTO SUD

Figure 7.2 Towers at Nocera. (Drawing after Johannowsky 1994, fig. 7.)

Pompeian type is missing only because the invention of concrete that would have allowed for the construction of such a vaulted space had not occurred. In the first century BCE these towers would receive an ornamental Doric frieze and pilasters with Corinthian capitals, indicating that similar decorative features on the Pompeiian and Nucerian buildings fit into a wider albeit elusive trend.[39]

The design and construction of the towers and double trace system at Pompeii required the technical expertise of skilled architects educated in the tactics of siege warfare. With the cities in close alliance with Rome and the existence of the Nucerian league questionable, Roman engineers were probably involved in the designs of both construction episodes. Given the rapid expansion of Rome throughout the Mediterranean at the time, architects must have adopted designs and defensive tactics encountered in the east. The adaptation of Samnite, Greek, and Roman ideals and construction techniques is a remarkable example of syncretism that must have affected notions of identity and self-determination at Pompeii and other settlements in the region.

Masonry, fortifications, and the definition of a city

The web of design transmission and architectural influences becomes more complex when considering how architects used construction materials and masonry to achieve aesthetic aims such as the deliberate use of tuff and travertine in the Pompeian enceinte. A few basic techniques stand out to include finishing masonry in visible areas, embossing/rustication of blocks, and the application of divergent types of construction materials. They are present to varying degrees on the Italian peninsula as part of a growing trend. Instances such as Pompeii where the entire circuit has two distinct construction materials are rare. However, unlike Pompeii, relatively few centers had access to extensive quarries of two different construction materials such as the travertine and tuff present in the Sarno River valley. Instead, cities and towns would emphasize their circuit by taking care to finish the masonry or leave it rusticated and by building sections of wall with elegant tight fitting joins. On occasion, patrons would import material, such as tuff stone in prevalently limestone areas, thereby adding to the expense and therefore the prestige of the monument. These examples foreshadow the copious use of marble in the monuments of the early Empire. Given the expenses of transporting the material, the Republican examples tend to focus on areas of high visibility, such as city gates and adjacent wall sections, where they created political statements or expressed a patron's munificence on local communities.

The practice of ornamenting fortifications came in a variety of manners. The most obvious date to as early as the Bronze Age Hittite and Mycenaean circuits, which included elaborate reliefs of lions and other carvings at gates. A more subtle approach was to display finished masonry in areas of high visibility as opposed to unworked blocks in remote tracts.[40] Dressing masonry was a laborious task. The appearance of dressed masonry in any enceinte is an indicator of the sophistication of the wall. The reason is clear: A community would put its best foot forward to viewers passing through the walls. Another technique that carried an equal ornamental value was to leave the blocks unfinished in otherwise tightfitting masonry in a technique known as a rustication, or *bugnato*. In the fortress of Eleutherae, built in the fourth century BCE, the architect retained rusticated blocks for an ornamental effect in an otherwise impeccably built wall. Worked and unworked blocks were also set out in deliberate ornamental contrasts. The earliest examples are in Greece where fortified circuits would first become synonymous with the image of cities. As early as the ninth century BCE, a tower in the circuit of Old Smyrna employed a purposeful contrast between worked and unworked masonry. The same technique would reappear in the early fifth century BCE when the city of Thassos built a circuit where courses of unworked ashlar blocks surmount a socle composed of smooth orthostats.[41] The circuit at Eleusis, built in the fourth century BCE, displays a more subtle effect where unworked blocks serve to emphasize a decorative socle for the upper wall.[42] These practices visually defined curtain walls into two sections: a socle supporting an upper section, differentiated through the application of a divergent construction material or technique. Although difficult to prove, they may find an origin as a petrification for ornamental purposes of a staple construction method where a stone socle acted as foundation for a mudbrick wall. The Greek examples prefigure in many ways the development of the Pompeian circuit.

On the Italian peninsula, separate regional building traditions would develop similar trends toward the ornamentation of enceintes. The construction of fortified circuits divides broadly into three main types strongly influenced by the availability of local materials and cost: (1) Circuits could be built in perishable materials. Etruscan sites at Felsina and Spina in the Po River plain – an area poor in stone – had a typical *agger* system with an earth embankment strengthened with a surmounting palisade, an exterior wooden revetment, and an exterior ditch.[43] In other examples, engineers would employ a (low) stone socle to support a mudbrick superstructure to create a freestanding wall or battlements, as in Velia and Gela.[44] Very few remains of these circuits survive today because the materials have since degraded. (2) Areas rich in travertine, sandstone, and particularly the tuff landscapes of southern Etruria, Lazio, and Campania have a prevalence of ashlar masonry because these materials lend themselves to regular cutting. The great Etruscan circuits of Veii (tuff) and Fiesole (sandstone) are just two of many examples. (3) In central Italy and along the Apennine range where limestone is prevalent, many circuits employ polygonal masonry because the stone naturally breaks at oblique angles and is far more difficult to cut in regular sizes. Both ashlar and polygonal masonry are common construction techniques throughout the Mediterranean, as is the use of perishable materials.[45] However, regional phenomena are prevalent in the design of fortifications.[46] The so-called Italic tradition is one that features containment walls for terraces or the revetment of an earthen bank – hence the *agger* type of defenses. Within this tradition, regional variations and the availability of materials must have been as evident in antiquity as they are today, establishing identifiable visual cues in the landscape. Such distinct construction manners must have operated as markers of cultural and ethnic identity in a similar fashion as Onians has proposed for the Greek architectural orders (i.e., Doric, Ionic, and Corinthian).[47] The masonry manners (types), because of their use of specific materials and techniques, must have fostered local and regional identities, such as the Latin communities south and east of Rome, the Samnite hinterland, and cities in northern and southern Etruria.

The ornamentation of fortifications and in particular of the curtain walls responded to their scale and position in the landscape. Rather than minute details, it is the masonry and its finish that was a key component because of its visibility. The nature of the materials employed allowed for different approaches. In the areas of Etruria where we find an abundance of tuff and travertine, the curtain walls sometimes received subtle embellishments in the form of a *bugnato* (rusticated) socle starting in the fourth century BCE. The western gate at Vulci is a case in point (see Figure 7.3). As a response to the increasingly complex military tactics of the fourth century BCE, the architect built a sophisticated gate that included a separate forward bastion shaped as a triangle. The design forced the attacking enemy into deadly corridors where the defenders could pelt them from three sides. The contemporary adjacent curtain wall, built entirely in tuff, still displays a decorative bossed socle in its three lowest courses. A similar treatment is present farther inland at Chiusi where the scant remains of its fourth-century circuit, built in local travertine, displays traces of an ornamental rusticated socle in the three lowest courses.[48] Subtle traces of rustication also survive in the Servian Wall in Rome, although its presence on the entire circuit is somewhat elusive. A surviving section on the modern via Carducci displays hints of surviving rustication on the lowest four courses.[49]

In contrast to the tuff structures, which are usually relatively homogeneous, limestone polygonal masonry often comes in variety of degrees of finish. Giuseppe

Figure 7.3 Decorative embossing surviving in the lowest two courses adjacent to the western gate of Vulci.

Lugli classified polygonal masonry into four manners, or *maniere*, divided chronologically according to their grade of workmanship (see Figure 7.4). The first two *maniere*, prevalent between the sixth and fourth centuries BCE, are present in the circuit at Roselle where builders stacked rough stones in a more or less random order. The third type featured finished masonry with precise joins between the blocks. It appeared in the fourth century BCE in places with a long tradition in applying the technique, such as the Sacco valley where the citadel of Alatri is perhaps the best known example. On occasion, in regions such as modern Formia, this tight-fitting masonry displays rustication as a further means of ornamentation.[50] The fourth manner, developed contemporaneously to the third and used well into the Imperial period, featured irregularly sized trapezoidal blocks arranged in almost regular but discontinuous courses. Although a refinement of polygonal masonry techniques certainly occurred, a major caveat to Lugli's chronology is assigning dates to any given the circuit or even sections of walls based on the degree of finished masonry. Once masons perfected a technique like the third *maniera*, for example, it does not mean that they abandoned and never reapplied the other two.[51] Instead, the degree of finish on wall faces also relied on factors such as patronage, resources, and workmanship. In terms of the presentation of town walls, factors such as a higher degree of finish in key areas of visibility remains a little considered variable that further complicates their chronology.

The area of the Sacco valley and Lepini Mountains was a region of ancient Latium south of Rome particularly rich in limestone that developed a strong tradition of polygonal construction techniques. It preserves the remains of some

Figure 7.4 The polygonal manners as illustrated by Lugli (1968, fig. 1).

twenty-eight dated circuits.[52] Architects would often incorporate subtle embellishments in the masonry that have remained virtually unnoticed. The town of Norba,
perched on the Lepini Mountains overlooking the Pontine plain, became a Roman
colony in 492 BCE as a bastion against the Volsci who inhabited the area.[53] Its ruins
preserve an enceinte, about 2 kilometers long and in places 13 meters high, built in
the fourth century BCE. The fortifications were constructed in the form of large terracing walls, creating a broad platform that leveled the terrain in a similar fashion
to the southern ridge at Pompeii. The circuit is one of the most homogeneous of
central Italy in terms of the application of construction techniques and design.[54] A
striking aspect is the degree to which workers finished the polygonal masonry and
the location of the tracts. The four gates, as well as the long tract of walls on the
eastern side of the city, invariably display carefully finished polygonal masonry
according to Lugli's third *maniera* (see Figure 7.5.) By contrast, the western side of
the circuit, which covered no more than the cliff edge, displays entirely unworked
blocks with rough surfaces and open joins between the stones. This circumstance
is hardly accidental; the areas with the best masonry also mark the approach routes
and entrances to the city.

The Latin town of Cori, perched on a hilltop nearby, preserves the remains of a
contemporary circuit reaching roughly 2.2 kilometers in length that protected the
lower town. Two smaller terraces, also in polygonal masonry, which doubled as
inner circuits, defended the citadel. Such terraces were of vital importance in the
defense and layout of the city. They gave Cori the level space necessary for construction and an in-depth defensive strategy with multiple lines of fortifications.
These grand terraces also carried a monumental element: They resonated with the

Figure 7.5 The main southern gate of Norba.

outer fortifications, providing visual unity to the city in both the urban and the rural landscapes. The intermediary terrace dating to the third century BCE provides an eloquent example of masonry display. Today, it lines the modern via della Libertá, which in antiquity was part the main road leading from the Porta Romana in the lower city up to the *arx*, which held the Forum and sanctuaries such as the Temple of Hercules.[55] The terrace was a place of high visibility where a viewer would have to pass the wall on the commute between the upper and the lower cities. Figure 7.6 shows the masonry on the corner of the terrace where one side faces the main road and the other a minor street. The very same corner blocks are highly finished on the side facing the regional road, whereas those on the street side are unfinished, indicating a clear preoccupation with the proper display of the masonry.

Figure 7.6 Worked and unworked masonry at a street corner in Cori.

By the end of the third century BCE, polygonal construction techniques would carry ideological weight. When Roman colonists laid the foundations of the new settlement at Cosa in 273 BCE, they did so on a strategic hilltop in a sparsely occupied and hostile landscape wrested from the Etruscan city of Vulci. The settlers were Latins from the area to the south of Rome near the Sacco valley and Lepini Mountains. They brought with them the polygonal techniques to build new fortifications.[56] At about 1,500 meters long, the walls exclusively feature highly finished polygonal masonry. They also include eighteen towers, most of which are concentrated on the higher side of the hilltop where they have little strategic use against siege machinery. Instead, they offered views of the sea against enemy ships and a clear delineation of the colony in the landscape for many kilometers. The colonists would also build a Capitolium temple on top of the *arx* in the early second century BCE with a podium built of polygonal masonry as a reference to the enceinte protecting the city. The fortifications were thus not only defensive. The masonry was familiar to the Latin colonists, and it made a political statement for an outpost founded in newly conquered territory.

Ideological associations with polygonal masonry are also manifest in the Sanctuary of Fortuna Primigenia. Located east of Rome at Palestrina, the site was a gateway to the Sacco valley and its traditions of construction using the polygonal technique. Built around 120 BCE, the sanctuary would become an important interregional oracular center.[57] It is famous for the manner in which architects used a hillside to manipulate visitors by means of dramatic vistas, terraces, ramps, tunnels, and open and closed spaces as part of their religious epiphany. Despite the copious use of concrete throughout the complex, the lowest terrace displays a polygonal masonry in the third manner, built using local limestone. It served as a colossal podium in the landscape for the complex of artificial terraces leading up to the temple at the summit. The terrace was a distinct regional marker at an important interregional sanctuary that drew pilgrims from far and wide. For a pilgrim, the masonry carried deeper symbolism, marking the end of the journey to get to the sanctuary and the start of the religious trek up to the inner oracle above.

Polychrome masonry and city walls

A more emphatic approach to the ornamentation of city walls and terraces is the use of polychrome masonry, which I define broadly as the use of different construction materials to achieve a deliberate visual effect. The concept is clear for the Pompeian fortifications, but it is not an isolated phenomenon. Instead, Pompeii is part of a trend that Rome would export during the Second Punic War. The application of different stone types in construction finds its roots in the Italic building tradition for practical reasons. Already in the sixth century BCE, builders in Rome used different types of tuff and, later, other stone types rationally for their inherent properties and load-bearing capacities.[58] In the Greek city of Paestum, the contemporary Doric frieze of the Athena Temple displays yellow travertine metopes alternating with gray sandstone triglyphs. The metopes lack any further sculpture, but they must have had a rich coat of paint like the rest of temple. The use of the two materials was probably a practical choice where sandstone lends itself better to carving the triglyphs.[59] Nevertheless, the presence of the two materials in such a visually prominent location on the building implies an aesthetic choice. In other words, although the measured use of different stone types in construction is certainly practical, architects must have recognized the ornamental value of contrasting colors.

The earliest use of polychrome masonry for ornamental purposes is in monumental funerary architecture. The pyramid of Menkaure at Gizeh dating to 2500 BCE had an outer casing that covered its inner core. It was composed of sixteen lower courses in red granite that acted as an ornamental socle for the upper yellow Tura limestone masonry above. Most pyramids featured a similar but monochrome casing. For Menkaure, the architect chose construction materials to grab a viewer's attention perhaps as measure to compensate for the fact that this was the smallest of the three great pyramids.[60]

Centuries later, the technique appears in Etruscan funerary monuments, particularly in the Banditaccia necropolis outside of the ancient city of Caere (Cerveteri). Tumuli were large circular earthen burial mounds on drums cut out of the living rock with low containment walls carved as a cornice. As monuments to the dead, they were venues for ostentatious display with the largest belonging to wealthy citizens competing for prominence in the funerary landscape. The living rock and the quarried masonry that compose the cornice often have slightly different hues or values, creating a contrast that becomes all the more evident when exposed to moisture. Rather than elaborate and detailed decoration, this type of ornamentation reflected the function of the tumulus as a type of large-scale funerary monument: Both the embellishments and the structure had to stand out in the landscape.

The approach adopted for the tumuli becomes more emphatic in the Tumulo delle Cornici di Macco – also known as the Polychrome Tumulus, or the Tomba Policroma (see plate 27) – and in the standardized designs of the *tombe a dado*. Dating to the second half of the sixth century BCE, the Polychrome Tumulus is the earliest surviving example in ancient Italy where an architect employed construction materials for their ornamental value. As opposed to its neighbors, with their smooth drums hewn from the living rock, this drum is composed of regular ashlar masonry set in bands of red tuff, gray tuff, and macco – a whitish calcareous stone typical of the area.[61] Although much smaller than the large tumuli that characterize the period, its design clearly stands out against its competitors. Soon after its construction, sumptuary laws would seek to limit such flamboyance into the standard shape of the *tombe a dado*. Tombs would be built as cubes along an urban orthogonal layout – a virtual city of the dead.[62] Many tombs display a smooth base carved from the living rock, supporting ashlar masonry in red tuff, and cornices composed of a lower band in white macco and another in gray tuff.[63] The tradition of subtle embellishments by means of construction techniques and masonry established in the larger tumuli continued despite the restrictions. The smaller size of the later tombs may very well have augmented the practice as a means to make the new tombs stand out in the funerary landscape. The endurance of the materials and the lasting ornamental effect they created were a tribute to the dead and to the role of the tombs as their final resting place.

Similar notions of permanence and strength associated with stone construction, as well as the fact that fortifications, like tumuli, were large-scale monuments in the landscape, must have led to the adoption of polychrome masonry in city walls. However, its application did not occur until the fourth century BCE, perhaps because social and economic crisis in Etruria led to a hiatus in the construction of large-scale fortifications in the fifth century BCE.[64] Rome's own Servian circuit, in its mid-Republican phase (4th century BCE), preserves traces of polychrome masonry. Built as a classic *agger*, the fortifications included a tall exterior and a low interior containment wall to hold the earth embankment. In the section near the Termini railway station, engineers built the exterior wall using a yellowish tuff known as *tufo giallo* and used a grayish tuff called *tufo del palatino* for the interior.[65] Such a conscious decision to mark each wall with its own construction material must have carried aesthetic as well as practical considerations. When builders first adopted *tufo giallo* it conveyed symbolic associations with Rome's territorial reach. The material came from quarries in the territory of Veii that became available to Roman architects after Rome conquered the Etruscan city in 396 BCE.

A somewhat later stretch in the saddle between the peaks of the Aventine Hill preserve courses of a yellow tuff, superimposed on a socle – roughly nine courses high – of gray tuff. The superimposition has occasioned a debate on whether the types of stone represent two different building events or a single construction episode. The yellow tuff was a revetment of a cement core that may indicate a slightly later construction date than the gray tuff socle built in the fourth century BCE.[66] Regardless of the chronology, the choice to use such a revetment must have had a measure of aesthetic consideration. The yellow tuff revetment of the cement core obviated the necessity of a plaster coat to protect the masonry, and the stone would have been far more expensive to procure and install. Instead, the stone revetment would have acted as a functional marker highlighting the renewed strength of the defenses, the extent of the renovation, and the associated patronage. Similar notions apply to the Pompeiian fortifications irrespective of whether the travertine/tuff juxtaposition of the curtain wall represents differing construction events.

A contemporary comparison for Pompeii exists at Ferentinum, a Hernician town located in the Sacco valley near Rome. As at Pompeii, the town built its walls in the early third century BCE as part of a new orthogonal layout of the city.[67] The new fortifications display an ornamental superimposition of construction techniques: a lower socle composed of polygonal limestone masonry laid according to Lugli's fourth manner, surmounted by a pseudo-isodomic framework in gray travertine. Some have ascribed this superimposition of techniques to successive construction events where the travertine masonry represents a later intervention dating to the first century BCE. An equally plausible hypothesis is that the two types represent a single construction event aimed at reaching a deliberate ornamental effect.[68] The masonry of the Porta Sanguinara – a main entrance into the city – and its adjacent curtain wall highlight the sophistication of the ornamental emphasis (see Figure 7.7). The gate is composed of a single large opening set in tightly fitting polygonal masonry that acts as a socle for a radial arch built with ashlar travertine blocks. A measure of the care taken to build the gate is already evident in the polygonal socle where the tightly fitting joins are in stark contrast to looser gaps in the masonry farther south. The use of the arch is partly a necessity because the opening was too wide for a single lintel. Builders rarely applied arches in polygonal circuits such as Segni, Alatri, and Arpino, which typically employ lintels or corbel techniques to span spaces. The lesser posterns in the circuit employ corbel openings set in the polygonal masonry, resulting in a visual hierarchy between the main gate and the minor entrances.[69] These factors indicate that the architect used the arch and the differentiation in the masonry to monumentalize the gate as well as the circuit.

Figure 7.7 The Porta Sanguinara at Ferentino and the adjacent curtain.

Aesthetic considerations must have influenced the decision to use travertine in the upper part of the curtain wall as a contrast to the lower polygonal masonry. Workers quarried most of the limestone for the polygonal masonry on site or very nearby. By contrast, they had to retrieve the travertine from the lower Sacco valley in a process that would have required considerably more time and effort. There must be a reason why builders went through the trouble to procure and transport the material. In practical terms, the travertine tends to produce smaller regular ashlar masonry when quarried as opposed to the limestone, which is carved into large blocks. Their size and use in the socle suggests that builders used the material in the immediate vicinity of the circuit for its load-bearing capacity.[70] In aesthetic terms, just as is the case in Pompeii, Rome, and Cerverti, Ferentino is one of the few examples where the town had ample quarries of divergent construction materials within its territorial reach. Even if the masonry represents two different construction events, the resulting visual effect in the defenses evoked territorial hegemony, patronage, and architectural and construction techniques in the image of an independent city.

The ornamental arrangement of the fortifications would become a hallmark of Ferentinum in the same manner that Pompeii's town walls would dialogue with its own urban layout. In the early first century BCE the local censors Aulus Hirtius and Marcus Lollius expanded the town's citadel with a new bastion designed to accommodate a temple. An inscription on the base still highlights their achievement:

> A(ulus) Hirtius A(uli) f(ilius) M(arcus) Lollius C(ai) f(ilius) ce(n)s(ores) fundamenta murosque af solo faciunda coeravere eidemque probavere/in terram fundamentum est pedes altum XXXIII in terram ad idem exemplum quod supra terra[m silici][71]

Translated the inscription roughly reads:

> Hirtus and Lollius, censors, supervised and completed the construction of the walls; the foundations reached 33 feet deep on the same plan as above in concrete.

In identical fashion to the city's main circuit, the structure stands on a limestone polygonal socle with an upper wall composed of ashlar travertine masonry. The upper ashlars are the revetment of a concrete core that is otherwise entirely absent in the fortifications, implying that Hirtius and Lollius commissioned the extension after the original lower circuit.[72] As an extension of the *arx*, the bastion would have accommodated a temple, resulting once again in close relationship between the patronage of both religious and defensive structures. By means of divergent construction materials and techniques, the two censors consciously referenced the ornamental emphasis present on the fortifications as a reflection of a local civic image.

Elsewhere in Republican Italy, the application of divergent construction materials for ornamental effect finds a place primarily at city gates rather than on the entire circuit. It is a reflection of targeted patronage where single patrons paid for the upgrade of individual gates, towers, or stretches of curtain wall. The use of divergent materials for an ornamental effect is particularly evident where patrons chose to import construction material rather than use local stone, thereby adding cost but also prestige to the project. Some of the most overt examples are again south of Rome where polygonal circuits built in local limestone preserve gates that display tuff ashlar masonry as part of refurbishments carried out in the third and second centuries BCE, as Rome consolidated its grip on the peninsula. The gates as places of high visibility were the natural venues for ostentatious display, which would also carry political associations.

The town of Segni displays examples of both original and later modifications to the embellishments in its fortifications. At about 5 kilometers in length, the it preserves one of the most complete circuits surviving since antiquity. Built in the fourth century BCE, the fortifications hug the rough topography of a strategic hilltop overlooking the Sacco valley. Local quarries where construction crews retrieved the limestone to build the polygonal masonry of the circuit still dot the perimeter.[73] The surviving circuit displays sections in two of Lugli's manners: rough second and highly polished third manner polygonal masonry. Their incidence seems to be more a matter of visibility rather than a result of different construction episodes.[74] The gate known as the Porta Saracena is perhaps the best preserved example of a gate built in polygonal masonry dating to Republican Italy. Erected using the corbel technique and a heavy lintel, the gate is a simple opening with an ogival shape. The masonry on the exterior curtain leading up to the entrance is in a highly polished third polygonal manner, whereas the wall transitions into a rough facade to mark a viewer's passage into the city. These subtle details indicate that builders paid particular attention to present the exterior and interior curtain walls to those entering the city.

The concept is more emphatic at Segni's Porta Romana, located slightly farther downhill. As the most important gate into the city, it marked the arrival of the main provincial road into Segni from the via Latina passing through the valley below. In the first half of the first century BCE, the gate received an overhaul that included a large stretch (about 500 meters) of the adjacent curtain.[75] For the curtain wall, workers carefully razed the old polygonal limestone masonry to create a socle upon which they built a wall of tuff ashlar masonry (see plate 28). For the gate, the architect applied a grand concept that obliterated any trace of the previous entrance. As opposed to the old ogival design, the new gate featured two parallel vaulted entrances, each flanked by a tower. To build it, workers employed the same brown tuff ashlar masonry they had used for the curtain. The tuff was an imported material. Its use in the fortifications was a deliberate effort to create a contrast with the polygonal white limestone masonry of the original circuit.[76]

Such a deliberate juxtaposition of materials and techniques carried much more significance than a simple reconstruction. A similar development occurred at the Temple of Juno Moneta on Segni's citadel. First built in the fifth century BCE, the temple sat on top of the *arx* with wide views over the Sacco valley below. The contemporary defensive circuit doubled as the terracing structure for the temple, once again creating visual relationship between the defenses and a protective deity. At the turn of the first century BCE, the temple received a refurbishment that was roughly contemporary with the Porta Romana. Although it later became a church, much of what stands today is part of this ancient refurbishment. The church stands on the original tiered podium of the first sanctuary composed of tightly fitting polygonal limestone masonry. The building has mostly disappeared, but its cella, composed of brown tuff ashlars, survives as the exterior walls of the church. Admittedly, there is no way of telling whether the patrons wanted tuff masonry left bare without a plaster coat to emphasize the color of the material. Nevertheless, the two materials would have contrasted dramatically through the difference in construction techniques: polygonal vs. regular isodomic ashlars. Perhaps a reason to imagine bare tuff masonry on these public structures is the fact that the areas of the Monti Lepini and the Sacco valley are rich in limestone but poor in tuff. Patrons had to import it from quarries close to Rome, where architects used it extensively. The application of these two construction techniques and materials in Segni's temple is a reflection of a broader dialogue between polygonal masonry – a mostly local building tradition – and tuff ashlars, a construction technique associated with Rome. Their appearance together conveys a message of association between local and Roman building traditions.[77] At Segni, this interchange extended to the fortifications where viewers entered the city. The result was an architectural and visual dialogue between sacred and military structures that reinforced the social and political bonds with Rome.

The nearby settlement of Cori presents similar developments, implying that they are part of a wider trend. In a parallel event to Segni, the city built its first enceinte in the fifth century BCE using the rough polygonal second manner. In the third century BCE – roughly contemporary with events at Pompeii and Ferentino – Cori rebuilt at least two of its main gates in the lower enceinte: the Porta Ninfina and the Porta Romana. The gates are heavily damaged, but enough survives to discern the main elements of their reconstruction. The refurbishment included replacing the rough second manner of the adjacent curtain with highly polished third manner polygonal masonry, as well as applying tuff ashlars for the gates.[78] Once again, this arrangement of imported brown tuff ashlars is a clear juxtaposition against the adjacent white masonry, which here finds further distinction in the division between the second and third polygonal manners in the adjacent curtain. The patron used the masonry to emphasize his euergetism and improve his city as well as signal his political association through local and Roman building traditions.[79]

If the enceintes of Ferentino, Cori, and Segni set out a discernible trend in Latium, the ancient town of Trebula Balliensis lies within the Samnite sphere of influence where the fortifications parallel the layout of the Pompeian gates. The site is located in the Monti Trebulani near Capua. Founded somewhere in the sixth century BCE, it was an independent settlement that would side with Hannibal until it fell to Quintus Fabius Maximus in 215 BCE.[80] Despite its Punic allegiance, Trebula would continue to flourish as an independent city within Roman orbit. Much of the site is unexcavated, with the exception of its western gate and part of the adjacent circuit. The tract dates to the fourth century BCE and is built in a rough polygonal limestone masonry. In front of it are the remains of a contemporary *proteich-isma* built in contrasting dark tuff ashlars that workers removed in a later phase. The architect who designed the gate had clear military considerations in mind. It opens at the apex of a natural topographical reentrance, creating a deep funnel that enhanced the gate's defensive capability. The entrance itself is reminiscent of the Porta Saracena at Segni in its ogival corbel arch capped with a heavy lintel.[81] A long corridor, built later in the third century BCE, stretches through the *agger* to provide access into Trebula proper. The corridor displays a revetment of brown tuff ashlar masonry that contrasts sharply with the adjacent white polygonal walls (see plate 29).[82] As at Pompeii, limestone and tuff were readily available in the immediate hinterland. Trebula would similarly exploit these materials to achieve a deliberate ornamental effect: first to create a contrast in the exterior curtain and later to emphasize a viewer's transition into and out of the city.

As a testament to the ideological framework associated with the ornamenta-tion of defenses in Latium and Samnium, Roman forces would export it outside of Italy. The east coast of the Iberian Peninsula saw heavy fighting during the Second Punic War, and the hinterland would struggle against Rome long after its end. In 218 BCE, in an effort to challenge Carthaginian forces in Spain, Gnaeus Cornelius Scipio Calvus made Tarragona (Tarraco) one of his bases after landing with his fleet in nearby Empúries. The two cities are the first Roman settlements of Hispania Citerior, the province that would later become Hispania Tarraconensis. They were at the vanguard of Roman efforts in the war, and both would remain important active agents in the following 200-year struggle against the Celtiberians to conquer the peninsula.[83] They also preserve the closest similarities to central and southern Italy in terms of ornamentation in their town walls, suggesting the display of a deliberate ideological statement.

When Scipio arrived in Tarraco, he first overran Cissus, a local Iberian *oppidum* located in the area of the later port. Leaving much of the early settlement intact, the Romans would build a separate fortified camp, or *castrum*, on a nearby hill, transforming it into a regional administrative centre. The two settlements would coexist for the remainder of the Republic.[84] In what is arguably the first set of for-tifications that Rome erected outside Italy, engineers built an enceinte in limestone polygonal masonry roughly 6 meters tall. A stretch of about 1 kilometer survives complete with three towers.[85] (see Figure 7.8). The towers reached higher than the curtain wall with a superstructure composed of neatly worked and embossed ashlar masonry made of local sandstone. A low decorative cornice further signaled the transition between the two construction techniques. The contrast in techniques is so stark that a common assumption erroneously attributed the sandstone ashlars to a Roman upgrade on a preexisting enceinte instead of seeing it as the result of aesthetic choice. The Torre de San Magin, also known as the Torre de Minerva, exemplifies the layout. It first functioned as one of two flanking towers of a main gate into the city that engineers closed in a later upgrade.[86] Further embellishments included some minor apotropaic busts on the corners and a prominent sandstone relief depicting Minerva. Below the Minerva relief, a masonry plaque was reserved for an inscription that has since disappeared or was never carved. The intended

Figure 7.8 The development of the circuit at Tarraco, showing the original gate (left) and its later modification (right). (After Hauschild 2006, figs. 11 and 20.)

text must have referenced the patron(s), the name of the city, or both. Although the polygonal masonry was ornamental in its own right, the ashlars on the towers, the sandstone relief, and the inscription were part of an intentional juxtaposition of masonry reminiscent of the circuits at Pompeii and Ferentinum. Such an emphasis not only served to highlight the towers; it also accentuated a main entrance into the new fortified enclosure.

In the mid-second century BCE, Tarraco would extend its circuit to include the lower city in a manner reminiscent of the developments in Italic circuits. This was the time of the Numantine War, when Rome decisively defeated the Celtiberian tribes along the River Ebro.[87] Engineers first raised the curtain of the existing wall with several courses of sandstone ashlars to reach the level of the towers. In a reference to the earlier manner of the walls, the new extension included a low socle composed of megalithic polygonal masonry with a surmounting 12-meter-high facade of sandstone ashlars complete with decorative embossing. Some of the ashlars display mason marks similar to those recovered at Pompeii and elsewhere on the Italian peninsula, further implying that the architects and builders were working within a well established tradition.[88] As is the case with Ferentinum, architects even assigned a visual hierarchy to the gates: low posterns were composed of megalithic masonry, whereas main gates, such as the Portal del Roser, received a megalithic socle with a surmounting arch in isodomic masonry. With the new extension, the fortifications continued to display an ornamental emphasis appropriate for the importance of a provincial administrative center such as Tarraco. The city walls exemplify how authorities could use construction techniques and materials to add ideological weight to urban fortifications. They established and created an image for Rome far away from the actual center of power.

The nearby settlement of Empuriés, located to the north of Tarraco, displays a similar development in its walls. The city was in origin a Greek foundation. Scipio chose the port to disembark his troops, and it would remain an important base throughout the Second Punic War. Tensions between local Spanish tribes and the Greek community persisted at the time of the Roman arrival, eventually leading to an uprising in 195 BCE. A Roman camp would spring up on a nearby high plateau outside the city in an effort to control both the Greek settlement and the local tribes.[89] The Roman camp would develop into a formalized settlement according to the orthogonal layout of colonial foundations. In the early first century BCE, the city would rebuild its walls. Builders carefully razed the previous second-century enceinte, in the process making sure to leave behind a few courses that would function as a socle for the new wall. The remaining socle is composed of large white blocks arranged in pseudo-isodomic masonry further embellished with bossing.[90] Above it, workers then added a wall of *opus incertum* to create the new defenses. As is typical of the material, the concrete would have featured some sort of plaster embellishment perhaps in the First Style, as proper for public buildings. The effect on the walls thus created seems to include an architectural nod toward the defenses of Tarraco, thereby creating clear colonial associations.

Towers and the ornamentation of the city

After the construction of the main circuit, the towers at Pompeii would be the focus of ornamentation in the subsequent upgrade of the late second century BCE. Early ornamented towers go as far back as the Geometric circuit of Old Smyrna and late archaic towers of Larissa (Buruncuk) in Greece. However, it is in the Hellenistic period that towers began to receive more embellishments because of their prominence as tall landmarks in the urban landscape. On occasion, stringcourses on the facades highlighted internal floors. More often, towers displayed highly polished, or deliberately rusticated masonry, or a combination of the two such as at the late Classical round tower at Eleusis.[91] These elements would develop into purely ornamental architectural elements, such as the Doric frieze documented on the Hellenistic towers of Sillyon and Selge. At Perge, the two round towers flanking the Hellenistic South Gate carried a decorative Doric frieze and pilasters carved in low relief.[92] They are the remnants of a pincer-shaped gate built in the second century BCE, which the local patron Plancia Magna elaborated further in honor of Emperor Hadrian centuries later.[93] Although the full extent of the second-century design is now lost, its surviving towers indicate that they were part of an elaborate structure built to promote the city and its image.

In Italy, the towers of Pompeii and Nocera displayed the closest affinities in their First Style ornamentation. Further parallels are difficult to find, although they must have existed. Contemporary towers built in *opus incertum* in places such as Cori and Telesia probably carried a similar ornamentation to cover the masonry. It has since disintegrated due to exposure. The towers of Paestum form an exception because workers used travertine to build and ornament the buildings (see Figure 7.9). In a refurbishment carried out in the early decades of the first century BCE, the towers on the north side of the Paestan circuit, as well as those flanking the western Porta Marina, received new ornamentations. They included a carved Doric frieze and pilasters with surmounting Corinthian capitals featuring central busts. These protective busts remain unidentified, but they found architectural and religious resonance in the Temple of Bona Mens (the Doric-Corinthian Temple) located in the Forum of the city, which displayed almost identical capitals.[94] Despite the paucity of evidence, ornamentation on the towers of Paestum, Nocera, and Pompeii signals a widespread tradition on the Italian peninsula where cities would decorate their towers in the late second and first centuries BCE.

Figure 7.9 Reconstruction of a Tower at Paestum. (After Krischen 1941, pl. 8.)

Patronage and the concept of the city

The strong ideological associations with fortifications means that assuring the patronage of their construction, maintenance, and upgrades was an important element to the social layout of a city. In early (pre-) Roman Italy enceinte construction was more of a communal affair, but that would change when, as public monuments, circuits would become the object of individual eurgetism. Such benevolence was a fundamental part of the Roman *cursus honorum*, where the financing of public structures was an integral duty to the community. Magistrates would advertise their contributions by means of inscriptions celebrating their accomplishments in construction projects. About twenty-seven dedicatory inscriptions survive, of which fifteen register repairs or restorations of fortifications. Some tell us of magistrates who financed entire walls, such as is the case of Quinticius Valgus and Magius Surus at Aeclanum.[95] Others mention patrons who would pay for upgrades, new tracts of walls, or single towers and gates such as the Porta Nola at Pompeii. These inscriptions begin to appear in the late third century BCE, with the earliest attested at Luceria and Paestum, as concepts of urbanization and cities in Italy came into further contact with the Hellenistic world.[96]

The settlement of Telesia preserves the most complex example of the process at work, revealing some of the dynamics of design transmission and patronage. Located in the open valley known as Valle Telesina, midway between Benevento and Capua, Telesia was a Samnite settlement that Sulla transformed into a Roman colony populated with veterans from his easterns campaigns. Its defensive system, dating to the early first century BCE, is exceptionally preserved. Built entirely in *opus incertum*, it included regularly spaced towers protruding out from concave curtain walls, designed as deadly cul-de-sacs of flanking defensive fire. Such a design was unprecedented in Italy, but it was common in the Hellenistic world in places such as Side; Philon described it in his treatise on siege warfare. Sulla's veterans must have applied this layout after witnessing its effectiveness in the east.[97] Given the recent unrest on the Italian peninsula, the drive to build such a sophisticated enceinte must have responded to defensive necessities.[98] Six surviving inscriptions inform us of separate instances where local patrons built individual towers in the circuit.[99] Although any ornamentation on these structures has long since disappeared, the uniform remains indicate a remarkable unity in their original design. Patrons clearly operated under an overall conceptual presentation for the walls as a public monument that defined a communal identity. Their aim was to evoke Roman colonial ideals as well as Hellenistic notions where fortifications were an integral part of an urban image.

The euergetism of individuals on fortified structures and the appearance of inscriptions commemorating such interventions seem to echo an exposure to the Greek east. These were announcements of personal aggrandizement that departed from the previous more anonymous efforts to fortify settlements in Italy. The notion that city walls are part of the civic image was already prevalent in the classical Greek *polis*, when the earliest announcements of elite patronage appear on fortifications. Once independent *poleis* in the Greek east adapted to be part of kingdoms in the Hellenistic world, cities and rulers would maintain an emphasis on their fortifications.[100] Starting in the late fourth and continuing into the first centuries BCE, many Greek enceintes became obsolete in the face of new siege techniques. Most cities were unable to maintain large fortifications or even upgrade or build new ones to the extent required by the new military realities. Many patrons, therefore, would act locally to upgrade fortifications with towers or spend money to upgrade and embellish gates, often displaying their achievements in prominent inscriptions. Only the richest settlements could afford full upgrades by building a series of new towers or defensive walls. On occasion, outside elements would aid in the (re)construction effort, as is the case of the Attalids who helped finance the circuits of Perge and Side in Asia minor in the second century BCE. Although these cities were part of a larger royal institution, their walls would still symbolize local autonomy. The idea of independent classical *poleis* would give way to Hellenistic notions of the "fortified city" where urban centers would strongly identify with their fortifications.[101] Ideals of civic image eventually found an embodiment in Tyche, the Greek goddess of fortune who became a tutelary deity of cities. In sculptures where she is displayed as a personification of cities and/or countries, Tyche symbolically wore a crown of walls. This model seems to function in a similar manner for the Italian context where many cities kept building fortifications despite the heavy influence of Rome.

The city of Velia located to the south of Paestum offers an example of how such ideals found ready acceptance in Italy. Like Paestum and Pompeii, it witnessed a complex web of influences that would shape its identity. Originally a Greek Phocean foundation, the city had built a strong enceinte by the end of the fourth century BCE. It remained loyal to Rome in the years of the Second Punic War, entering a *foedus* that allowed Velia to retain a degree of self-governance and open trade contacts with the east. After the end of the war, the city built extra towers and gates in the most vulnerable areas of the circuit. The Porta Rosa, dating to the early second century BCE, still displays a deliberate ornamental rustication and two surmounting arches in a construction technique common to centers such as Ferentino and Perugia but absent in the Greek world[102] (see Figure 7.10). A similar rustication exists on the seawall on the southern side of the city, where sculptors carved the edges of the blocks smooth but left a rough center for aesthetic effect. These architectural developments point to Velia's self-awareness as a proper city in both the Hellenistic and the Italic traditions. The later bust of a female deity, perhaps Cybele or Velia herself, wearing a turreted crown encapsulates the concept. Excavators recovered it, together with statues of members of the Julio-Claudian dynasty in a public *caesareum*-type sanctuary dating to the first century CE.[103] In this sanctuary, elite patronage – now the imperial family – and fortifications came together in the conceptual/divine manifestation of Velia.

The development of an ornamental element in the city walls of Republican Italy that began in the fourth century BCE carried with it ideological connotations of civic identity. It would lead to its export to places such as Spain where Roman forces struggled for control of the region. The associated patronage that paid for these structures would continue throughout the late Republic to become a tradition adopted by emperors like Augustus, Tiberius, and Claudius in places such as Verona, Fano, Rimini, and Spello in Italy, and Nimes and Autun in Gaul in the Roman west. The tradition would continue into the second century CE, when local patrons in cities in Asia Minor, such as Nicaea and Attaleia, would build monumental city gates in honor of the emperor. In the end, Pompeii, Velia, Paestum, Nocera, and Telesia are but a few examples in Republican Italy where communities attempted to propose themselves as a proper city by means of construction. Such

Figure 7.10 Porta Rosa, Velia.

an effort is most compelling considering that the application of the First Style on the towers at Pompeii and Nocera is symptomatic of the local Samnite community modeling itself as an ideal city. Such ideological connotations would intertwine with the divine, where religious associations with fortifications would protect the community and the city.

Notes

1 Liv. *ab. ur. con.* I.7.2–3.
2 For more on these dual concepts, see Beard 2015, 57–60.
3 See Jouffroy 1986, 24–26.
4 Winter 1971, 79–91.
5 Onians 1988, 24.
6 Chiaramonte 2007, 145.
7 Strabo *Geo.* V.4.8; Cristofani 1991, 17; De Caro 1992b, 74; Guzzo 2007, 41–43.
8 Lawrence 1979, 39.
9 Goldsworthy 2003, 22. This development is especially true for Greece and probably Italy too.
10 Garlan 1968, 252; Gat 2002, 135; van Daele 2004, 30.
11 Torelli 2008, 265–278.
12 Vitr. *De arch.* 1.V.5.

13 Frigerio 1935, fig. 11; Adam 1982, 39.
14 Sewell 2010, 35–36.
15 Säflund 1932, 115–118, 170–174.
16 In this case, the plethron measurement was 36 meters; see Barbera and Magnani Cianetti 2008, 21–23.
17 Tréziny 1986, 197.
18 Cifani 2008, 239; Bernard 2012, 10–11.
19 The earliest examples at Veii date to the eighth century BCE; see Boitani et al. 2016, 19–32. Miller insists that the construction techniques used to build using ashlar masonry in Etruria come from Greek and Phoenician traditions but that the development of the *fossa-agger* system is typically italic; see Miller 1995, 190–195.
20 For the quarry marks, see Fitzgerald Marriot 1895, 62–86; Sogliano 1898a, 68–69; Maiuri 1929, 280; Tréziny 1986, 197.
21 Bonucci 1830, 81–82.
22 Pesando 2010b, 47–75.
23 E.g. the *meddix pumpeianus* for Pompeii; see De Caro 1991, 32; Cornell 1995, 124.
24 Senatore 2001, 185–265. Carafa sees the league dissolved in 304 BCE at the end of the Second Samnite War. He is unclear how magistrates continued to exercise power; Carafa 2011, 93.
25 Rowlands 1972, 452; Lawrence 1979, 123.
26 Johannowsky 1982, 841.
27 Colonna 1962, 86–99; Conta Haller 1978, 78; Fontaine 2013, 268–272.
28 Richardson 1988, 49. Johannowsky, in search of Greek precedents, suggests that Neapolis (Naples) may have supplied a model for the Samnite double trace system when it added an exterior wall to a previous fortification in the fourth century BCE. The reality is that the double trace system is more common in the Samnite sphere of influence (note 27), suggesting that it developed there rather than in Neapolis; Johannowsky 2009, 13–15.
29 Philon *polior.* I.84.
30 See Adam 1982, 112–113; Greco et al. 2014, 1271–1281.
31 Coarelli 1995, 23.
32 Winter 1971, 120–121.
33 Liv. *ab ur. cond.* XXIII.15.6.
34 Johannowsky 1994, 123.
35 See Zanker 1998, 61–78; Zanker 2000, 126–140.
36 Johannowsky 1982, 842; See also Johannowsky 1994, 123–131.
37 Winter 1971, 188–191; Johannowsky 1994, 132; Lawrence and Tomlinson 1996, 174, 176.
38 D'Ambrosio 1990, 88.
39 Krischen proposes two reconstructions; Krischen 1941, pl. 7 and 8; D'Ambrosio 1990, 74–75.
40 Winter 1971, 79.
41 Hellmann 2010, 294–295.
42 Winter 1971, 79.
43 Malnati and Sassatelli 2008, 429–471.
44 Tocco Sciarelli 2009, 20. On mudbrick, see Winter 1971, 69–71.
45 For a survey, see Brasse and Müth 2016, 75–100.
46 Müth and Ruppe 2016, 231–248.
47 Onians 1988, 8–22.
48 Bianchi and Gargiani 1988, 82–85; Borghi 2002, 92.
49 Cifani and Fogagnolo 1998, 369.
50 Cifarelli 2012, 295–301; Ciccone 2015, 81–96.
51 See Lugli 1966, 65–69; debated by Miller 1995, 78; Becker 2007, 99–105; Brasse and Müth 2016, 75–76.

52 Jouffroy 1986, 368.
53 Liv. *Ab ur. con.* II.34.6.
54 Quilici and Quilici Gigli 2001, 181–244; De Haas and Attema 2016, 251–262.
55 Brandizzi Vitucci 1968, 22–38.
56 Fentress 2000, 13.
57 The sanctuary and its phases are notoriously difficult to date, but the traditional view puts its construction at about 120 BCE; see Coarelli 1987, 35–84.
58 See Jackson et al. 2005, 485–510; Jackson and Marra 2006, 403–436; Bernard 2012, 9.
59 See Greco and Longo 2008, 24 for the painting notice. For instance, sculptors used sandstone to carve the metopes of the sanctuary at Foce Sele.
60 See Müller-Römer 2008, 132 and in particular note 37.
61 Proietti 1986, 152; Naso 1996, 53.
62 Proietti 1986, 183; Prayon 2001, 335–343.
63 Scholars have often equated these Etruscan types as a reflection of actual domestic architecture. It would be interesting if we could project a similar ornamentation to contemporary housing. Unfortunately, we know too little of ancient Cerveteri to make such a link.
64 Miller 1995, 90–91. This is not to say that polychrome masonry did not occur in Greece. The late classical curtain wall at Larissa (Buruncuk) displays randomly placed stones of different coloration for an ornamental effect. The stringcourse of the surviving tower in the same circuit is red, as opposed to the white masonry of the tower;see Müth et al. 2016, 140.
65 Bernard 2012, 9.
66 Bernard 2012, 20.
67 Quilici and Quilici Gigli 1995, 243; Valchera 2009, 160.
68 Quilici and Quilici Gigli 1995, 234–235; Spaziani 2012, 231–234.
69 Spaziani 2015, 33–39 suggests that the Porta Pentagonale could be a later addition to the circuit and that it might represent a large sewer. As part of this theory, she reconsiders that the upper ashlar masonry of the walls may represent a later addition (see previous note).
70 De Rossi 2009b, 61, Spaziani 2012, 231–234.
71 CIL X 5837–5840.
72 In the past, the two construction techniques present on the bastion led to a scholarly debate on their construction sequence; see Agiletti and Frasca 2015, 21–32.
73 De Rossi 2009a, 52.
74 Cifarelli 1992, 10 and 18–22.
75 Despite refurbishments carried out in the Middle Ages, much of this intervention is still visible today; see Colaiacomo 2015, 61–72.
76 On the development of the gate, see Cifarelli 1992, 35.
77 Wallace-Hadrill 2008, 123; Cifarelli 2013, 48–52.
78 Palombi et al. 2013, 525–528.
79 These same aspects would apply when the town built its towers and carried out refurbishments in *opus incertum* in the second century BCE.
80 Liv. *Ab ur. con.* XXIII, 39, 6.
81 Conta Haller 1978, 103.
82 See Caiazza 2009, 7–38 and 55–82.
83 Liv. *Ab ur. con.* XXI, 60–61.
84 See Ruiz de Arbulo 2006, 33–37.
85 On the chronology of the fortifications, see Serra Villaró 1949, 221–236; Hauschild 1979, 204–250; Sánchez Real 1985, 91–117.
86 See Hauschild 2006, 153–171.
87 Ruiz de Arbulo 2006, 38.
88 See Balil 1983, 231–235.
89 Liv. *Ab ur. con.* XXXIV, 9–16.

 90 See Aquilué et al. 2006, 24–29.
 91 Winter 1971, 79–91; Müth et al. 2016, 140.
 92 Winter 1971, 86–91.
 93 McNicoll 2007, 129; Caliò 2012, 197.
 94 See chapter 8 of this volume for further discussion.
 95 CIL IX 1140.
 96 See CIL IX 800; Jouffroy 1986, 18–23.
 97 See Quilici 1966, 85–99.
 98 Also Gabba 1972, 96 and 108; Pocetti 1988, 317. For the personification of cities, see
      Pollitt 2009, 2–3.
 99 CIL XI 2230, 2233, 2235; Ramanius 2012, 117.
100 Caliò 2012, 190–213.
101 Maier 1961, 55–68; Caliò 2012, 169–221.
102 Tocco Sciarelli 2009, 77–79.
103 Tocco Sciarelli 2009, 19–74; also Greco 2014, 30.

8   City walls and the gods

The concept of tutelary deities defending a settlement was widespread in antiquity and an integral part of the political and social identity of cities. Their origin stems from the lion gates of Mycenae and Hattusa and the Lamassu guardians at Assyrian palaces and traces through to the Greek and Roman worlds in famous examples such Athena at Athens, Juno Regina at Veii, and Jupiter, Juno, and Minerva at Rome. Fortifications and the divinities protected each other in a reciprocal process. Fortified enclosures built by the populace protected both the community and sanctuaries, whereas tutelary divinities, in turn, shielded defenses and the population. City walls were dangerous liminal spaces vulnerable to real military as well as elusive threats such as disease, misfortune, and the evil eye. Gates and circuits featured apotropaic devices to keep such invisible uncontrollable forces from entering the city. It was thus critical for a community to maintain the sanctuaries and the defensive enclosures as part of the divine and military elements protecting the community. A complex symbolism would emerge from this basic relationship to include social, religious, and political significance.

One of the common themes throughout the history of Pompeii is the religious symbolism associated with its city walls.[1] It certainly is not an isolated example, but Pompeii preserves exceptionally abundant evidence. Yet because of its piecemeal recovery, much of its significance is neglected. Gates such as the Porta Stabia and Nola indicate the presence of a cult dedicated to Minerva that found a wider resonance in the city. The cult would retain a presence at the gates, but its importance would change overtime, together with the political and social layout of Pompeii. The discussion concerning the religious associations of the Pompeian fortifications has otherwise focused on whether the defenses followed the line of a *pomerium* as a marker of a possible Etruscan *sulcus primigenius* foundation ritual. It is in part an attempt to project the *pomerium* attested in the literature for Rome where it was a religious as well as a legal boundary defining the actual city as traced by Romulus upon the foundation of the city. Already in antiquity, authors sparred as to whether the fortifications of colonies and Etruscan cities followed the course marked during the foundation ritual.[2] The Servian Wall in Rome, for example, most certainly did not do so. It is doubtful whether Pompeii ever had a *pomerium*, and, if it did, its potential course need not have followed the line of fortifications (*Neustadt*) or that the *Altstadt*.

The issue of the *pomerium* at Pompeii is distracting since it is limited to one particular ritual. Roman grammarian Sextus Pompeius Festus writing in the second century CE recounts how specific Etruscan rituals existed to build walls and more specifically to sanctify gates that seem disconnected from the *pomerium*.[3] Similarly, Cicero states that priests declared fortifications as sacred even though nothing could be as powerful as the divine protection of the city.[4] Roman law codified city walls into a special category known as *res sanctae*, or holy things, protected by but not consecrated to the gods. Scaling or crossing them illegally was a sacrilege punishable by death.[5] Regardless of the *pomerium* issue, then, the implication is that city gates and defenses often received special religious protection and carried apotropaic devices to protect the community.[6] Irrespective of its Etruscan origin, the Pompeian enceinte must have carried cultural and religious notions of self-determination, identification, and political dominance.

Perhaps the most recognized religious symbol on the Pompeian fortifications is the keystone on the city side of the Porta Nola (see Figure 8.1). Next to the bust, excavators recovered the inscription describing how Vibius Popidius constructed and dedicated the gate in the late second century BCE.[7] A certain amount

Figure 8.1 The bust on the Porta Nola.

of controversy has surrounded its translation, where an early erroneous interpretation led to the identification of the bust with the Egyptian goddess Isis. Although
weathering has damaged the bust almost beyond recognition, a few identifiable
locks of curled hair and the base of a helmet on the neck identify her as Minerva.
Presumably, such busts decorated more than just one of the gates at Pompeii, but
no others were found in situ. In addition to its apotropaic role, the bust that Vibius
dedicated together with the Porta Nola carried multivalent religious, social, and
political associations. Vibius was a wine merchant and a member of the powerful elite *gens* Popidia who was trying to promote his reputation.[8] The careful
placement of the inscription next to the bust of Minerva is a clear statement of
association with the goddess that advertised their relationship as benefactors and
protectors of the community.

Minerva was also present at the Porta Stabia (see Figure 8.2). The gate still carries two small niches carved into the eastern wall of the court, meant to evoke a
deity protecting the walls as well as travelers traversing the liminal space.[9] At the
time of their discovery, the upper niche retained traces of a plaster coating with
a graffito on the back wall that read PATRVA, a reference to Minerva Patrua.[10]
Set about 1 meter apart, the two niches have a spatial correlation implying either
some kind of functional association or that one, the larger niche above, replaced
the other. Excavations in 2006 and 2007 by the Pompeii Archaeological Research
Project: Porta Stabia have identified a continuity of worship stretching to the early
third century BCE. The lower niche belonged to a shrine in the first version of the
gate, which saw the addition of an altar in a second phase. In a third phase, workers
partially buried the altar after raising the sidewalk in the late second or early first
century BCE.[11] A new sidewalk laid in the first century BCE completely buried the
altar. It was at this time that workers must have cut the second, larger and higher
niche, in an effort to maintain a continuity of cult.

The partial burial of the altar in the late second century BCE included a ritual
deposition placed along its south side. The date is roughly contemporaneous with
Vibius's dedication at the Porta Nola, suggesting that it was an offering to Minerva.
Among the votive deposits were small votive cups, burnt remains of a sheep/goat
vertebra and a pig mandible, and three broken parts of a small terracotta figure (see
Figure 8.3). Although the remains of a bull are missing, the animal bones indicate
a dedication similar to a *suovetaurilia*. This ritual offering of a bull, sheep, and a
pig, meant as a purifying rite and intended to repel evil spirits, seems particularly
apt at the gate of the city.[12] The figurine may have been purposefully broken as
part of the ritual. All that survived was the head as well as the left and right sides
of her torso. A slight flanging at the edge of each arm suggests that it was once part
of a bowl-shaped *bruciaprofumo* (incense burner). The lack of defining attributes
makes any further identification difficult, but her frontal pose is strikingly similar
to a statue of Minerva recovered at the Porta Marina, as well as votive figurines
related to a regional cult.[13]

The Oscan inscription located on the inside of the western bastion of the Porta
Stabia indicates further religious associations with Minerva. To briefly recapitu-
late, the inscription mentions how the aediles M. Sittius and N. Pontius repaired

Figure 8.2 The niches and altar in the Porta Stabia. (Courtesy Pompeii Archaeological
   Research Project: Porta Stabia.)

the road from the gate up to the *pons Stabianus*, crossing the Sarno River and
the *via Pumpaiiana* up to the Temple of Jupiter Meilichios. The *via Pumpaiiana*
mentioned in the inscription must have passed through or reached the gate. A
common consensus identifies it as the modern via Stabiana heading from the gate
up to and including its side street, the via del Tempio d'Iside, to reach the Temple

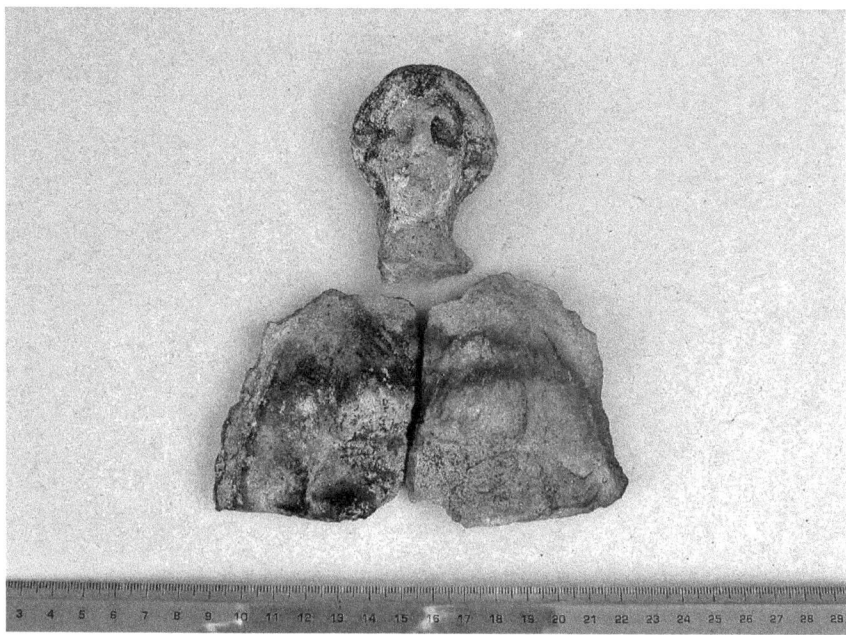

Figure 8.3 Remains of the terracotta figurine excavated at the Porta Stabia. (Courtesy Pompeii Archaeological Research Project: Porta Stabia.)

of Minerva in the Triangular Forum.[14] Given this itinerary, the *via Pumpaiiana* may have acted as a sacred processional route to honor Hercules and Minerva. Ancient sources recount in mythological terms how Hercules passed through the Bay of Naples as he led Geryon's cattle to Eurystheus from Spain. The drive was a veritable triumph, or *pompé* (πομπή) in Greek, which, according to some, gave Pompeii its name.[15] During the drive, Hercules rested the cattle at the future location of the Doric Temple in the Triangular Forum, which was then dedicated to Minerva and Hercules. Admittedly, the problems with these sources are many, not least that most postdate the destruction of Pompeii by several hundred years. However, this background provided the *via Pumpaiiana* with its name: The road traced the mythical route that Hercules took and more simply marked a yearly procession honoring Hercules and Minerva as protectors of the city. The Porta Stabia, as a critical element to this ritual, was an important religious landmark in the city.

The Porta Marina preserves a large niche embedded in its southern exterior bastion (see Figure 8.4). The niche is now devoid of embellishments, but a nineteenth-century print indicates that it once featured a plaster coating similar to the niches of the Porta Stabia. The excavation report of 1861 describes the recovery of three large terracotta statue fragments and two smaller pieces of a hand associated with the shrine. Hermann Von Rohden reunited the fragments in a drawing published in 1880, but they subsequently vanished into obscurity and considered lost.[16] In the

Figure 8.4 Niche in the Porta Marina.

summer of 2010, some research and the aid of the site director at Pompeii, Dott. ssa Grete Stefani, led to the rediscovery of the fragments, which were stored in a corner of one of the *magazzini* in Pompeii. Although somewhat battered, the pieces unmistakably join into Von Rohden's reconstruction (see plate 30). The figure stands on her left leg in a gentle contrapposto pose with her right hand resting on her hip. She wears a sleeveless *chiton*, and a *chlamys* draped over her right shoulder and arm. The head is gone as is almost the entirety of her left side above the knee. A stylized rock next to her left foot preserves the remnant of a shield that once stood next to her leg, indicating that the statue represents the same Minerva present at the other Pompeian gates. The fragmentary state makes any dating based on stylistic grounds problematic. The niche in its current version is almost certainly associated with postcolonial refurbishment of the gate.[17] However, given the continuity of cult recovered at the Porta Stabia, the statue may be a relic of an earlier phase. The fact that excavators recovered only about half of the statue indicates that already in antiquity it was an important artifact. The context indicates that the statue broke at some point before the eruption. Although broken and battered, the remaining pieces were put back into place where they continued to evoke her presence.

The religious references from other Pompeian gates are somewhat vaguer. The Porta Vesuvio displays the remains of two altars in the cul-de-sac on the exterior northwest tip of the gate. Upon excavation in the early twentieth century, it preserved the remains of a plaster coating and a fresco associated with the altars. Any identifiable signs had already deteriorated beyond recognition and have now completely disappeared. Antonio Sogliano identified the shrine as a *lararium*, and the fresco presumably depicted the public *lares* protecting the city.[18] However, given the prevalence of Minerva on the other gates, it seems plausible to assume that she played a role here as well. At the Porta Ercolano, during the excavations next to the Tomb of M. Cerrinus Restitutus, Maiuri recovered a stratum filled with votive materials including small terracotta vases as well as fragments of figurines. Two of them carried an undefined headgear, whereas other slightly more identifiable figures represent *erotes*.[19] Maiuri ascribed the deposit to a vanished religious *sacellum* dedicated to Venus or Minerva that was associated with the Samnite gate. It would have served the gate throughout its history right up to its reconstruction after the earthquake. This is when the protective role of Minerva at the Porta Ercolano seems to have waned, presumably in favor of whichever emperor financed its reconstruction.[20] The semidivine status of the emperor would have changed the dynamic. The relationship between divinity and patron, so neatly set out previously by Vibius at the Porta Nola, conflated into one.

Further religious references may lie in the adornment of Pompeii's fortified towers. The First Style ornamentation is part of an architectural language usually associated with temples and public buildings to stress their status as religious or public structures. However, the specific use of the Doric frieze may also represent an allusion to Minerva. Vitruvius notes that the strong proportions of the Doric order reflect the martial strength of Mars, Minerva, and Hercules, specifically recommending the use of the style on their temples.[21] The Doric sanctuary dedicated to Minerva in the Triangular Forum is one of many examples. For Pompeii, the Triangular Forum was a particularly charged location due to the presence of a *mundus* and a *heroon* dedicated to the mythical founder of the city.[22] Although the structure warrants a systematic investigation, most authors place Tower I at the eastern tip of the Triangular Forum. Here the tower was arguably part of the

Menerv[ium], the sanctuary to Minerva mentioned in an *eituns* inscription on the facade of House VIII.5.19–20.[23] The tower, the fortifications, and the adjacent Temple of Minerva would have created an immediate visual relationship visible in the surrounding countryside and to those visiting the sanctuary.[24]

The fact that the topography in this part of the city barely merits any height-ened military defense further underscores the association between cult and defense. Tower I is the only such structure on the cliff edge that marks the southwestern part of the city. The visibility of this location naturally served as the starting point for the counterclockwise numbering of the towers, which ended at Tower XII near the Porta Ercolano. Such an emphasis for counterclockwise movement in Italic cities has its origins in religious and superstitious practices, which would influence daily life from the organization of Roman cart traffic, to pathways through domestic and religious spaces, and to the configuration of shop fronts and their entry.[25] For the towers, the counterclockwise direction perhaps reflected the course of religious cer-emonies partly associated with the act of tracing the *sulcus primigenius* furrow, as well as *lustratio* processions designed to purify the city and keep evil spirits away. Instructions for such a *lustratio* come from Iguvium, modern Gubbio. Priests led a procession around the *arx* and conducted a sacrifice at each of its three gates, which represented the weak openings in the protection of the city.[26] In Rome, a collection of priests, Vestals, and magistrates would participate in processions to designated sacred spaces in the city, known as the *Argei*, to collect fake altars and images, which they then threw into the Tiber from the *pons Sublicius*.[27] These notions explain why almost all of the religious elements at the Pompeian gates are on the right where they further emphasized the counterclockwise direction as one approached the city.

The Triangular Forum was thus a node for the divine and military elements pro-tecting the city.[28] The near simultaneous construction and renovations events for the fortifications and the temple seem deliberate. Minerva's cult was one of the oldest in the city and a critical part of Pompeian identity throughout its history. Built in the early sixth century BCE at the very inception of the city, the temple served as a psychological landmark binding the community and the landscape.[29] It maintained this role when Pompeii built its new enceinte and refurbished the temple in the late fourth century BCE. The surviving metopes and antefixes representing Hercules and Athena/Minerva Phrygia, as well as the associated votive deposits, indicate that the temple housed the Phrygian version of the goddess.[30] Rather aptly, this Minerva, among her traditional roles promoting fertility and artisanship, also protected navi-gation, a crucial activity for a port city like Pompeii.[31] The new enceinte functioned as a terrace platform for the temple, thereby renewing both the military and divine protection of the city and its hinterland.[32] These statements were of critical impor-tance to those who had financed the construction of both monuments.

These observations imply that the Phrygian Minerva had patent associations with Pompeian cultural identity. It was August Mau, inspired by the bust carved on the Porta Nola, who believed that Minerva's presence at the city gates was the result of an overt Greek influence on the city and her role as Athena Polias. Juno would have been a more apt Roman cult since the Aeneid describes her role as the protectress of city gates.[33] However, a number of elements imply that Pompeii's relationship with Minerva was more complicated and perhaps more Italic and even local in meaning. After all, Cicero provides a measure of her role in Roman thinking when he describes Minerva as *custos urbis*, or the protectress of the city.[34] Minerva, together with Apollo, had a similar role in an Etruscan context at the Portonaccio Temple outside of Veii, where, as a tutelary divinity of the city, she required worship by those entering its limits.[35] The cult of the Phrygian Minerva had a Trojan connection that associated it with wider series of foundation myths and Greek identity in Magna Graecia and Sicily. Recent evaluations have also recognized it as a particularly pro-Roman cult. Minerva's appellative Phrygia and hence her Trojan background result from her association with the Palladium and its transport to Rome by Aeneas.[36] Although many cities claimed ownership of the Palladium, its status as one of Rome's protective relics, the relationship with Aeneas, and its foundation myth suggest that the Phrygian Minerva had a wider role in the protection of cities and the projection of Roman power. Rome may have purposefully adopted the Phrygian Minerva in southern Italy after the cult's revitalization by Hellenistic monarchs in order to promote the unity of the Greek Italiot league against the threat posed by Apennine populations.[37] At Pompeii, the strong association between cult and fortifications seems explicit. Perhaps more difficult to disentangle, given the ambiguous origin of the cult and its wider regional implications, is what it means for Pompeian civic identity and image.

The archaeological evidence indicates that the cult of the Phrygian Minerva was part of an orchestrated political introduction in the Bay of Naples (see Figure 8.5).

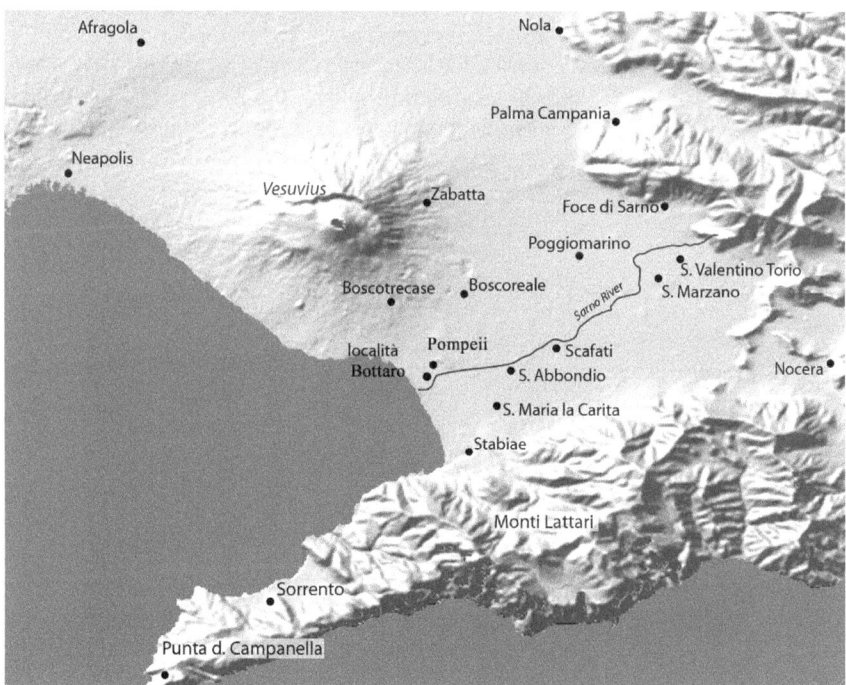

Figure 8.5 Map with some of the locations described in the chapter. (Courtesy S.J.R. Ellis.)

Along with Pompeii's Doric Temple, sanctuaries in nearby Stabia (Località Privati), Punta della Campanella, Località Bottaro, Fratte di Salerno, and Paestum, featured antefixes depicting the Phrygian Minerva and Hercules fashioned from the same molds.[38] Some have identified the Nucerian league, or a renascent Neapolis (Naples), as financing the sanctuaries outside Pompeii, motivated by either their new alliance with Rome or a final assertion of local identity during the Samnite Wars.[39] For the Punta della Campanella sanctuary, the most important in the bay, this effort may represent an attempt to take over or to reestablish a previous cult and therefore the region. Although Strabo claimed that Odysseus founded the sanctuary to Athena, the archaeology indicates that its foundation occurred in the sixth century BCE in an act that likely expressed political associations between Neapolis and Athens.[40] Such a date correlates with the foundation of the first Doric Temple at Pompeii. At least initially, the construction of the sanctuaries may therefore represent an attempt by Neapolis to take over, by cultural means, the southern Bay of Naples.[41] The early political connection between the sanctuaries is difficult to trace with any further certainty.

Terracotta figurines depicting Minerva Phrygia recovered in votive pits at each location display remarkable similarities. The most important examples, found also at Punta della Campanella and Pompeii, show Minerva standing frontally and wearing a Phrygian helmet; she braces a shield against her left leg, and in her right hand she holds a *patera* resting on a colonnette (see Figure 8.6).[42] Its popularity implies that the figure represents a copy of the cult statue at Punta della Campanella.[43] A similar depiction also appears in a fresco on the facade of the Shop of the Carpenters' Procession (VI.7.8–11), where carpenters carry an effigy of Minerva to honor her role as protectress of their trade (see plate 31). Only a fragment of the statue is visible, but enough remains to identify the goddess standing in a similar frontal pose as the votive figurines, including the *patera* and shield. The image may well reflect the Punta della Campanella or the Temple of Minerva (Doric Temple) cult

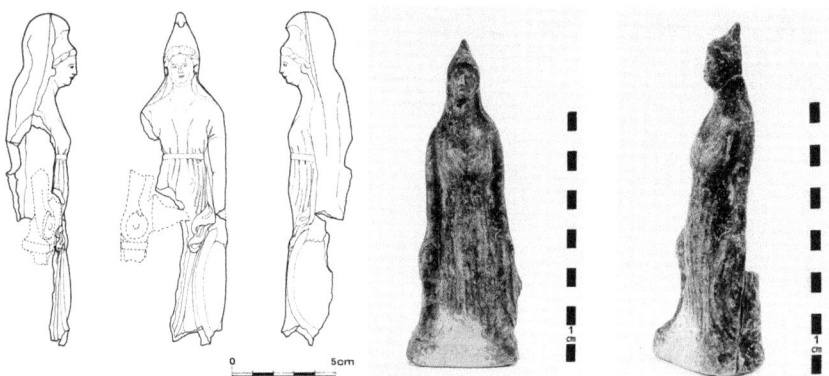

Figure 8.6 Figure from the Punta Campanella sanctuary (left) (after Russo 1990 fig. 15) and from Pompeii (right) (after D'Alessio 2001, pl. 16).

statues, considering that the lowered *patera* on the figurines is a practical artistic shorthand for her extended arm.[44] It suggests a continuity of the cult right up to the final days of the city.

The archaeological evidence attests an affinity between the Phrygian Minerva and the Pompeian enceinte. A variant figurine of the Campanella type, recovered in votive pits from the sanctuaries dating to the third/second centuries BCE at Paestum and Stabia, displays the goddess holding a shield and standing in a similar chiastic pose to the statue fragments of Minerva recovered at the Porta Marina.[45] Although the image on the *bruciaprofumo* fragment from the Porta Stabia example remains ambiguous, the figure's marked frontality and recovery context, along with the graffito in the upper niche, indicate a similar association. Assuming a continuity of worship and the reach of Minerva's cult, the niches and the altar at the Porta Stabia imply a connection between her temple and the city walls stretching back at least to the construction of the first Samnite enceinte. Given the pro-Roman association with the cult, these factors indicate a heavy Roman influence on Pompeii during its

Figure 8.7 Figurines recovered from the *Località Privati* (Stabia) (left) (after Miniero et al. 1997, fig. 15) and from Paestum (right) (after Tocco Sciarelli 1988, pl. 53). (Image courtesy Museo Archeologico Nazionale Paestum.)

Samnite period. Such a Roman involvement, perhaps military and financial, may have extended to the construction of the first Samnite enceinte. Nevertheless, given Minerva's role in Pompeii since the inception of the city, a local component, or even a Greek or Etruscan element for that matter, also seems at play considering her association with the city's identity. One may envision the projection of a Roman aspect onto a cult of particular importance to local identity.

A measure of how defenses and religious factors must have worked in the image of Pompeii comes from neighboring Paestum. Its circuit has traditionally formed a point of comparison for Pompeii and for good reason. The first phase of its enceinte, built in the last decades of the fourth century BCE, is broadly contemporary. The design of the main gates is roughly similar: Both have forecourt gates with the exception that the examples at Paestum are built into a wall erected in the Greek manner, namely a wide freestanding wall (i.e., no *agger*) with an internal rubble fill. The two settlements also share a similar history. They would fall to populations – the Lucani for Paestum and the Samnites for Pompeii – coming down from the Apennines to conquer the cities in the fifth century BCE. Unlike Pompeii, however, which would remain independent until 88 BCE, Paestum would side with King Pyrrhus against Rome, leading to the installation of a Latin colony after its fall in 273 BCE.[46]

These sociopolitical developments provide the background for the remains of shrines at the gates of Paestum. The Porta Marina on the western side of the city preserves a niche that is similar in size to the upper niche preserved at the Porta Stabia in Pompeii. Carved into the rear right wall of the gate, the one at the Porta Marina is more elaborate with two Ionic pilasters and an entablature to create an aedicule (see Figure 8.8).[47] Its date and the deity evoked here are matters of speculation, but the gate was part of the first defenses built in the fourth century BCE. On the opposite side of the city, the Porta Sirena has yielded the remains of a *sacellum* dedicated to Athena.[48] Once again located on the right-hand side as one entered the city, the sanctuary was no more than a miniature temple-like structure set within a defined precinct. A series of associated votive deposits nearby contained the same figurine types as those recovered at the Temple of Athena inside Paestum, the Porta Marina and the Doric Temple at Pompeii, and near Stabia.

The religious connection between the shrine at the Porta Sirena and the Temple of Athena – one of the most recognizable buildings of Paestum – created a deep interconnection with the image of the city and the gates. Athena presided at important rituals of transition, including the passage into adulthood. Her presence at the Porta Sirena is the result of her role as protectress of liminal spaces, in this case the boundary between the city and its *chora*. The *sacellum* was destroyed, and the offerings of figurines at the temple ceased following the foundation of the Latin colony and the subsequent expansion of the defensive circuit. In its third phase, the gate received new apotropaic devices in the form of two keystones that are still visible today: a carved dolphin (city side) and a Siren or Nereid (extramural side).[49] Both symbols referred to Poseidon and the past when Greek settlers first established the city as Poseidonia. Although the Temple of Athena would continue to function, the relationship between cult and fortifications seems deliberately broken with the arrival of the colonists who changed the focus to a less explicit association.

Figure 8.8 Aedicule in the Porta Marina at Paestum.

At the turn of the first century BCE, the citizens of Paestum would reinvigorate the connection between their enceinte and the cults protecting the city. At the time, Paestum embarked on the (re)construction of the Temple of Bona Mens, which, along with the *capitolium*, was the only sanctuary built in the Colonial period. Located at the very heart of the Forum, the temple was a landmark that symbolically covered the old *comitium* that had fallen into disuse. The phasing of the structure and its attribution is not without controversy, but it is the final version of the temple that preserves an ornamental reference to the city walls.[50] In the absence of a defining cult, the building is known as the Doric-Corinthian Temple, a name that points to its unique local blend of architectural styles (see Figure 8.9). It incorporated the Doric and Corinthian architectural orders with a frontal Italic-style temple standing on a tall podium. Its syncretic design, as well as its position at the heart of Paestum, made the temple a potent symbol for the city. The Corinthian capitals display female busts placed centrally between the volutes. Each capital had different effigies, and it is unclear which divinities the busts represent.[51] Nevertheless, the busts found an immediate resonance in the defensive towers that the city refurbished in the same period on the north side of the enceinte. Each of these new towers included a Doric frieze and ornamental pilasters that featured similar Corinthian capitals with a central female bust (see Figure 7.9).[52] Their location on the exterior of the defenses must have carried an apotropaic function

Figure 8.9 Surviving capital from the Doric-Corinthian Temple at Paestum.

as well as a direct association between the cult at the temple and the fortifications in a fashion similar to that in Pompeii. In both cities, patrons and architects used such landmarks to transform the image of their cities, in the process presenting themselves as protectors and benefactors of the community.

The development of Venus Pompeiana

Parallel to Minerva, the cult of Venus had a direct correlation to the concept of Pompeii and its fortifications. Unlike Minerva, she did not feature prominently at the gates of the city. The cult of Venus Fisica Pompeiana was an overt stamp of Sullan conquest on Pompeii. The dedication of her temple directly onto the city walls was a mark of the new colonial reality and a symbol of Pompeii's loss of independence.[53] Pompeii's refoundation as *Colonia Cornelia Veneria Pompeianorum* provides a measure of the cult's importance.[54] *Cornelia* evoked Sulla's nephew and cofounder of the colony, P. Cornelius Sulla, and the Sullan *gens* Cornelia, whereas *Veneria* alluded to Venus who acted as the dictator's personal protective goddess.[55] Although disputed, the cult of Venus Fisica Pompeiana may be an appropriation of the Samnite Mefitis Fisica whose main sanctuary was in the territory of the Hirpini in the D'Ansanto valley.[56] After her transformation, Venus would remain popular right up to the eruption. In the surviving depictions, she usually carries a standard set of attributes: a mural crown or diadem, a scepter, an olive or myrtle branch, and a ship's rudder. She is often heavily bejeweled and wears a blue mantle covering a purple *chiton*, occasionally decorated with stars.[57] Her attributes are similar to those of Cybele and Fortuna, sometimes leading to debates concerning her identification. The mural crown reflects those of the Greek Tychai personifying cities, whereas the rudder refers to Fortuna as guider of lives and carried further meaning for Pompeii as a port city.[58] Her overall iconography indicates her close ties with the civic image or at least the conceptualization of Pompeii.

Located upon the fortifications a little farther westward along the ridge from the Doric Temple, the Temple of Venus fulfilled a similar role in the landscape as Minerva's sanctuary. Cultic traces stretch back to the Archaic period and indicate the presence of preceding religious structures roughly contemporaneous to the first Samnite enceinte and even the Pappamonte circuit. Construction of the main sanctuary began in the last quarter of the second century BCE, well before the foundation of the colony.[59] The temple was a grand concept imitating the terraced sanctuaries of Praeneste, Tivoli, and Terracina. Albeit on a smaller scale, it signaled Pompeii's ambition to join the architectural developments occurring on the Italian peninsula.[60] The sanctuary featured a double terrace supporting three porticoes that surrounded a central temple with an open view toward the sea and the Monti Lattari. A tuff wall, matching the adjacent curtain wall, acted as a terracing structure and fortification throughout the Samnite period.[61] The material referenced the Pompeian territory where workers quarried it and the houses of the elite who used it in the facades of their houses to advertise their wealth.[62] From a distance, the display of construction material was a marker of visual unity for the city, creating an urban image as well as an architectural and symbolic dialogue between the sanctuary's patrons and the divinities protecting Pompeii.

With the arrival of the colonists, the Temple of Venus received modifications that included a new decorative program and the reduction of the double terrace supporting it into a single platform (see Figure 5.7).[63] Although less dramatic than building an entirely new temple, this modification, or completion depending on the point of view, still provided a powerful example of Roman appropriation of both the cult and the city.[64] The contemporaneous refurbishment of the nearby Porta Marina acted in concert with these developments. The gate emphasized the monumentality of the area, linking the sanctuary with the mouth of the Sarno River and the newly built *Navalia*.[65] An expansion of the sanctuary in the Augustan period, this time using marble in a move reminiscent of the grand *aurea templa* built by the emperor in Rome, continued the political association with the emperor's protective goddess and the city.[66] By the time of the eruption, the temple was under reconstruction, initially sponsored by Nero after the earthquakes of the early 60s CE had damaged the building. Nero even dedicated a golden lamp to the goddess as a symbol of his sponsorship and perhaps because his wife Poppaea and her *gens* were from Pompeii.[67] The high-level political sponsorship explains why, by the time of the eruption, the Temple of Venus played a more prominent role in protecting Pompeii, whereas the Doric Temple dedicated to Minerva lay in ruin.

Venus/Mefitis and Minerva played distinct roles in projecting Pompeii's image and protecting the city and its fortifications. Both occupied temples that – in differing degrees in the city's history – emphatically displayed the goddesses' authority as well as that of the elite individuals who paid for the construction of the sanctuaries and the defenses. They also carried implicit political associations and connected to the image of the city and the identity of its population. Although the Triangular Forum saw some refurbishments in the Augustan and postearth-quake periods,[68] the new emphasis on Venus Pompeiana led to the ruinous state of the Temple of Minerva, reducing it to a small cella housing the goddess.[69] Whether the introduction of the Phrygian Minerva is a specifically Roman cult or whether it is a Roman or a Samnite attempt at association with a Greek foundation myth is unclear.[70] Nevertheless, the cult continued to play a role in the protection of the city gates, following a tradition associated with the precolonial aspects of the cult. Although the cult declined, it never completely disappeared, nor did the colonists appropriate it. This continuity is in stark contrast to the development of Venus Pompeiana out of the original Samnite goddess Mefitis. Her transformation into Venus, her association with the *gens* Cornelia, and the refurbishment of the temple provided a powerful reminder and testament of allegiance to the new social order dominating the city. At any rate, two of the most prominent Pompeian deities found resonance in its city walls where they created a strong sense of devotion, pride, security, and community among the populace. Whether in Greek, Oscan, or Roman form, they were the clear markers of Pompeii's civic identity and its development throughout the history of the city.

Religion, fortifications, and identity outside Pompeii

The connection between fortifications and religion evident at Pompeii finds parallels elsewhere in ancient Italy. On occasion, the presence of apotropaic devices on circuits in could take on primarily superstitious forms without much further political significance. There are about two dozen examples of one or multiple phalli or ithyphallic carvings meant to protect gates.[71] The town of Alatri preserves a few on its citadel and lower circuit built in the late fourth century BCE. Three phalli sculpted on a lintel still protect the Porta Minore, which was a small postern offering access to the citadel. The ancient Porta Bellona, a main gate into the city, displays three vague separate carvings: a standing figure, a seated figure (perhaps a deity), and a bearded dancing *silenos*.[72] Each image was ithyphallic and functioned most likely in an apotropaic role to scare off and defend the community against evil spirits that were also the bringers of disease and misfortune. The concern with protecting liminal spaces was ubiquitous, reaching into the domestic sphere. Tertullian attests that four separate divinities protected the liminal space of doors: Janus as the overarching guardian, Ferculus who watched over the door leaves, Limentius who protected the threshold and lintel, and finally Cardea who attended the hinges.[73]

These deep-seated social and religious beliefs meant that protective symbols, divinities, and shrines on gates must have acted as group identifiers that could carry further political associations. Gates such as the Porta Stabia were landmarks for the beginning of religious processions. A measure of their importance in the urban landscape comes from the Dipylon gate in Athens. The gate was the starting point of the sacred parade, known as the Panathenanic procession, that reached the Acropolis to honor Athena in her role as *Polias*, or protectress of the city. The architecture of the Dipylon gate was an elaborate affair with a double opening, a deep corridor, and four flanking towers on each corner. It also featured a small altar located in the city side of the gate between its two main passageways. It was dedicated to Zeus Herkeios (protector of forecourts), Hermes as the patron of travelers, and Acamas. Considering the role of Athena in the city, this trifecta of protectors certainly complicates which dieties and figures the population evoked at the city gates. However, Acamas stands out as son of Theseus, hero of the Trojan War, and patron of the Akamantis tribe that occupied this part of the city.[74] His presence at the gate points to a measure of factional identity within a quarter of a city that translated to its associated gate. Only hints of such relationships remain at Pompeii, where the ancient names of the gates and the altar of Caecilius Jucundus indicate similar group dynamics that may have extended to lost ornamental programs. Through their religious and social roles, city gates were landmarks to single urban districts and could become identifiers for inhabitants and neighborhoods.

The visual relationship between defenses and temples also helped define the image of an ideal city in the Greek world where such notions developed together with the growth of urban centers. The Acropolis in Athens, for instance, still displays this relationship through the remains of the first Parthenon that were purposefully embedded into its defenses. In Sicily, colonists from Gela founded the city of Acragas in 580 BCE and ensured that its defenses and shrines would foster the creation of a civic identity. In earlier colonial foundations on the island, aristocrats typically controlled most religious sanctuaries, particularly those that functioned as boundary markers on the fringes of a city's territory. Those at Acragas – many of which the city monumentalized with temples in the fifth century BCE – display a different setup. The Temple of Athena stood on the Acropolis, whereas all the other cults were laid out as a ring around the city on top of the outer fortifications or near the gates. Rather than falling under the control of the elite, this layout indicates that the temples were in the public domain as part of a more egalitarian political system. Although the identification of the deities that occupied these shrines remains contentious, well acquainted viewers certainly could identify each temple as they passed through the fortifications or approached the city.[75]

The relationship among tutelary gods, patronage, and the city present in the Greek world is far more tangible and politically driven on the Italian peninsula. A sign of its strength comes from rituals known as the *evocatio* and its counterpart the *devotio*, which Roman generals applied in war. The *devotio* was more common and divided further into two: the *devotio ducis* and the *devotio hostium*. In a *devotio ducis*, a commander would devote his life to the gods to ensure victory. In a *devotio hostium*, he would offer the enemy city for destruction: All of its possessions were sacrifices to the gods of the underworld – the fates of Veii, Carthage, and Corinth are the most famous examples.[76] In an *evocatio*, the commander would invite the god(s) to abandon their efforts to protect the enemy city and come to Rome. The best known example of an *evocatio* is that of Veii in 396 BCE, where after a ten-year siege Marcus Furius Camillus invited Juno Regina to abandon Veii and come to Rome where he promised her a new temple on the Aventine.[77] Despite the prominence of the cities where generals conducted these rituals, the use of the *evocatio* is rare. Commanders reserved it for extreme cases and only on those cities that had a ritual foundation. Consequently, most cases occurred during the Roman expansion in Italy where rituals associating tutelary gods, the city, and its walls were the strongest.[78]

The visual relationship between sanctuaries and fortifications on the Italian peninsula served to strengthen the associations between cities, political leaders, and their tutelary divinities. As a separate space often removed from the lower city, the *arx* would perform as the center of political as well as religious life in Etruscan and Latin cities.[79] From the hilltops, the augurs (priests), who were usually members of the local elite, would perform the ritual known as the *augurium* to take the *auspicia*, or the divination by flight of birds. Starting in the late Bronze Age, settlements would develop below these highly charged locations – the *arx* was the actual nucleus from which the urbanization process often began. As a measure of this separation in the development of urbanism in Italy, Vitruvius specifically prescribed that temples dedicated to the protective gods should be in eminent places with commanding views of the city.[80] His comment reflects a tradition that was almost ubiquitous. The towns of Segni and Ferentino are just two of many examples where the Italic *arx* carried a defensive as well as a religious function. Perhaps less obvious is that patrons would continue to exploit this relationship in subsequent reconstructions. This was a recurring factor for Pompeii, where the temples and fortifications on the southwestern tip of the plateau would receive nearly simultaneous upgrades.

The Temple of Jupiter Optimus Maximus (JOM) on the Capitoline Hill in Rome delineates how a sacred structure was a central element of the political, cultural, and religious identity of a community.[81] In the late Republic, Sulla would exploit its legacy by engaging in a program of reconstruction on three critical buildings on the Capitoline and the adjacent Forum Romanum. His efforts seem almost programmatic in nature.[82] In 83 BCE, a devastating fire swept over the Capitoline Hill, destroying the Temple of Juno Moneta as well as the Temple of JOM, which had overlooked the city for some 400 years as the single most important temple at the heart of Roman religion. Sulla would soon rebuild, including in the process the reconstruction of the Senate house that stood nearby in the Forum below. As the heart of Roman politics, Sulla clearly sent out a political message of restoring the Republic. In the area between these two buildings, Sulla and his ally Quintus Lutatius Catulus began construction of the large multistory building known as the *Tabularium*, parts of which functioned as the state archives. It also performed as the foundations of a sanctuary similar to the Temple of Hercules at Tivoli and the Temple of Jupiter Anxur in Terracina that had also inspired the Temple of Venus at Pompeii. Authors spar over whether the substructure supported the Temple of Juno Moneta[83] or acted as a podium for the Temple to the Genius of the Roman People, The Temple to Venus Victrix, and the Temple to Faustae Felicitas.[84] The first hypothesis seems stronger, since Sulla as dictator presumably chose to rebuild the Temple of Juno Moneta to evoke a connection with the gens *furia* who in the past had supplied two illustrious dictators that had saved Rome from disaster. Marcus Furius Camillus was the dictator who had conquered Veii and drove the Gauls away after the sack of Rome, and his son Lucius Furius Camillus, who had built the first temple to Juno Moneta as fulfillment of a vow he made while battling the Aurunci.[85] The triple temple hypothesis is equally ideological: In addition to the genius of the Roman people, Venus Victrix referred to Sulla's personal protectress, whereas the Temple of Faustae Felicitas referenced his nickname Felix and his twins Fausta and Faustus. Irrespective of the layout, the new building carried patent political associations.

Sulla's inspiration to fashion the foundations of the Tabularium as a terrace to the Capitoline Hill may well have carried further associations with his military successes overseas. The new temple of JOM perched nearby proudly displayed columns from the Olympieion in Athens, where he had recently won a victory in the war against Mithridates. Sulla witnessed firsthand the strength of the Athenian defenses during a bitter siege, even going as far as demolishing the long walls that

Figure 8.10 The foundations of the so-called Tabularium in Rome. (Photo by N.A. Van der Graaff.)

connected the city with its harbor after the capitulation.[86] He must have been aware of the symbolism associated with the walls of the Athenian Acropolis that preserves one of the few Greek monuments commemorating a historical event. In the aftermath of the Persian attack of 480 BCE, the Athenians would erect a new Parthenon and, along with other important shrines such as the Erechtheion, build the northern fortification wall as a terrace for the Acropolis. As a memorial of the Persian sack, engineers arranged the column drums, frieze, and cornice blocks of the old Parthenon destroyed during the raid into the new fortification for viewers to admire from below. The resulting monument had political and religious associations, evoking a struggle that Athens and the Greeks had won against almost overwhelming odds.[87]

The foundations of the Tabularium would act as a new symbolic fortification for the Capitoline Hill, complete with accompanying temples. In Rome, the destruction of the monuments on the Capitoline Hill during the Social and the Civil Wars – many of which had survived for centuries – must have been a similar traumatic event. In terms of its construction technique, the foundations of the Tabularium, a solid mass of tuff ashlars pierced with arches, referenced the Servian wall built in similar fashion in the early fourth century BCE.[88] The foundations acted as a new citadel wall, visually connecting the Temple of JOM and the Senate house, in the process creating a Sullan facade defining the western edge of the Forum. Together, the three monuments imparted the same associations among patronage, tutelary gods, and fortifications found elsewhere in the late Republic – in this case in the very center of Rome.

Patronage, religion, and power at the gates

The relationship between patronage, the city, and its tutelary deities is most apparent in city gates. Once again Greece supplies a precedent to Italy. Many fortifications had niches or shrines in or near gates where a protective deity, perhaps the tutelary civic gods, presided over the liminal space.[89] However, unlike the Roman and Etruscan tradition where fortifications were among the sacred things, the religious connection for the Greeks was limited to the protection of the liminal space. On occasion, the deities worshipped at these shrines, and their significance would tie into the image of the city. The walls of Thassos dating to the early fifth century BCE preserve reliefs that give the gates their name: the Gate of Herakles, the Satyr Gate, the Gate of Zeus and Hera, the Gate of Hermes, and the Gate of the Chariot. As some of the earliest examples in the Greek world, inspiration for their design may have come from elaborate types in the eastern Mediterranean.[90] Although the reliefs have slightly different carving dates, each had lavish ornamentation, including Doric and Ionic architectural elements, a faux balustrade, a pediment and carved doorjambs with the images of the divinities. These divinities must have had a more complex protective and political relationship with the population and the patrons who commissioned the gates.[91] The Satyr Gate featured the relief of a dancing Silenus associated with Dionysus (see Figure 8.11). The Gate of the Chariot preserves the fragmentary remains of a figure in a chariot, identified as Artemis led by Hermes. The Gate of Zeus and Hera displayed the divine couple on opposite jams in a seated position facing the city. They are in the company of their messengers,

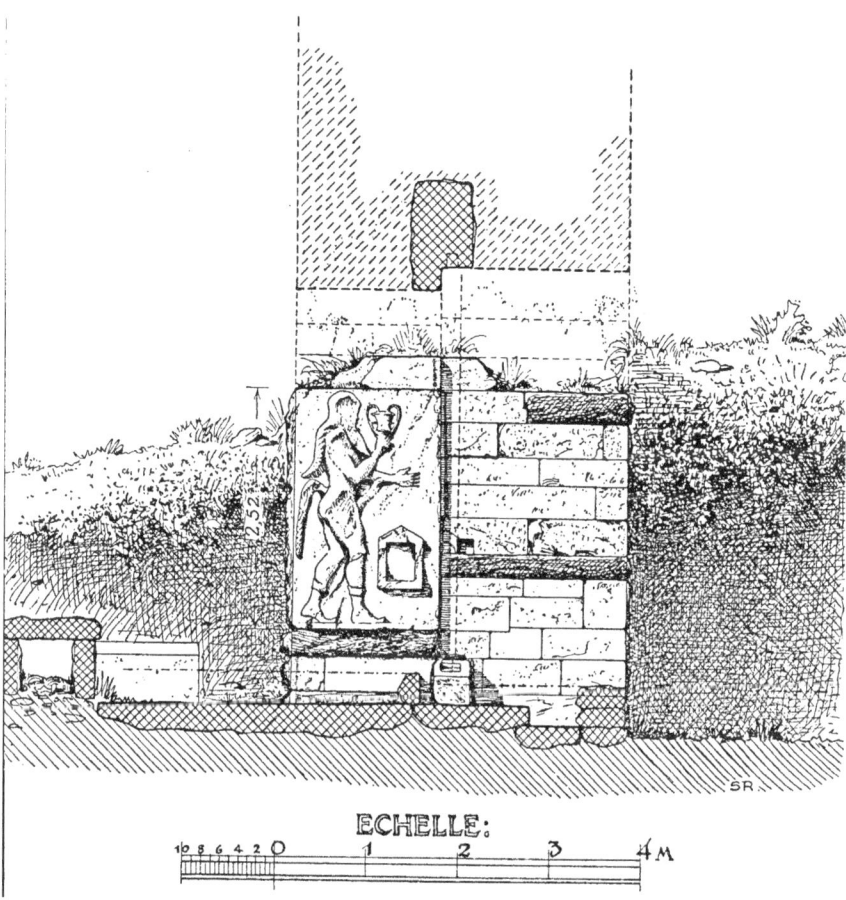

PORTE DV SILÈNE
COVPE A-B,
CÔTÉ OVEST

ECHELLE:

Figure 8.11 Drawing of the Selinus relief in Thassos. (After Picard 1962, pl. 13.)

Hermes for Zeus and Iris for Hera. This gate is also the earliest preserved structure presenting a gallery carved in low relief above the entrance. The form would inspire a similar layout to the entrance of the Great Tomb at Leukadia (Mieza), Macedonia, dating to 300 BCE and continued in Italy with the Porta Marzia in Perugia.[92]

In Italy, the earliest examples of ornamented gates survive in the Etruscan cities of Volterra, Perugia, and the Roman town of Falerii Novi. The Porta Marzia in Perugia survives because the Renaissance architect Antonio Sangallo preserved the upper portion of the gate when he designed the new walls. It previously spanned the via Amerina to honor the city's alliance with Rome signed around 241 BCE.[93] The surviving decoration is composed of two flanking pilasters in the Corinthian order that hold up a balustrade carved in low relief. Four smaller pilasters create five windows displaying horses on the two outer windows, whereas three male figures stand in the center. Below the balustrade, three protective busts decorate the keystone and the spandrels. Time has heavily damaged the sculptures, making their interpretation difficult. One view describes them as members of the local elite, perhaps deified ancestors.[94] Another interpretation views the central figure as Jupiter flanked by the foundational twins of Perugia, Oncus and Aulestes with their horses. As protectors of Perugia, the deities referenced the Roman relationship with the Dioscuri and its foundational twins Romulus and Remus.[95] At Volterra, the outward facade of the Porta all'Arco displays three heavily eroded heads at the springing and keystone of the arch (see Figure 8.12.) Their identity remains

Figure 8.12 Porta all'Arco, Volterra.

debated, but they may represent Jupiter and the Dioscuri or the Capitoline triad, symbolically embracing the city after its Roman takeover.[96] A clear political and religious reference resides in the Porta di Giove at Falerii Novi, where a head of Jupiter still decorates the keystone. Its placement here references Rome in its role as founder of the settlement in 273 BCE as a measure to relocate and control the recalcitrant inhabitants of Falerii Vetres.[97]

Elsewhere, but still relevant because of the strong aesthetic and political parallels to Italic enceintes such as Pompeii and Ferentino, Roman forces would carry Minerva to protect the defenses they would build at Tarragona during the Second Punic War. The Torre de San Magín preserves a series of unidentifiable apotropaic busts, as well as the bottom half of the relief dedicated to Minerva (see Figure 8.13). Enough survives to identify her holding a lance and leaning on a

Figure 8.13 Drawing of the relief in Tarragona. (After Grünhagen 1976, fig. 1.)

shield decorated with a wolf's head.[98] The tower was one of the most prominent in the Roman city, functioning originally as part of its main gate built in the late third century BCE (see Figure 7.8). Minerva would have fulfilled her role as protectress of fortifications and projected a measure of *romanitas* associated with the city.[99] An accompanying dedicatory plaque once carried a vanished inscription that must have referenced the patron and the city. Although the Tarragonian example is almost a century older, its context is reminiscent of the bust and inscription adorning the Porta Nola at Pompeii.[100] Both examples show how patrons associated themselves with the fortifications and Minerva as the divine protectress of the city.

With the advent of Augustus, the relationship between patronage and the divine protection of cities shifted from local patrons to the figure of the emperor. This was particularly the case for the western provinces, where the Roman conquest brought with it new urban centers, and Italy where the collapse of the Republic had led to the neglect of infrastructural works critical to the self-image of cities.[101] The Augustan program of urban renewal concentrated on basic civic needs such as sewage systems and roads, as well as the ex novo foundation and upgrade of cities to colonial status to accommodate the veterans of the civil wars. In provinces such as Gaul where Roman cities and infrastructure were almost nonexistent, local patrons and the emperor or his family intervened with financial aid to establish colonies with coherent civic images.[102] City walls and gates would be part of the buildings that projected the image of the ideal Roman city. Their ornamentation is just as much a symbol for the city as it is a reflection of the political and social realities of the empire.

With this emphasis on the city and the development of new construction materials such as concrete, a new type of gate appeared in the first century BCE that would become typical of Roman fortifications. Architects designed multistoried gates featuring two or more lower openings with surmounting arcaded galleries (see Plate 32). Tall hexagonal or semicircular towers framed the gates, adding to the architectural symbolism. An inner gate on the city side created a courtyard known as a *cavaedium* that gives the design its name. Albeit with slight variations in plan and finish, surviving examples abound in the colonies of northern Italy, such as Fano, Torino, Spello, and Verona, and in Autun, Saintes, and Nimes in France.[103] The origin of the upper galleries typical of these designs may derive from lost Greek prototypes later sculpted as ornaments on the Gate of Zeus and Hera at Thassos, for example. However, the interior courtyard of the *cavaedium* design seems to be a natural continuation of the forecourt gate types typical of the *agger* system, such as the Porta Stabia in Pompeii.[104] Such forms typically included some sort of bridge spanning the exterior bastions and a deep court that carried a similar military advantage of being able to trap the enemy inside the gate. A continuing debate surrounds the military effectiveness of *cavaedium* gates due to the relative vulnerability of the upper galleries although the flanking towers and interior court must have negated the flaw. The emphasis on size and scale in the design carried an unmistakable monumentality.[105] The gates contributed to the urban decor and evoked the military discipline needed to defend the *virtus* of the state in the tradition established in Republican circuits.[106] Similar concepts must have pervaded the refurbishments carried out at the Pompeian gates during the Augustan period. The miniature merlons present at the Augustan Palaestra and tomb precincts are just a further reflection of these ideals.

The separation between the figure of the emperor, civic image, and the protective roles of the gates would conflate further under Augustus.[107] Existing fortifications and the construction of new gates would come to be associated with imperial power. As a reward for siding with Augustus during the Perusine War, the colony of Spello (Hispellum) founded as the Colonia Julia Hispellum, received a further contingent of veterans and a new set of city walls.[108] The Porta Venere was a crown jewel of the new enceinte, announcing the city to any travelers passing on the via Flaminia below (see Plate 32). It survives today as one of the prime examples of *cavaedium*-type gates. The visible remains are the result of a restoration effort carried out in the 1930s where workers rebuilt the facade using impressions left in the concrete as well as a surviving section of the exterior decoration.[109] The gate features large octagonal towers that frame a triple gateway and that once had a surmounting loggia. The entrance is composed of a large central arch and two flanking passageways in a design reminiscent of a three bay triumphal arch. The concrete core featured a white travertine revetment sculpted with Doric pilasters and horizontal cornices that served to emphasize each of the passageways against the adjacent masonry. Any further sculptural details and inscriptions are irrevocably lost, but they must have evoked the new emperor and his donation to the city. Visible for miles through the surrounding Valle Umbra, the gate projected an image of the colony and an element of its *romanitas* throughout the countryside.[110]

The conflation between emperor, patronage, religion, fortifications, and urban image would reach triumphal overtones in the so-called Arch of Augustus in Rimini (see Figure 8.14). Voted on by the Roman Senate in 27 BCE, the arch honored

Figure 8.14 Arch of Augustus, Rimini. (Photo by N.A. Van der Graaff.)

Augustus and his completion of the repair works to the via Flaminia, which arrived at this gate from Rome.[111] The surviving arch is now a freestanding structure composed of travertine blocks, but tall fortified towers flanked either side in antiquity. The attic was refurbished in the Middle Ages and still incorporates parts of the inscription honoring Augustus. Presumably, it also supported a triumphal statue displaying the emperor in a *quadriga* that has since disappeared.[112] Framing the entrance are two engaged Corinthian columns that stand on high socles reminiscent of temple architecture. The shafts support a carved entablature and faux pediment rich with sculpted details such as gorgons, floral designs, eagles, weapons, dolphins, and hippocamps. The interior (city side) and exterior facades each preserve two *imagines clipeatae* (bust in shields) on either side of the fornix. Those that face the city represent Neptune (left) and the goddess Roma, or Minerva (right), with a trophy of armor, whereas those on the exterior display Jupiter (left) and Apollo with a lyre (right). The image of a bull decorates the keystone of the arch on either side.

The imagery is overwhelmingly one of triumph and victory but also of tradition and renewal. The bull on the keystone, which also finds a parallel in the Arch of Augustus opening into the city walls of Fano to the south, referenced the sacrificial consecration and colonial foundation of Rimini. The subsidiary ornamentation of gorgons, eagles, and the like performed a traditional apotropaic function that is now conflated with the new imagery of empire. The *imagines clipeatae* connected the gate to the earlier Etruscan examples such as Perugia and Volterra. Jupiter protected the walls and gate, whereas Apollo, as the patron of the new emperor, had played an instrumental role in the victory at Actium and the reunification of the east and west.[113] On the inner facade, Neptune alluded to Actium and the role of Rimini as a port city, whereas Minerva (or Roma), in her guise as protectress of the gates, referenced the eternal city. Through this imagery, the gate was a crowning symbol of the new city, which would stand as a model for the Augustan ideals of empire. Rimini was Caesar's first major allied city in his march on Rome. Augustus showered it with privilege, refounding it as Colonia Augusta Ariminensis and conducting a comprehensive reconstruction campaign that included repaving the roads, installing a sewage network, building new bridges, erecting a triumphal arch, and constructing a new theater.[114] Although the arch was a monument to the emperor, it combined the imagery and function of city gates to relay a new message of empire, one intimately related to Augustus as patron and protector of the colony.

The new arch also acted on a regional level, alluding to Rimini's historic strategic role. As a colony founded in 268 BCE, Rimini was a significant harbor on the Adriatic as well as a gateway to the north, connecting the via Flaminia crossing the Apennines and the via Aemilia reaching into the Po River valley.[115] In the same vote that authorized the monumental gate, the Senate also ordered construction of its twin arch to act as its symbolic counterpart at the beginning of the via Flaminia on the Milvian Bridge outside of Rome.[116] The twin arch has disappeared, but the two monuments must have made an indelible impression on anyone traveling the distance between the cities and crossing the thresholds marking urban vs. non-urban contexts of Roman protection.[117] For such viewers, the *imagines clipeatae* on the facades of the arch had further meaning: Neptune and Rome reminded one of final destination of the road, whereas travelers entering the city would fall under the protection of Augustus (via Apollo) and Jupiter. The arch triumphantly commemorated the completion of a large infrastructural project, as well as the benevolence of the emperor and empire. Augustus had redeployed the explicitly monumental shape of the triumphal arch into city gates.[118] Through this integration and their associated symbolic function, the gate in Rimini marks the transition of the triumphal arch from an isolated victory monument into a structure celebrating imperial patronage, protection, empire, and civic image.

The associations and meaning that Augustus placed in the gates and circuits would continue to develop in the following Julio-Claudian period, when triumphal arches would integrate further with city gates.[119] Triumphant overtones are present in the walls of Saepinum, built and financed by the future emperor and successor of Augustus, Tiberius, between 2 and 4 CE. The Porta Bojano is the best preserved of four almost identical gates (see Figure 8.15). It features two towers flanking a single fornix (opening) faced with elegant travertine blocks. The jams support statues of captured Dalmatian prisoners who frame the dedicatory inscription. It recounts how Tiberius financed the construction of the city walls presumably from the booty acquired during his recent Dalmatian campaigns. These military operations were part of the renewed impetus of the empire after decades of civil war.[120] The only deity present on the gate is Hercules, who decorates the exterior keystone. The ornamentation on the gate otherwise carries an unmistakable imperial overtone replacing the traditional imagery associated with deities and local patrons. This climate of urban renewal and meaning invested in fortifications must have applied as well to the Pompeiian gates refurbished in the Augustan period.

The gates of Verona are the ideological bridge for the postearthquake reconstruction of the Porta Ercolano at Pompeii. The city built them first under Emperor

Figure 8.15 Porta Bojano in Sepino.

Augustus and refurbished the facades during the reign of Claudius.[121] The Porta Borsari originally opened onto the main *decumanus*, whereas the Porta Leoni straddled the *cardo* of the city (see Figure 8.16). Each featured the *cavaedium* design with a three-story court and tall flanking towers. Only one of the original two arches that composed the Porta Leoni survives. Inscriptions recount how the local *quattroviri* P. Valerius, Q. Caecilius, Q. Servilius, and P. Cornelius oversaw its first construction. This version of the gate featured an elaborate brick facade embellished with cornices and a crowning Doric frieze sculpted in tuff stone. A further inscription dating to the time of Emperor Claudius (41–54 AD) highlights how Ti. Flavius, P. F. Noricus, and perhaps two more *quattroviri*, added new facades on either side of the gate. The surviving remains display about half of the deliberately ornamental structure. An aedicule featuring Corinthian columns frames the lower gateway. Three arched windows survive above each, framed with protruding columns that support a tall superstructure with a central alcove flanked by a spiral fluted Corinthian column. The complete facade would have featured two lower gateways, an arched gallery with six windows, and an upper alcove, each framed with columns in an arrangement that would resemble a theatrical *scaenae frons*. As paralleled in the Porta Marzia in Perugia and as was common in theaters, the alcove must have displayed some sort of statuary, perhaps an image of Claudius or a member of the imperial family.[122]

Almost nothing survives of the first phase of the Porta Borsari. The standing white limestone facade is primarily what remains of its second phase. Any record of the original patron(s) has vanished, but they must have operated as part

Figure 8.16 The Porta Leoni (left) and the Porta Borsari (right) in Verona. (Photos by N.A. Van der Graaff.)

of the reconstruction of the city carried out under Claudius. The facade has two lower arched gateways framed with a Corinthian *aedicula* in a formal element reminiscent of the *Tabularium* in Rome and the Augustan arch in Rimini. The frieze displays an inscription commemorating the restorations that Emperor Gallienus carried out on the circuit in 265 CE.[123] As opposed to the Porta Leoni, the superstructure of the Porta Borsari is quite different, with two superimposed galleries of six windows framed with various smaller *aediculae*. It lacks the upper alcove, resulting in much less of a theatrical presence. The difference in design is attributable to their position in the gates: The surviving facade of the Porta Leoni originally faced the city, whereas that of the Porta Borsari opened onto the exterior. The theatricality of the Porta Leoni must be the result of the city side carrying less military importance than the exterior of the gate.

The monumentality of the Veronese gates is a referential nod toward Emperor Claudius and the Julio-Claudian dynasty. The scant remains of a third gate across the Adige River near the theater district yielded an associated inscription.[124] In an act of civic loyalty, it honored Claudius, as the emperor and protector of the state, and his wife Messalina. The inscription implies that this gate and the others in Verona experienced a refurbishment specifically tied to the figure of the emperor. Claudius received a very similar monument in a gate known as the Porta Aurea in nearby Ravenna – today completely destroyed – after he visited the city on his way back from his British campaigns.[125] Although less labor-intensive than building entirely new gates, the application of a second veneer on the Veronese gates would have achieved a similar effect and essentially dedicated the fortifications to the emperor and his protection.

The examples of Verona, Spello, Saepinum, and Rimini illustrate how gates had become overt Imperial monuments associated with a proper civic image. These same elements would eventually find a place in the reconstruction of the Porta Ercolano at Pompeii after the earthquake of 62 CE. Its monumental three-bay arch spanning a main regional road out of the city was a *cavaedium* gate design without the flanking towers. The religious and military protective aspects of fortifications previously connected with the individual(s) financing the defenses would now meet in the figure of the emperor who in his semidivine status was ideally suited to fulfill such an overarching protective role. It was the culmination of the association among religion, patronage, and city gates, which began in pre-Republican Italy.

In the end, the city walls at Pompeii were as rich and full of symbolism as any other existing monument. Their significance changed in unison with the political and social landscape. The Roman conquest immediately appropriated the fortifications, a strong symbol of independence in the Samnite period, and ushered in the new social order of the city. The tensions of appropriation and continuity inherent with the establishment of the colony are evident in the religious landscape with the development of Venus Pompeiana and the preservation of Minerva's cult in the city gates. Despite the eventual decline in the military role of the defenses, the symbolic protective association would remain constant. It would resonate throughout the Augustan program of civic renewal and even find a place in the *lararium* of Caecilius Jucundus in the years before the eruption. The imperial message would find its final voice in the Porta Ercolano as the crown jewel of the reconstruction effort. It was the culmination of a dynamic process of changing significance that made the Pompeian fortifications far more symbolically complex than their monolithic appearance tends to suggest.

Notes

1 A portion of this chapter has appeared as an abbreviated article coauthored with S.J.R. Ellis. See Van der Graaff and Ellis 2017, 283–300.
2 See discussion in chapter 2 of this volume.
3 Festus *De ver. sig.* 358 L.
4 Cicero *De nat. deo.* 3.49.
5 Just. *inst.* I.2,1,10; Gai. 2, 8-9; Theoph. *par. inst.* 2,1,10; Seston 1966, 1489–1498; Smith and Tassi Scandone 2013, 455–473.
6 See Turner 1977, 94–98.
7 The most complete engraving of the bust is on the cover of De Clarac and Mori 1813; Mazois 1824b, 27 and 53; Nissen 1877, 511; Overbeck and Mau 1884, 46 and 52; Conway 1897, 45; Maiuri 1929, 213. Another example, depicting either Dionysus or a satyr, still decorates the western *dromos* in the main theater; see Overbeck and Mau 1884, 128. Olga Elia describes another bust, but it was not found in situ; see Elia 1975, 121 and fig. 13.
8 Maiuri 1929, 258–259; Vetter 1953, Ve 71; Castrén 1975, 44, 207.
9 The upper, larger niche measured 44 centimeters wide, 63 centimeters tall, and 30 centimeters deep. The lower, smaller niche measured 23 centimeters wide, 30 centimeters tall, and 20 centimeters deep.
10 See CIL IV 5384; Fiorelli 1875, 29; Calderini 1924, 87.
11 Ellis and Devore 2006, 11–15; Devore and Ellis 2008, 317–319. For the final publication of these excavations, see Van der Graaff forthcoming.
12 Cato *De agri.* 14.
13 Devore and Ellis 2008, 11–14. Also Ellis and Devore 2006, 1–15. De Caro, Baldassare, Coarelli, and Anniboletti discuss the identification of the figurine in Guzzo and Guidobaldi 2008, 512–513.
14 Sogliano 1937, 34–35; Sgobbo 1942, 26–27.
15 Virgil *Ad aen.* VII, 662; Solinus *Collec.* II, 3; Martianus Capella *De nup.* VII, 642; Isidorus *Ety.* XV, 1, 51.
16 Fiorelli 1861, 370; Breton 1870, 80; Von Rohden 1880, 44; Eschebach and Müller-Trollius 1993, 12.
17 Von Rohden 1880, 44.
18 Sogliano 1906, 99–100.
19 Maiuri 1929, 239 and fig. 39.
20 As if to emphasize a communal need for protection, someone carved a graffito in the shape of a flower – a common apotropaic device found on the house facades of Pompeii – on the northern exterior threshold into the city.
21 Vitruvius *De arch.* I.2.5.
22 De Caro 2007, 74.
23 Vetter 1953, Ve 27; Antonini 2007, 48; Varone and Stefani 2009, 368; and Tav. XXXc.
24 The area is still buried today; Van der Poel 5 1981, 88.
25 On the organization of traffic, see Poehler 2006 and Poehler 2017. On the configuration of retail landscapes and movement through urban spaces more generally, see Ellis 2011 and Laurence and Newsome 2011.
26 See Fowler 1966, 183; Pocetti 1988, 324–327; Coarelli 2000, 103; Guzzo 2007, 116.
27 Palmer 2009, 84.

28 On the continuity of cult at the Triangular Forum, see Carafa 2011, 98. Osanna has argued for different iterations between the first cult to Minerva and its later Phrygian version that began in the fourth century BCE. The early version is enigmatic, and much of this analysis focuses on the fourth-century cult; Osanna 2016, 179–201.
29 De Caro 2007, 74.
30 Bernabei 2007, 20–24; Carafa 2011, 95.
31 Breglia Pulci Doria 1998, 101–108; D'Alessio 2005, 541.
32 Guzzo 2007, 40.
33 Serv. Com. in Verg. *Ad aen.* II, 620; Mau 1902, 531.
34 Cic. *Fam.* XII 25,1; Plutarch *Cic.* 31, 6; Von Hesberg 1998, 374; Pina Polo 2003, 111.
35 Neil 2016, 21.
36 Apollod. *Biblio.* 3.12.3; Ov. *Met.* xiii. 337; De Caro 1992a, 175–178; Bernabei 2007, 20–24.
37 See Scatozza Höricht 2005, 662–666.
38 See D'Ambrosio 1984, 11–17, 153–207; Miniero et al. 1997, 11–16; Scatozza 1997, 189. On the distribution of sites between Sorrento and Pompei, see Jacobelli 1994, 65–77; Russo 1998, 23–98.
39 Scatozza Höricht 2001, 224–227; Pesando 2010a, 242. For an in-depth discussion, see De Caro 1992a, 173–178; Scatozza Höricht 2005, 662–666. For the early connection with Greek Neapolis, see Russo 1992, 207–213 and Russo 1998, 24; Scatozza 1997, 192–197.
40 Strabo *Geo.* V.4.8. On the debate, see Breglia Pulci Doria 1998, 101–108.
41 Morel 1982, 147–153.
42 Russo 1990, 57, fig. 16; D'Alessio 2001, pl. 3, 16.
43 De Caro 1992a, 175–176; Carafa and D'Alessio 1999, 34–39; Carafa 2008, 13.
44 Clarke 2003, 85–89.
45 Tocco Sciarelli 1988, 36, pl. 53. Also Miniero et al. 1997, 24, fig. 15.
46 Strabo *Geo.* V.4.13.
47 Schläger 1957, 9; Brands 1988, 151.
48 See Cipriani and Pontrandolfo Greco 2010, 219–229 and 369–374.
49 The identification is controversial; see Cipriani and Pontrandolfo Greco 2010, 39–44 and 325–326.
50 For the identification of the cult, see Torelli and Cipriani 1999, 64–68; Greco and Longo 2008, 39. For a single construction event dating to the first century BCE phase, see Torelli and Cipriani 1999, 64–68. For the identification of the two phases, see Krauss and Herbig 1939; Denti 2004, 665–697.
51 Krauss and Herbig 1939, 27–29, 41.
52 Krauss and Herbig 1939, 42; Krischen 1941, 23–25; Von Mercklin 1962, 67, cat. #176; Brands 1988, 157.
53 Zanker 1998, 61–70; Wolf 2004, 193–194; Tybout 2007, 408; Carroll 2008, 64–74, 92–97.
54 Lo Cascio 1991, 122.
55 Descoeudres 2007, 16.
56 See Sogliano 1932, 359–361; Lejeune 1964, 393–394; Coarelli 1998, 185–190; Lepone 2004, 162–167. Carroll contests this view, suggesting that Venus already had a place in Pompeii before the colony; see Carroll 2008, 96–97.
57 Della Corte lists fourteen representations; Della Corte 1921a, 68–70, 85–87; see also Clarke 2003, 89.
58 Bastet 1975, 65. On the Greek Tychai wearing a turreted crown, see Pollitt 2009, 2–3.
59 Curti 2008, 50; Coletti et al. 2010, 189–211. Carroll contests this early phase; see Carroll 2008, 64–74.
60 Zanker 1998, 61–70; Wolf 2009, 322–323.
61 Curti 2008, 52–54; Coletti et al. 2010, 204–207.
62 Richardson 1988, 371; Zanker 1998, 61–70.
63 Curti 2008, 52–54.

64 Lepone 2004, 163–167. See also Bernabei 2007, 46; Curti 2008, 72. Wolf 2009, 269 suggests the Temple of Venus at Pompeii was finished under Sulla.
65 See Curti 2005, 59 and Coarelli 2010, 437–450, who counter the previous theory, suggesting that the quay wall was defensive in nature, on which see Descoeudres 1998, 216.
66 Jacobelli and Pensabene 1995–1996, 45–53. Wolf 2009, 269. On the *aurea templa*, see Zanker 1988, 104–110.
67 De Caro 1998, 239–244. See also Zevi 2003, 856–864.
68 See Pesando 2000, 260–163; D'Agostino et Al, 2001, 337; Pesando and Guidobaldi 2006, 56; Carafa 2011, 99.
69 Laurence 2007, 23.
70 De Caro 1992a, 177.
71 Lugli 1968, 96–97.
72 See Lugli 1968, 132; later, during the Christian era, crowds would pelt these "profane" images with stones on the second day of Easter, resulting in extensive damage.
73 Tert. *De idol.* XV.
74 Fields 2010, 25; Greco et al. 2014, 1277–1280; Steffelbauer and Ruggeri 2013, 58.
75 Those at gates I, III–V are associated with Demeter, Hera, Heracles, and a chthonian god respectively; Holloway 2000, 60–61.
76 Macrobius 3.9; Gustafsson 2000, 42–62.
77 Livy *Ab urb. con.* V.21.1–3, V.22.3–7, V.23.7; Plutarch *Cam.* 5.1. Other cases may have been Volsinii, Falerii and Carthage.
78 Gustafsson 2000, 71–82.
79 Torelli 2000, 195–199.
80 Vitr. 1.7.1.
81 For the relationship between religion and the elite in Rome, see Stamper 2008, 32–33, 38–40.
82 Gisborne 2005, 119–121; Sulla also erected an equestrian togate statue of himself on the Rostra, expanded the Lacus Curtius, and began repaving the Forum; see Van Deman 1922, 1–31; Purcell 1995, 332; Giuliani 1996, 166.
83 Here I follow Tucci's arguments; see Tucci 2005, 6–33. Sulla was the principal force behind the construction of the Tabularium complex. Quintus Lutatius Catulus completed it shortly after Sulla's death in 78 BCE.
84 Coarelli 2010, 125–128.
85 Liv. *Ab ur. con.* V.49.; VII.28.1-8
86 Appian *Mith.* 30.121; Strabo *Geo.* IX.1.15. Platner and Ashby 1929, 299; Conwell 2008, 194–195.
87 Rhodes 1998, 32–33; Müth et al. 2016, 136–137.
88 The defenses of Rome featured similar arched openings to accommodate defensive artillery, which would have added further architectural reference between the two. It is possible that the tuff on the Tabularium substructure had a plaster coating. However, the masonry technique displaying rows of stretcher and header blocks, similar to the masonry on the city walls, would still be evident.
89 Geis 2006, 18–20; Hellmann 2010, 294–295.
90 Picard and Risom 1962; Adam 1982, 90; Müth et al. 2016, 126–137. Geis 2006, 101–104 suggests that these reliefs are the results of contacts with the east.
91 On the apotropaic role, see Picard and Risom 1962, 13–41; Geis 2006, 102–104.
92 Rebecchi 1978, 153–166; Bacchielli 1984, 79–87.
93 Defosse 1980, 760–819.
94 Warden 2012, 359.
95 Coarelli 2004, 85–87.
96 Pasquinucci and Menichelli 2000, 49.
97 Torelli 2008, 277.
98 Grünhagen 1976, 212 and 225; Hauschild 1985, 75–90.

99   Pina Polo 2003, 117. A winged Minerva, dating to Domitian's reign is also prominent at the Porta Romana in Ostia, see Von Hesberg 1998.
100 See Mazois 1824a, 52; Maiuri 1929, 213; Pesando and Guidobaldi 2006, 33.
101 Gros 1996, 35.
102 Gabba 1972, 73–112; Zanker 1988, 327; Gros 1992, 217.
103 Ward-Perkins 1994, 171–179.
104 Kähler 1942, 10; Rosada 2000, 181, 191–192, who also argues that the Porta Stabia is a Greek as well as a protohistoric design. I tend to disagree with the purely Greek assessment of the Porta Stabia because the design is a practical solution to a passage-way through an *agger*.
105 Kähler 1942, 86–88; Rebecchi 1987, 130–150; Van Tilburg 2008, 136–137.
106 Zanker 1988, 328.
107 Rosada 1992, 129–139; Stevens 2016, 288–299.
108 Kähler 1942, 100–101.
109 Rebecchi 1987, 136–139; Moretti 2014, 229–237. Enough remained to replicate the friezes and Doric pilasters following standard proportions. A similarly austere facade survives at Fano.
110 Richmond 1932, 52–62; Baiolini 2002, 61–101.
111 De Maria 1991, 260.
112 Cassius Dio *Rom. hist.* 53.22–2; Mansuelli 1960, 374.
113 G. Mansuelli 1960, 33–37 argues that the pediments and columns suggest a religious association with gates. See also De Maria 1979, 59–91; Ortalli 1999, 15–26.
114 Ortalli 1995, 469–529.
115 Galsterer 2006, 11–17.
116 Cassius Dio *Rom. hist.* 53.22–2.
117 Ortalli 1995, 469–529.
118 Richmond 1933, 160; Mansuelli 1957, 167–171; Mansuelli 1960, 374; Macdonald 1986, 137–148.
119 Richmond 1933, 149–174.
120 Gros 1996, 40.
121 On the ideological associations of the Veronese gates, Richmond 1933, 164–170; Cavalieri Manasse 1987, 3–57; Rosada 1990, 388–402; Cavalieri Manasse 1992, 9–42.
122 For a detailed description, see Kähler 1935, 138–197; Forlati Tamaro 1965, 12–34; Cavalieri Manasse 1987, 30; Cavalieri Manasse 1992, 13–22.
123 Kähler 1935, 143–158.
124 Cavalieri Manasse 1992, 13–22.
125 Rebecchi 1987; Cavalieri Manasse 1992, 37; La Rocca 1992, 265–314.

Glossary

Acropolis	Upper fortified citadel of an ancient city.
Aedicula	Small niche or structure used as a shrine.
Aedile	Elected magistrate in charge of public works, grain supply, and public games.
Agger	Earthen rampart or bulwark, sometimes reinforced with an external ditch and a masonry revetment.
Agrimensor	Roman land surveyor.
Album	In politics, a blank tablet used to record edicts.
Altstadt	German for "old city." At Pompeii, it correlates with the group of irregular streets in the southwestern part of the city.
Antefix	Edge ornament marking the join between roof tiles.
Antiquarium	The museum of Pompeii destroyed during the air raids of 1943.
Apotropaic device	Purpose of a device or mechanism intended to avert or turn away evil.
Argei	Designated sacred spaces in Rome that were the focus of yearly processions designed to purify the city.
Arx	Upper fortified citadel of a city in the Etruscan and Latin tradition.
Ashlar	Regularly shaped square or rectangular hewn stone.
Atriolum	Small hall or antechamber.
Atrium	In a Roman house, the central space with an open roof and surrounded with rooms.
Augur	Roman or Etruscan priest and diviner.
Augurium	The observation and interpretation of omens carried out by an *augur*.
Augustales	A college of priests selected from freedmen that was dedicated to the worship of the *lares* and Augustus in cities other than Rome.
Aurea templa	Group of temples (re)built using marble under Augustus in Rome.
Auspicia	The interpretation of signs by birds used to predict future events.
Bastion	Fortified projection from a defensive line.

Battlement	Parapet surmounting a wall with regularly spaced openings.
Biga	Roman chariot drawn by two horses.
Bruciaprofumo	Incense burner.
Bugnato	Unfinished or roughly surfaced stone blocks used in an ornamental fashion in architecture.
Caesareum	Shrine dedicated to the imperial cult.
Campus	Open field used for (military) training or other events.
Capitolium	Temple dedicated to the Capitoline triad (Jupiter, Juno, and Minerva), located at the heart of Roman settlements.
Cardo	Main road(s) in a Roman city aligned in a north–south direction.
Castellum (aquae)	Water distribution and storage tank of a Roman aqueduct.
Castrum	Roman military camp.
Cavaedium (gate)	City gate with a central square or rectangular court and flanking towers.
Cella	Inner sanctuary of a Roman or Greek temple.
Centuriation	Land division into a grid for urban planning and land distribution to settlers.
Chiton	Ancient garment worn full length by women and short by men.
Chlamys	Cloak pinned on one shoulder.
Chora	Territory associated with an ancient city.
Cippus	Small pillar or carved stone used as land or grave marker.
Cisarii	Drivers of the *cissum*.
Cissum	Light two-wheeled cart.
Civitas foederata	Free city or settlement closely allied with Rome.
Colonia	Outpost of Roman citizens.
Comitia centuriata	Assembly (centuriate) composed of soldiers. One of the three voting assemblies of the Roman Republic.
Comitium	Space or building used for political assembly.
Construction seam	Join or meeting point of two different types of masonry.
Corbel arch	A technique where architects span a space by offsetting successive courses of stone toward the center.
Cubiculum	Small private room in a Roman house often used as a bedroom.
Cul-de-sac	Dead end.
Cursus honorum	The ladder of political offices that aspiring politicians were expected to climb during the Republic.
Curtain wall	Fortified defensive wall linking gates and towers together.
Decumanus	Main road(s) in a Roman city aligned in an east–west direction.
Devotio	Vow where a Roman general consecrated himself and the enemy to the chthonic gods.
Dodecapoli	Loose federation of twelve Etruscan cities.

Dressing (stone)	Final act of surfacing a quarried block.
Duoviri (duumviri)	Joint magistrates.
Duovir iure dicundo	One of the two chief elected magistrates of a Roman city.
Duumvir quinquennalis	One of two magistrates elected every fifth year to exercise the censorship.
Eituns	Samnite word for "militia" or "troops."
Emblema	Central panel of a Roman mosaic.
Emplecton	Construction technique where ashlars compose two outer faces of a wall and rubble fills in the intervening space between the faces.
Emporium	Designated space in the territory of a city for trade with foreign merchants, often under the protection of a tutelary deity.
Enceinte	Fortified circuit enclosing a stronghold or settlement.
Erotes	Cupids.
Euergetism	Expected practice where Roman and Greek members of the elite used their wealth for public benefaction.
Evocatio	Ritual carried out during war in which a Roman general invited a divinity protecting a besieged enemy city to Rome.
Garum	Fish sauce.
Gens	Family or group of families who claimed a common ancestor and shared the same name.
Facadism	Practice in architecture where the facade of a building is left in place and a new interior is built behind it.
Fauces	The corridor between the main door and the atrium in a Roman house.
First Style	Style in Roman painting where plaster applied on walls imitates slabs of marble.
Foedus	Treaty.
Forecourt gate	City gate with a forward court.
Fossa	Ditch.
Fossa-agger	Fortification system composed of an exterior ditch and a rear earthen embankment.
Header-stretcher	Masonry technique where rectangular ashlars or bricks are laid down in alternating patterns of short sides (headers) and long sides (stretchers).
Heroon	Shrine dedicated to the (mythical) founder of a city.
Hippocamps	Mythical creatures, half horse and half fish.
Hippodamian plan	Urban plan arranged along a rectangular grid as proposed by Hippodamus of Miletus.
Imago clipeata	Portrait of individuals or image of a divinity set in a round shield.
Imperium	Authority, command.
Impluvium	Shallow basin in a Roman atrium used to collect rainwater into an underground cistern.
Insula	City block.
Insula occidentalis	City block on the western side of Pompeii.
Isodomic masonry	Masonry where blocks measure an equal size.

Iuvenalia	Festivals given by the *collegia iuvenum* (iuventus) in honor of the goddess *Iuventus* as protectress of youth.
Iuventus	Youth corps organized in cities across the Roman world in groups known as the *collegia iuvenum*. Also goddess of youth.
Lararium	Shrine dedicated to the guardian deities (*lares*) often inside a Roman house.
Lares	Guardian deities commonly worshipped inside a household.
Libertini	Social class of men freed from slavery.
Lustratio	Purification ceremony.
Macco	Calcareous stone found near the town of Cerveteri.
Magazzini	Warehouses.
Maniera (manner)	Used in art and architecture to define particular stylistic characteristics.
Maniple	Unit of a Roman legion.
Meddix	A Samnite magistrate (pl. *meddices*).
Meddix tuticus	Chief Samnite magistrate.
Merlon	Vertical solid section of a crenellated parapet.
Metope	Element of a Doric frieze set in between the triglyphs. Often carved or painted.
Mores	Moral customs or attitudes.
Mundus	The symbolic center of an Etruscan city, often marked with a pit containing a stone from which surveyors plotted out the city.
Municipium	Settlement with Roman citizenship that retained a measure of cultural independence.
Murus(m)	Wall.
Nativa praesidia	Natural topographical elements defending a settlement or stronghold
Navalia	Port (river) or dockyard.
Neustadt	German for "new city." At Pompeii, it correlates with the area of the city arranged in an orthogonal plan.
Opus africanum	Type of masonry with a frame of vertical and horizontal ashlars. Rubble or mudbrick fills the intervening spaces of the framework.
Opus caementicium	Wall built in concrete with facing of rough undressed stones.
Opus incertum	Wall built in concrete with a facing of fist–sized, irregularly shaped stones.
Opus quadratum	Square or rectangular ashlars laid in regular courses.
Opus reticulatum	Concrete wall with a facing of square stones or bricks set diagonally in diamond shapes.
Opus scutulatum	Marble or stone paneled floor arranged in geometric patterns.
Opus vittatum mixtum	Concrete wall with a facing of alternating bands of small tuff blocks and terra-cotta bricks.
Ordo decurionum	Municipal town council of 100 men.

Orthostat	Thin slabs of stone set vertically in masonry or as revetment.
Oscan	Ancient Italic language spoken in southern Italy.
Pagus	A district or region outside of a city.
Palaestra	A Roman gymnasium.
Palisade	Defensive line composed of (wooden) stakes.
Palma Campania	A Bronze Age culture of Campania that takes its name from the modern town of Palma Campania where it was found first.
Pappamonte	Friable tuff stone typical of the Pompeian plateau.
Parapet	Low wall protecting the wall-walk on top of the fortifications.
Parilia	Annual festival of pastoral origin celebrating the foundation day of Rome.
Patera	Saucer-shaped vessel used for sacred libations.
Patronus municipii	One or more patrons chosen to defend and develop municipal interests.
Pax augusta	Period of peace that began with the reign of Augustus.
Peripteral	A building surrounded with a row of columns on all sides.
Pietas	Religious duty or proper behavior; piety.
Plethron	Greek unit of measure composed of 100 feet (30 meters).
Plumam	Undefined, probably a parapet.
Polygonal masonry (manners)	Construction technique where the stones composing a wall have irregular polygonal shapes.
Pomerial street	Street tracing the boundary of a city wall.
Pomerium	Religious boundary traced at the foundation of a city.
Porticus	A roofed open corridor supported by a colonnade or arcade.
Postern	A small (hidden) entrance in a fortified circuit about the size of a domestic door.
Postmurum	Area beyond the wall into the city.
Pozzolana	Volcanic ash found in the area of the Bay of Naples used to make concrete.
Praetorian Guard	Elite imperial guard unit of the Roman army.
Proteichisma	Defensive outer fortified line running in front of the main defensive wall.
Protome	Ornamental element composed of a bust or animal head in frontal view.
Pseudo isodomic	Masonry composed of blocks measuring almost equal size.
Quadriga	Chariot drawn by four abreast.
Quadriporticus	Open square surrounded on each side with a portico.
Quattroviri	Four magistrates in charge of a city.
Rampart	Defensive enclosure or wall surrounding a city.
Regio(nes)	Modern organization of Pompeii into regions as defined by Fiorelli.
Romanitas	Loosely translated as "Romaness." Collection of ideals (political and cultural) defining Roman spirit.
Roma Quadrata	Square or rectangular enclosure on the Palatine hill associated with the beginning of Rome.

Rustication	*See* ***Bugnato.***
Sacellum	A small religious shrine.
Samnites	Ancient group of peoples from the central and southern Apennine Mountains in Italy.
Scaean gate	A gate design where the attacking enemy are forced to expose their right unshielded side. Named after a gate at Troy.
Scaenae frons	The scene building set behind the stage in a Roman theater.
Schola (tomb)	Tomb type shaped as a semicircular bench.
Second Style	Style in Roman wall painting where painters open up the wall plane using illusionistic architectural landscapes.
Securitas	Safety, security.
Spolia	Reused building materials from previous structures.
Stylobate	Stone platform supporting a colonnade.
Sulcus primigenius	Ritual furrow plowed to delineate the extent of a city at its foundation.
Summa honoraria	Sum of money that magistrates and priests paid upon entering office.
Suovetaurilia	Purification ritual involving the sacrifice of a sheep, a pig, and a goat.
Synoecism	The merging of cultures or villages into a new settlement.
Tablinum	Room behind the atrium of a Roman house used to receive clients and store documents.
Tabularium	Record office in ancient Rome.
Templum	Sacred space, shrine, or building.
Terminus	End point.
Terminus ante quem	"Limit before which," e.g., the last possible date for something
Tombe a dado	Etruscan tomb type built in rows and carved in standard cube shapes.
Travertine	Limestone deposited by (hot) springs
Triclinium	Dining space that accommodates three couches used for reclining during meals.
Triglyph	Element of a Doric frieze in the shape of a tablet with three grooves. Separates the metopes.
Tuff	Volcanic ash consolidated into a soft and porous rock.
Urbs	(Walled) town.
Urbs quadrata	Geometrically planned city. In Roman colonies, it referenced *Roma quadrata* and the foundation of Rome.
Velum	Sail covering offering shade during gladiatorial or theatrical performances.
Vicus	Small settlement or village.
Virtus	Combination of values and virtues (strength, valor, courage, character, excellence, military prowess) associated with masculinity.
Wall-walk	Space on top of the fortifications used for patrols and communications.

Bibliography

Adam, J.P. 1982. *L'Architecture militaire grecque*. Paris: Picard.

———. 1999. *Roman Building: Materials and Techniques*. Translated by A. Mathews. London: Routledge.

———. 2007. "Building Materials, Construction Techniques and Chronologies." In *The World of Pompeii*, edited by P. Foss and J.J. Dobbins, 98–114. New York: Routledge.

Adamo Muscettola, S. 1992. "La trasformazione della città tra Silla e Augusto." In *Pompei*, edited by F. Zevi, 2:75–114. Naples: Banco di Napoli.

Adams, W.H.D. 1873. *The Buried Cities of Campania; or, Pompeii and Herculaneum: Their History, Their Destruction, and Their Remains*. London: Nelson and Sons.

Agiletti, S., and R. Frasca. 2015. "Fundamenta murosque af solo faciunda coeravere. La 'questione' dell'avancorpo dell'acropoli di Ferentino nell'ottocento." In *Studi sulle mura poligonali. Alatri atti del quinto seminario 30–31 ottobre 2010*, edited by L. Attenni, 21–31. Naples: Valtrend Editore.

Aineas. 2001. *Aineias the Tactician: How to Survive under Siege: A Historical Commentary, with Translation and Introduction*. Translated by D. Whitehead. London: Bristol Classical Press.

Albertson, F.C. 1990. "The Basilica Aemilia Frieze: Religion and Politics in Late Republican Rome." *Latomus* 49 (4): 801–815.

Amery, C., and B. Curran Jr. 2002. *The Lost World of Pompeii*. London: Frances Lincoln.

Amodio, G. 1996. "Sui 'vici' e le circoscrizioni elettorali di Pompei." *Athenaeum* 84: 457–478.

Amoroso, A. 2007. *L'insula VII, 10 di Pompei. Analisi stratigrafica e proposte di ricostruzione*. Rome: L'Erma di Bretschneider.

Andreau, J. 1980. "Pompéi: mais où sont les vétérans de Sylla ?" *Revue des Études Anciennes* 82: 183–199.

Anonymous. 1889. "Pompei giornale dei soprastanti." *Notizie degli Scavi di Antichità*, 278–281.

———. 1899. "Giornale redatto dai soprastanti." *Notizie degli Scavi di Antichità*, 406–407.

Antonini, R. 2004. "Eítuns a Pompei. Un frammento di DNA italico." In *Pompei, Capri e la Penisola Sorrentina. Atti del quinto ciclo di conferenze di geologia, storia e archeologia. Pompei, Anacapri, Scafati, Castellammare di Stabia, ottobre 2002–aprile 2003*, edited by F. Senatore, 273–321. Capri: Oebalus.

———. 2007. "Contributi Pompeiani II-IV." *Quaderni di Studi Pompeiani* 1: 47–113.

Aoyagi, M., and U. Pappalardo. 2006. *Pompei. Regiones VI–VII, Insula Occidentalis, 1*. Naples: Valtrend.

Apollodorus. 2002. *Apollodorus: The Library: with an English Translation*. Translated by J.G. Frazer. Cambridge, MA: Harvard University Press.

Apollodorus. 2010. *Apollodorus Mechanicus, Siege-Matters: Poliorketika*. Translated by D. Whitehead. Stuttgart: Franz Steiner Verlag.

Appian. 1912. *Roman History, Volume I: Books 1–8.1*. Translated by H. White. Cambridge, MA: Harvard University Press.

Aquilué Abadias, J. 2006. "Noves troballes a les excavacions de la cuidad romana d'Empúries." *Cota Zero: Revista d'Arqueologia i Ciència* 21: 17–22.

Aquilué, X., P. Castanyer, M. Santos, and J. Tremoleda. 2006. "Greek Emporoin and Its Relationship to Roman Republican Empúries." In *Early Roman Towns in Hispania Tarraconensis*, edited by L. Abad Casal, S. Keay, and S. Ramallo Asensio, 18–32. Portsmouth, RI: Journal of Roman Archaeology.

Aristotle. 1944. *Politics: In Twenty-Three Volumes XXI*. Translated by H. Rackham. Cambridge, MA: Harvard University Press.

Arthurs, P. 1986. "Problems of the Urbanization of Pompeii. Excavations 1980–1981." *The Antiquaries Journal* 66: 29–44.

Attenni, L. (ed.). 2015. *Studi sulle mura poligonali. Alatri atti del quinto seminario 30–31 ottobre 2010*. Naples: Valtrend Editore.

Attenni, L., and Baldassarre, D. (ed.). 2012. *Quarto seminario internazionale di studi sulle mura poligonali, Palazzo Conti Gentili, 7–10 Ottobre 2009*. Ariccia: Aracne.

Azzena, G. 1987. *Atri: forma e urbanistica*. Rome: L'Erma di Bretschneider.

Bacchielli, L. 1984. "Le porte romane ad ordini sovrapposti e gli antecedenti greci." *Mitteilungen des Deutschen Archaeologischen Instituts, Römische Abteilung* 91: 79–87.

Baiolini, L. 2002. "La forma urbana dell'antica Spello." In *Città dell'Umbria*, edited by L. Quilici and S. Quilici Gigli, 61–101. Rome: L'Erma di Bretschneider.

Balil, A. 1983. "Segni di scalpellino sulle mura romane di Tarragona." *Epigraphica* 45: 231–236.

Barbera, M., and M. Magnani Cianetti. 2008. *Archeologia a Roma Termini: le mura serviane e l'area della stazione: scoperte, distruzioni e restauri*. Milan: Electa.

Barnabei, F. 1901. *La villa pompeiana di P. Fannio Sinistore scoperta presso Boscoreale*. Rome: Accademia dei Lincei.

Barré, L. 1841. *Ercolano e Pompei: raccolta generale de pitture, bronzi, mosaici, ec. fin ora scoperti e riprodotti dietro le antichità di Ercolano, il Museo Borbonico e le opere tutte pubblicate fin qui, accresciute de tavole inedite*. Vol. 5. Venice: Giuseppe Antonelli.

Bartoloni, G., and Michetti, L.M. (ed.). 2013. *Mura di legno, mura di terra, mura di pietra: fortificazioni nel Mediterraneo antico*. Rome: Quasar.

Bastet, F.L. 1975. "Venus in Pompei." *Hermeneus* 47: 63–72.

Beard, M. 2015. *SPQR: A History of Ancient Rome*. New York: Liverlight.

Becatti, G. 1971. *Mosaici e pavimenti marmorei*. Rome: Istituto Poligrafico dello Stato.

Bechi, G. 1851. "Sommario degli scavamenti di Pompei eseguiti nel corso del mese di agosto 1851." *Memorie della Regale Accademia Ercolanese di Archeologia* 7: 39–47.

———. 1852. "Relazione degli scavi di Pompei da agosto 1842 a gennaio 1852." *Reale Museo Borbonico* 14: 1–23.

Becker, J.A. 2007. "The Building Blocks of Empire: Civic Architecture, Central Italy, and the Roman Middle Republic." Ph.D. thesis, University of North Carolina at Chapel Hill, NC.

Bergmann, B. 1991. "Painted Perspectives of a Villa Visit. Landscape as Status and Metaphor." In *Roman Art in the Private Sphere: New Perspectives on the Architecture and Decor of the Domus, Villa, and Insula*, edited by E. Gazda, 49–70. Ann Arbor: University of Michigan Press.

Bernabei, L. 2007. *Contributi di archeologia vesuviana. Raccolta critica della documentazione III, i culti di Pompei.* Rome: L'Erma di Bretschneider.

Bernard, S. 2012. "Continuing the Debate on Rome's Earliest Circuit Walls." *Papers of the British School at Rome* 80: 1–44.

Bianchi, S., and B. Gargiani. 1988. "Chiusi. Orto del Vescovo: Alcune notizie preliminari sugli scavi 1985–1987." In *Archeologia in Valdichiana*, edited by G. Paolucci, 82–85. Rome: Multigrafica.

Blake, M.E. 1947. *Ancient Roman Construction in Italy from the Prehistoric Period to Augustus.* Ann Arbor: University of Michigan Press.

Blanc, N. 1997. "L'énigmatique Sacello Iliaco: contribution à l'étude des cultes domestiques." In *I temi figurativi nella pittura parietale antica (IV sec. A.c.- IV sec. D.c.): atti del VI convegno internazionale sulla pittura parietale antica*, edited by D. Scagliarini Corlàita, 37–42. Imola: University Press Bologna.

Boëthius, A. 1978. *Etruscan and Early Roman Architecture.* New Haven, CT; London: Yale University Press.

Boitani, F., F. Biagi, and S. Neri. 2016. "Le fortificazioni a Veio tra Porta Nord-Ovest e Porta Caere." In *Le fortificazioni arcaiche del Latium vetus e dell'Etruria meridionale (IX-VI Sec. A.c.): Stratigrafia, cronologia e urbanizzazione: Atti delle giornate di studio, Roma, academia belgica, 19–20 settembre 2013*, edited by P. Fontaine and S. Helas, 19–32. Bruxelles: Brepols.

Bonghi Jovino, M. 1984. *Ricerche a Pompei: l'Insula 5 della Regio VI dalle origini al 79 D.c. campagne di scavo 1976–1979.* Rome: L'Erma di Bretschneider.

Bonnet, P. 1980. "Pompéi Quartier des Théatres mémoire du P. Bonnet 1858." In *Pompéi travaux et envois des architects français au XIX siècle*, 309–330. Naples: Macchiaroli.

Bonucci, C. 1830. *Pompéi.* Naples: Imprimerie Française.

Borda, M. 1959. "Il fregio pittorico delle origini di Roma." *Capitolium* 34 (5): 3–10.

Borghi, R. 2002. *Chiusi.* Rome: L'Erma di Bretschneider.

Borrelli, L. 1937. *Le tombe di Pompei a schola semicircolare.* Naples: Torella.

Böttcher-Ebers, J. 2016. "Zur semantischen Funktion des Bogens im Stadttorbau. Ein Vergleich zwischen republikanischen und hellenistischen Stadttoren." In *Focus on Fortifications: New Research on Fortifications in the Ancient Mediterranean and the Near East*, edited by R. Frederiksen, S. Müth, P. Schneider, and M. Schnelle, 277–287. Oxford: Oxbow.

Bragantini, I. 1996a. "VII 1,40: Casa di M. Caesius Blandus." In *Pompei: Pitture e mosaici*, edited by G. Pugliese Carratelli and I. Baldassarre, 6: 380–458. Rome: Istituto della Enciclopedia Italiana.

———. 1996b. "VII 1,25.47 Casa di Sirico." In *Pompei: Pitture e mosaici*, edited by G. Pugliese Carratelli and I. Baldassarre, 6: 228–353. Rome: Istituto della Enciclopedia Italiana.

———. 1997. "VII 7,5 Casa di Trittolemo." In *Pompei: Pitture e mosaici*, edited by G. Pugliese Carratelli and I. Baldassarre, 7: 232–253. Rome: Istituto della Enciclopedia Italiana.

———. 1998. "VIII 3,8–9: Casa del Cinghiale I." In *Pompei: Pitture e mosaici*, edited by G. Pugliese Carratelli and I. Baldassarre, 8: 362–384. Rome: Istituto della Enciclopedia Italiana.

Brandizzi Vitucci, P. 1968. *Cora.* Rome; Florence: Leo S. Olschi.

Brands, G. 1988. *Republikanische Stadttore in Italien.* 458. Oxford: B.A.R.

Brasse, C. 2014. "Mauren im Wandel: Neue Forschungen zum Befestigungssytem von Pompeji." *Antike Welt* 45: 22–28.

Brasse, C., and S. Müth. 2016. "Mauerwerksformen und Mauertechniken." In *Ancient Fortifications: A Compendium of Theory and Practice*, edited by S. Müth, P. Schneider, M. Schnelle, and P. De Staebler, 75–100. Oxford: Oxbow.

Breglia Pulci Doria, L. 1998. "Atena e il mare. Problemi e ipotesi sull'Athenaion di Punta della Campanella." In *I culti della Campania antica. Atti del convegno internazionale di studi in ricordo di Nazarena Valenza Mele*, edited by S.A. Muscettola and G. Greco, 97–108. Rome: Giorgio Bretschneider.

Breton, E. 1855. *Pompeia décrite et dessinée*. Paris: Claye.

———. 1870. *Pompeia*. Paris: Claye.

Brilliant, R. 1984. *Visual Narratives: Storytelling in Etruscan and Roman Art*. Ithaca, NY; London: Cornell University Press.

Briquel, D. 2008. "La città murata: Aspetti religiosi." In *La città murata in Etruria: atti del XXV convegno di studi etruschi ed italici, Chianciano Terme, Sarteano, Chiusi, 30 marzo–3 aprile 2005: in memoria di Massimo Pallottino*, edited by G. Camporeale, 121–134. Pisa: Fabrizio Serra.

Bruno, V.J. 1969. "Antecedents of the Pompeian First Style." *American Journal of Archaeology* 73 (3): 305–317.

Burns, M. 2003. "Pompeii under Siege: A Missile Assemblage from the Social War." *Journal of Roman Military Equipment Studies* 14/15: 1–9.

Caiazza, D. 2009. *Trebula Balliensis: Notizia preliminare degli scavi e restauri 2007–2008–2009*. Alife: Arti Grafiche Grillo.

Calastri, C. 2004. "Una nuova villa con frontone a torrette dall'agro di Cosa." In *Viabilità e insediamenti nell'Italia antica*, edited by L. Quilici and S. Quilici, 173–186. Rome: L'Erma di Bretschneider.

Calderini, A. 1924. *Saggi e studi di antichità, escursioni pompeiane*. Milan: Vita e Pensiero.

Caliò, L.M. 2012. "Dalla polis alla città murata: l'immagine delle fortificazioni nella società ellenistica." *Archeologia Classica* 63 (2): 169–221.

Camporeale, G. 2008. "La città murata d'Etruria nella tradizione letteraria e figurativa." In *La città murata in Etruria: atti del XXV convegno di studi etruschi ed italici, Chianciano Terme, Sarteano, Chiusi, 30 marzo–3 aprile 2005: in memoria di Massimo Pallottino*, edited by G. Camporeale, 15–36. Pisa: Fabrizio Serra.

Cappelli, R. 1998. "The Painted Frieze of the Esquiline and the Augustan Propaganda of the Myth of the Origins of Rome." In *Palazzo Massimo alle Terme*, edited by A. La Regina, 51–58. Milan: Electa.

Carafa, P. 1997. "What was Pompeii before 200 B.C.? Excavations in the House of Joseph II, in the Triangular Forum and in the House of the Wedding of Hercules." In *Sequence and Space in Pompeii*, edited by S.E. Bon and R. Jones, 13–31. Oxford: Oxbow.

———. 2007. "Recent Work on Early Pompeii." In *The World of Pompeii*, edited by P. Foss and J.J. Dobbins, 63–73. New York: Routledge.

———. 2008. *Culti e santuari della Campania antica*. Rome: Istituto Poligrafico.

———. 2011. "Minervae et Marti et Herculi Aedes Doricae Fient (Vitr. 1.2.5). The Monumental History of the Sanctuary in Pompeii's So-Called Triangular Forum." In *The Making of Pompeii*, edited by S.J.R. Ellis, 89–111. Portsmouth, RI: Journal of Roman Archaeology.

Carafa, P., and M.T. D'Alessio. 1999. "Cercando la storia dei monumenti di Pompei. Le ricerche dell'università di Roma 'la Sapienza' nelle Regioni VII e VIII. Il santuario del tempio dorico a Pompei alla luce dei nuovi rinvenimenti." In *Pompei, il Vesuvio e la Penisola Sorrentina. Atti del secondo ciclo di conferenze di geologia, storia e archeologia, Pompei, ottobre 1997–febbraio 1998*, edited by F. Senatore, 17–43. Pompei: Edizioni Rufus.

Carrington, R.C. 1932. "The Etruscans and Pompeii." *Antiquity* 6: 5–23.

Carroll, M. 2008. "Nemus et Templum. Exploring the Sacred Grove at the Temple of Venus in Pompeii." In *Nuove ricerche archeologiche nell'area vesuviana (scavi 2003–2006). Atti del convegno internazionale, Roma 1–3 febbraio 2007*, edited by P.G. Guzzo and M.P. Guidobaldi, 37–45. Rome: L'Erma di Bretschneider.

Cassetta, R., and C. Costantino. 2006. "Parte V. La Casa del Naviglio (VI 10, 11) e le botteghe VI 10, 10 e, VI 10, 12." In *Rileggere Pompei. l'Insula 10 della Regio VI*, edited by F. Coarelli and F. Pesando, 243–333. Rome: L'Erma di Bretschneider.

———. 2008. "Vivere sulle mura. Il caso dell'Insula Occidentalis di Pompei." In *Nuove ricerche archeologiche nell'area vesuviana (scavi 2003–2006). Atti del convegno internazionale, Roma 1–3 febbraio 2007*, edited by P.G. Guzzo and M.P. Guidobaldi, 197–208. Rome: L'Erma di Bretschneider.

Castagnoli, F. 1971. *Orthogonal Town Planning in Antiquity*. Cambridge, MA; London: MIT Press.

———. 1981. "Politica urbanistica di Vespasiano in Roma." In *Atti del congresso internazionale di studi vespasianei: Rieti, settembre 1979*, edited by B. Riposati, 261–275. Rieti: Centro Studi Varroniani.

Castrén, P. 1975. *Ordo Populusque Pompeianus: Polity and Society in Roman Pompeii*. Rome: Bardi.

Castrén, P., R. Berg, and A. Tammisto. 2008. "In the Heart of Pompeii. Archaeological Studies in the Casa di Marco Lucrezio (IX,3, 5.24)." In *Nuove ricerche archeologiche nell'area vesuviana (scavi 2003–2006). Atti del convegno internazionale, Roma 1–3 febbraio 2007*, edited by P.G. Guzzo and M.P. Guidobaldi, 331–340. Rome: L'Erma di Bretschneider.

Cavalieri Manasse, G. 1987. "Verona." In *Il Veneto nell'età romana. Note di urbanistica e di archeologia del territorio*, edited by G. Cavalieri Manasse, 2:3–57. Verona: Banca Popolare di Verona.

———. 1992. "l'Imperatore Claudio e Verona." *Epigraphica* 54: 9–42.

Cerchiai, L. 1995. *I Campani*. Milan: Longanesi.

Cherici, A. 2008. "Mura di bronzo, di legno, di terra, di pietra. Aspetti politici, economici e militari del rapporto tra comunità urbane e territorio nella Grecia e nell'Italia antica." In *La città murata in Etruria: atti del XXV convegno di studi etruschi ed italici, Chianciano Terme, Sarteano, Chiusi, 30 marzo–3 aprile 2005: in memoria di Massimo Pallottino*, edited by G. Camporeale, 37–66. Pisa: Fabrizio Serra.

Chiaramonte, C. 2007. "The Walls and Gates." In *The World of Pompeii*, edited by J.J. Dobbins and P. Foss, 140–149. New York, NY: Routledge.

Chiaramonte Treré, C. 1986. *Nuovi contributi sulle fortificazioni pompeiane*. Milan: Cisalpino-Goliardica.

Ciccone, S. 2015. "La muratura poligonale nello sbocco marittimo di Formia: Criteri tecnici e ambiti d'impiego." In *Studi sulle mura poligonali, Alatri, atti del quinto seminario 30–31 ottobre 2010*, edited by L. Attenni, 81–96. Naples: Valtrend.

Cifani, G. 2008. *Architettura romana arcaica: Edilizia e società tra monarchia e repubblica*. Rome: L'Erma di Bretschneider.

———. 2010. "I grandi cantieri della Roma arcaica: Aspetti tecnici e organizzativi." In *Arqueología de la construcción. Los procesos constructivos en el mundo romano: Italia y provincias orientales (congreso, Certosa di Pontignano, Siena, 13–15 de noviembre de 2008)*, edited by S. Camporeale, H. Dessales, and A. Pizzo, II:35–49. Madrid: Consejo Superior de Investigaciones Científicas, Instituto de Arqueología.

————. 2016. "The Fortifications of Archaic Rome: Social and Political Significance." In *Focus on Fortifications: New Research on Fortifications in the Ancient Mediterranean and the Near East*, edited by R. Frederiksen, S. Müth, P. Schneider, and M. Schnelle, 82–91. Oxford: Oxbow.

Cifani, G., and S. Fogagnolo. 1998. "La documentazione archeologica delle mura arcaiche a Roma." *Mitteilungen des Deutschen Archäologischen Instituts, Römische Abteilung* 105: 359–389.

Cifarelli, F.M. 1992. "Il recinto urbano." In *Segni*, edited by G.M. De Rossi, 9–59. Salerno: Università di Salerno.

————. 2012. "Tecniche costruttive nel Lazio del tardo ellenismo: la cd. opera poligonale di IV maniera bugnata." In *Quarto seminario internazionale di studi sulle mura poligonali, Palazzo Conti Gentili, 7–10 Ottobre 2009*, edited by L. Attenni and D. Baldassarre, 295–301. Ariccia: Aracne.

————. 2013. "Tecniche costruttive del tardo ellenismo a Segni: verso una sintesi." In *Tecniche costruttive del tardo ellenismo nel Lazio e in Campania (Atti del convegno, Segni 2 dicembre 2011)*, edited by F.M. Cifarelli, 43–54. Rome: Espera.

Cipriani, M., and A. Pontrandolfo Greco. 2010. *Paestum: scavi, ricerche, restauri. I, le mura: il tratto da Porta Sirena alla Postierla 47*. Paestum: Pandemos.

Clark, W. 1831. *Pompei*. London: Charles Knight.

Clarke, J.R. 1979. *Roman Black-and-White Figural Mosaics*. New York: New York University Press.

————. 1996. "Landscape Paintings in the Villa of Oplontis." *Journal of Roman Archaeology* 9: 81–107.

————. 2003. *Art in the Lives of Ordinary Romans: Visual Representation and Non-Elite Viewers in Italy, 100 B.C.–A.D. 315*. Berkeley: California University Press.

Coarelli, F. 1987. *I santuari del Lazio in età repubblicana*. Rome: La Nuova Italia Scientifica.

————. 1995. "Le mura regie e repubblicane." In *Mura e porte di Roma antica*, edited by B. Brizzi, 8–38. Rome: Editore Colombo.

————. 1998. "Il culto di Mefitis in Campania e a Roma." In *I culti della Campania antica. Atti del convegno internazionale di studi in ricordo di Nazarena Valenza Mele, Napoli 15–17 maggio 1995*, edited by S.A. Muscettola and G. Greco, 185–190. Rome: Giorgio Bretschneider.

————. 2000. "Pompei. Il foro, le elezioni, le circoscrizioni elettorali." *Annali di Archeologia e Storia Antica. Istituto Universitario Orientale. Dipartimento di Studi del Mondo Classico e del Mediterraneo Antico* n.s.7: 87–111.

————. 2002. *Pompeii*. New York: Barnes & Noble.

————. 2004. "Le porte di Perusia." In *Stadttore: Bautyp und Kunstform; Akten der Tagung in Toledo vom 25. bis 27. September 2003*, edited by T.G. Schattner and F. Valdés Fernández, 79–87. Mainz am Rhein: Von Zabern.

————. 2008. "Il settore nord-occidentale di Pompei e lo sviluppo urbanistico della città dall'età arcaica al III secolo A.c." In *Nuove ricerche archeologiche nell'area vesuviana (scavi 2003–2006), atti del convegno internazionale, Roma 1–3 febbraio 2007*, edited by P.G. Guzzo and M.P. Guidobaldi, 173–176. Rome: L'Erma di Bretschneider.

————. 2010a. "Navalia pompeiana." In *Dall'immagine alla storia: Studi per ricordare Stefania Adamo Muscettola*, edited by C. Gasparri, G. Greco, and R. Pierobon, 437–450. Pozzuoli: Naus Editoria.

————. 2010b. "Substructio et Tabularium." *Papers of the British School at Rome* 78: 107–132.

Coarelli, F., and F. Pesando. 2011. "The Urban Development of NW Pompeii: Archaic Period to 3rd c.B.C." In *The Making of Pompeii*, edited by S.J.R. Ellis, 37–58. Portsmouth, RI: Journal of Roman Archaeology.

Coarelli, F., F. Pesando, M. Verzár-Bass, and F. Oriolo. 2006. *Rileggere Pompei*. Rome: L'Erma di Bretschneider.

Colaiacomo, F. 2015. "Il circuito murario di Segni in età medievale: Topografia e tecniche costruttive." In *Studi sulle mura poligonali. Alatri atti del quinto seminario 30–31 ottobre 2010*, edited by L. Attenni, 61–72. Naples: Valtrend.

Coletti, F., S. Prascina, H. Sterpa, and N. Witte. 2010. "Venus Pompeiana: Scelte progettuali procedimenti tecnici per la realizzazione di un grande edificio sacro tra tarda repubblica e primo impero." In *Arqueología de la construción. 2. Los procesos constructivos en el mundo romano: Italia y provincias orientales: (Certosa di Pontignano, Siena, 13–15 de noviembre 2008)*, edited by S. Camporeale, H. Dessales, and A. Pizzo, 189–211. Madrid: Consejo Superior de Investigaciones Científicas, Instituto de Arqueología.

Colin, M.G., ed. 1987. *Les enceintes augustéennes dans l'occident romain: (France, Italie, Espagne, Afrique du Nord): Actes du colloque de Nîmes (IIIe congrès archéologique de Gaule méridionale), 9–12*. Nîmes: Musée Archéologique.

Colonna, G. 1962. "Saepinum. Ricerche di topografia sannittica e medievale." *Archeologia Classica* 14 (1): 80–107.

Conta Haller, G. 1978. *Ricerche su alcuni centri fortificati in opera poligonale in area campano-sannitica*. Naples: Arte Tipografica.

Conticello De' Spagnolis, M. 1994. *Il Pons Sarni di Scafati e la via Nuceria-Pompeios*. Rome: L'Erma di Bretschneider.

Conway, R.S. 1897. *The Italic Dialects*. Cambridge: Clay and Sons.

Conwell, D.H. 2008. *Connecting a City to the Sea: The History of the Athenian Long Walls*. Leiden: Brill.

Cooley, A., and M.G.L. Cooley. 2004. *Pompeii: A Sourcebook*. London; New York: Routledge.

Coralini, A. 2001. "I pavimenti della Casa del Centenario a Pompei (IX 8, 3, 6, A). I temi figurati." In *Atti del VII colloquio dell'associazione italiana per lo studio e la conservazione del mosaico: Pompei, 22–25 marzo 2000*, edited by R. Paribeni, 45–60. Ravenna: Edizioni del Girasole.

Cormack, S. 2007. "The Tombs at Pompeii." In *The World of Pompeii*, edited by P. Foss and J.J. Dobbins, 585–606. New York: Routledge.

Cornell, T. 1995. "Warfare and Urbanization in Roman Italy." In *Urban Society in Roman Italy*, edited by T.J. Cornell and K. Lomas, 121–134. London: Routledge.

Cotugno, A., A. Maiuri, U. Pappalardo, and F. de Sanctis. 2009. *Il fondo bibliografico di Amedeo Maiuri: libri, carteggio e cimeli di un grande archeologo*. Naples: L'Orientale.

Cristofani, M. 1988. "Economia e società." In *Rasenna: Storia e civiltà degli Etruschi*, edited by M. Pallottino and G. Pugliese Carratelli, 83–138. Milan: Scheiwiller.

———. 1991. "La fase etrusca di Pompei." In *Pompei*, edited by F. Zevi, 1:9–22. Naples: Banco di Napoli.

Curti, E. 2005. "Le aree portuali di Pompei. Ipotesi di lavoro." In *Moregine: Suburbio "portuale" di Pompei*, edited by V. Scarano Ussani and P.G. Guzzo, 51–76. Naples: Loffredo.

———. 2007. "La Venere Fisica trionfante: Un nuovo ciclo di iscrizioni dal santuario di Venere a Pompei." In *Il filo e le perle. Studi per i 70 anni di Mario Torelli*, 67–77. Venosa: Osanna.

————. 2008. "Il tempio di Venere Fisica e il porto di Pompei." In *Nuove ricerche archeologiche nell'area vesuviana (scavi 2003–2006). Atti del convegno internazionale, Roma 1–3 febbraio 2007*, edited by P.G. Guzzo and M.P. Guidobaldi, 47–60. Rome: L'Erma di Bretschneider.

————. 2009. "Spazio sacro e politico nella Pompei preromana." In *Verso la città. Forme insediative in Lucania e nel mondo italico fra IV e III Sec. A.c. atti delle giornate di studio, Venosa, 13–14 Maggio 2006*, edited by M. Osanna, 497–511. Venosa: Osanna.

Curuni, S.A. 2012. "Le mura poligonali (megalitiche): I casi di del paleócastro di Nisyros (Grecia) e dell'acropoli di Butrinto (Albania)." In *Quarto seminario internazionale di studi sulle mura poligonali, Palazzo Conti Gentili, 7–10 Ottobre 2009*, edited by L. Attenni and D. Baldassarre, 9–19. Ariccia: Aracne.

D'Agostino, B. 2013. "Le fortificazioni di Cuma." In *Mura di legno, mura di terra, mura di pietra: Fortificazioni nel Mediterraneo antico*, edited by G. Bartoloni and L.M. Michetti, 207–227. Rome: Quasar.

D'Agostino, B., J. De Waele, and L. A. Scatozza Höricht. "La cronologia." In *Il tempio dorico del Foro Triangolare di Pompei*, edited by J.A.K.E. De Waele, 336–374. Rome: L'Erma di Bretschneider.

D'Agostino, B., F. Fratta, V. Malpede, M. Cuozzo, and L. Del Verme. 2005. *Cuma: le fortificazioni*. Naples: Università degli Studi di Napoli "L'Orientale."

D'Alessio, M.T. 2001. *Materiali votivi dal Foro Triangolare di Pompei*. Rome: Giorgio Bretschneider.

————. 2005. "Nuovi materiali votivi dal tempio dorico di Pompei." In *Depositi votivi e culti dell'Italia antica dell'età arcaica a quella tardo-repubblicana. Atti del convegno di studi, Perugia, 1–4 giugno 2000*, edited by A. Comella and S. Mele, 535–543. Bari: Edipuglia.

D'Ambrosio, A. 1984. *La stipe votiva in Località Bottaro (Pompei)*. Naples: Giannini.

D'Ambrosio, A., and S. De Caro. 1983. *Un impegno per Pompei. Fotopiano e documentazione della necropoli di Porta Nocera*. Milan: Touring Club Italiano.

D'Ambrosio, I. 1990. "Le fortificazioni di Poseidonia-Paestum. Problemi e prospettive di ricerca." *Annali di Archeologia e Storia Antica. Istituto Universitario Orientale. Dipartimento di Studi del Mondo Classico e del Mediterraneo Antico* 12: 71–101.

D'Auria, D. 2008. "Tratto dell'agger a nord dell'Insula VI 2." *Rivista di Studi Pompeiani* 19: 103–106.

De Caro, S. 1979a. "Nuovi rinvenimenti e vecchie scoperte nella necropoli sannitica di Porta Ercolano." *Cronache Pompeiane* 5: 179–190.

————. 1979b. "Scavi nell'area fuori Porta Nola a Pompei." *Cronache Pompeiane* 5: 61–101.

————. 1985. "Nuove indagini sulle fortificazioni di Pompei." *Annali di Archeologia e Storia Antica. Istituto Universitario Orientale. Dipartimento di Studi del Mondo Classico e del Mediterraneo Antico* 7: 75–114.

————. 1986. *Saggi nell'area del tempio di Apollo a Pompei. Scavi stratigrafici di A. Maiuri nel 1931–32 e 1942–43*. Naples: Istituto Universitario Orientale.

————. 1991. "La città sannitica. Urbanistica e architettura." In *Pompei*, edited by F. Zevi, 1:23–46. Naples: Banco di Napoli.

————. 1992a. "Appunti sull'Atena della Punta della Campanella." *Annali di Archeologia e Storia Antica. Istituto Universitario Orientale. Dipartimento di Studi del Mondo Classico e del Mediterraneo Antico* 14: 173–178.

————. 1992b. "Lo sviluppo urbanistico di Pompei." *Atti e Memorie della Società Magna Grecia* 3: 67–90.

252   *Bibliography*

———. 1998. "La lucerna d'oro di Pompei. Un dono di Nerone a Venere Pompeiana." In *I culti della Campania antica. Atti del convegno internazionale di studi in ricordo di Nazarena Valenza Mele, Napoli 15–17 Maggio 1995*, edited by S.A. Muscettola and G. Greco, 239–244. Rome: Giorgio Bretschneider.

———. 2007. "The First Sanctuaries." In *The World of Pompeii*, edited by P. Foss and J.J. Dobbins, 73–81. New York: Routledge.

De Caro, S., and D. Giampaola. 2008. "La circolazione stradale a Neapolis e nel suo territorio." In *Stadtverkehr in der Antiken Welt: Internationales Kolloquium zur 175-Jahrfeier des Deutschen Archäologischen Instituts Rom, 21. Bis 23. April 2004*, edited by M. Deiter, 107–124. Weisbaden: Riechert.

De Clarac, C.O.F.B., and F. Mori. 1813. *Fouille faite à Pompei en présence de s.m. la Reine des Deux Siciles, le 18 mars 1813*. Paris.

De Haas, T., and P. Attema. 2016. "The Pontine Region under the Early Republic." In *Focus on Fortifications: New Research on Fortifications in the Ancient Mediterranean and the Near East*, edited by R. Frederiksen, S. Müth, P. Schneider, and M. Schnelle, 251–262. Oxford: Oxbow.

De Jorio, A. 1828. *Plan de Pompei*. Naples: Imprimerie Française.

De Maria, S. 1979. "La Porta Augustea di Rimini nel quadro degli archi commemorativi coevi dati strutturali." In *Studi sull'arco onorario romano*, edited by G.A. Mansuelli, 59–91. Rome: L'Erma di Bretschneider.

———. 1991. *Gli archi onorari di Roma e dell'Italia romana*. Rome: L'Erma di Bretschneider.

———. 1996. "Mosaici di Suasa. Tipi, fasi, botteghe." In *Atti del III colloquio dell'Associazione italiana per lo studio e la conservazione del mosaico, Bordighera, 6–10 dicembre 1995*, edited by F. Guidobaldi and A. Guiglia Guidobaldi, 401–424. Bordighera: Istituto Internazionale di Studi Liguri.

De Rossi, G.M. 2009a. "Il divenire delle mura megalitiche." In *Le mura megalitiche: il Lazio meridionale tra storia e mito*, edited by A. Nicosia and M.C. Bettini, 41–53. Rome: Gangemi.

———. 2009b. "Il cantiere e la tecnica." In *Le mura megalitiche: il Lazio meridionale tra storia e mito*, edited by A. Nicosia and M.C. Bettini, 55–73. Rome: Gangemi.

De Vos, M. 1977. "Primo stile figurato e maturo quarto stile negli scarichi provenienti dalle macerie del terremoto del 62 D.c. a Pompei." *Mededelingen van het Nederlands Instituut te Rome* 39: 29–47.

De Waele, J.A.K.E. 1993. "The Doric Temple on the Forum Triangolare in Pompeii." *Opuscula pompeiana* 3: 105–118.

———. 2001. "La ricostruzione." In *Il tempio dorico del Foro Triangolare di Pompei*, edited by J.A.K.E. De Waele, 111–132. Rome: L'Erma di Bretschneider.

Defosse, P. 1980. "Les remparts de Perouse." *Mélanges de l'Ecole Française de Rome* 92: 760–819.

Degering, H. 1898. "Uber die Militarischen Wegweiser in Pompeji." *Mitteilungen des Deutschen Archäologischen Instituts, Römische Abteilung* 13: 124–146.

Delaunay, E. 1877. *Une promenade à Pompéi*. Scafati: Tipografia Pompeiana.

Della Corte, M. 1913. "Il pomerium di Pompei." *Rendiconti della reale accademia dei lincei. Classe di scienze morali, storiche e filologiche* 22: 261–308.

———. 1916. "La necropoli sannitico-romana scoperta fuori Porta Stabia." *Notizie degli scavi di antichità* 9: 287–305.

———. 1921a. "Dipinti pompeiani." *Ausonia* 10: 64–87.

————. 1921b. "Il Pagus Urbulanus ed i nomi antichi di alcune porte di Pompei." *Rivista Indo Greco Italica* 5: 87–88.

————. 1965. *Case ed abitanti di Pompei.* Naples: Fiorentino.

Denti, M. 2004. "Scultori greci a Poseidonia all'epoca di Alessandro il Molosso. Il tempio corinzio-dorico e i Lucani. Osservazioni preliminari." In *Alessandro il Molosso e i condottieri in Magna Grecia. Atti del quarantatreesimo convegno di studi sulla Magna Grecia, Taranto–Cosenza 26–30 settembre 2003*, edited by A. Stazio and S. Ceccoli, 665–697. Taranto: Istituto per la storia e l'archeologia della Magna Grecia.

Descoeudres, J.P. 1998. "The so-called Quay Wall North West of the Porta Marina." *Rivista di Studi Pompeiani* 9: 210–216.

————. 2007. "History and Historical Sources." In *The World of Pompeii*, edited by P. Foss and J.J. Dobbins, 9–27. New York: Routledge.

Devore, G., and S.J.R. Ellis. 2008. "The Third Season of Excavations at VIII.7.1–15 and the Porta Stabia at Pompeii. Preliminary Report." *Fastionline* 112: 1–15.

Dickmann, J.A., and F. Pirson. 2002. "Die Casa dei Postumii in Pompeji und ihre Insula." *Mitteilungen des Deutschen Archäologischen Instituts. Römische Abteilung* 109: 243–313.

————. 2005. "Il progetto Casa dei Postumii. Un complesso architettonico a Pompei come esemplificazione della storia dell'insediamento, del suo sviluppo e delle sue concezioni urbanistiche." In *Nuove ricerche archeologiche a Pompei ed Ercolano. Atti del convegno internazionale, Roma 28–30 novembre 2002*, edited by P.G. Guzzo and M.P. Guidobaldi, 10:156–169. Naples: Electa.

Di Maio, G. 2014. "Il paesaggio archeologico della costa di Oplonti." In *Oplontis: Villa A ("of Poppaea") at Torre Annunziata, Italy*, edited by J.R. Clarke and N. Muntasser, para. 693–702. Ann Arbor: University of Michigan Press.

Dio Cassius, Cocceianus. 1990. *Dio's Roman History: In Nine Volumes.* Translated by E. Cary. Cambridge, MA: Harvard University Press.

Diodorus Siculus. 1963. *Diodorus of Sicily.* Translated by C.H. Oldfather, C.L. Sherman, C. Bradford Welles, R.M. Geer, and F.R. Walton. Cambridge, MA: Harvard University Press.

Dionysios Halikarnasseus. 1937. *The Roman Antiquities.* Translated by E. Cary. Cambridge, MA: Harvard University Press.

Ducrey, P. 2016. "Defence, Attack and the Fate of the Defeated: Reappraising the Role of City-Walls." In *Focus on Fortifications: New Research on Fortifications in the Ancient Mediterranean and the Near East*, edited by R. Frederiksen, S. Müth, P. Schneider, and M. Schnelle, 332–337. Oxford: Oxbow.

Dyer, T.H. 1867. *Pompeii: Its History, Buildings, and Antiquities; an Account of the Destruction of the City with a Full Description of the Remains, and of the Recent Excavations, and also an Itinerary for Visitors.* London: Bell and Daldy.

Elia, O. 1975. "La scultura pompeiana in tufo." *Cronache Pompeiane* 1: 118–143.

Ellis, S.J.R. 2011. "The Rise and Re-Organization of the Pompeian Salted Fish Industry." In *The Making of Pompeii: Studies in the History and Urban Development of an Ancient Town*, edited by S.J.R. Ellis, 59–88. Portsmouth, RI: Journal of Roman Archaeology.

Ellis, S.J.R., and G. Devore. 2006. "Towards an Understanding of the Shape of Space at VIII.7.1–15, Pompeii. Preliminary Results from the 2006 Season." *Fastionline* 71: 1–15.

————. 2007. "Two Seasons of Excavations at VIII.7.1–15 and the Porta Stabia at Pompeii, 2005–2006." *Rivista di Studi Pompeiani* 18: 119–128.

————. 2010. "The Fifth Season of Excavations at VIII.7.1–15 and the Porta Stabia at Pompeii: Preliminary Report." *Fastionline* 202: 1–21.

Ellis, S.J.R., A. Emmerson, A. Pavlick, and K. Dicus. 2011. "The 2010 Field Season at I.1.1–10, Pompeii: Preliminary Report on the Excavations." *Fastionline* 220: 1–17.

Emmerson, A. 2010. "Reconstructing the Funerary Landscape at Pompeii's Porta Stabia." *Rivista di Studi Pompeiani* 21: 77–86.

Eschebach, H. 1970. *Die Städtebauliche Entwicklung des antiken Pompeji: mit einem Plan 1:1000 und einem Exkurs: Die Baugeschichte der Stabianer Thermen.* Heidelberg: Kerle.

Eschebach, H., and L. Eschebach. 1995. *Pompeji vom 7. Jahrhundert v.Chr. bis 79 n.Chr.* Cologne: Böhlau.

Eschebach, L., and J. Müller-Trollius. 1993. *Gebäudeverzeichnis und Stadtplan der antiken Stadt Pompeji.* Cologne: Böhlau.

Esposito, D. 2008. "Un contributo allo studio di Pompei arcaica. I saggi nella Regio V, Ins. 5 (Casa dei Gladiatori)." In *Nuove ricerche archeologiche nell'area vesuviana (scavi 2003–2006). Atti del convegno internazionale, Roma 1–3 febbraio 2007*, edited by P.G. Guzzo and M.P. Guidobaldi, 71–80. Rome: L'Erma di Bretschneider.

Etani, H. 2010. *Pompeii: Report of the Excavation at Porta Capua, 1993–2005.* Kyoto: Paleological Association of Japan.

Etani, H., and S. Sakai. 1995. "Preliminary Reports. Archaeological Investigation at Porta Capua, Pompeii. Second Season, September–December 1994." *Opuscula Pompeiana* 5: 53–67.

———. 1998. "Preliminary Reports. Archaeological Investigation at Porta Capua, Pompeii. Fifth Season, September–January 1997–98." *Opuscula Pompeiana* 8: 111–134.

———. 1999. "Rapporto preliminare. Indagine archeologica a Porta Capua, Pompei. Sesta campagna di Scavo, 26 Ottobre–11 Dicembre 1998." *Opuscula Pompeiana* 9: 120–136.

———. 2003. "Rapporto preliminare. Indagine archeologica a Porta Capua, Pompei. Settima campagna di scavo. 7 ott. 2002–27 genn. 2003." *Opuscula Pompeiana* 12: 123–137.

Facenna, D., C.F. Giuliani, C. Beltrame, P. Rockwell, and S. Agostini. 2001. "I rilievi Torlonia del Fucino." In *Il tesoro del lago: l'Archeologia del Fucino e la collezione Torlonia*, edited by A. Campanelli and S. Agostini, 34–45. Pescara: Carsa.

Favro, D. 2006a. "A City in Flux. The Animated Boundaries of Ancient Rome." In *Proceedings of the XVIth International Congress of Classical Archaeology: Boston, August 23–26, 2003*, 191–195. Oxford: Oxbow.

———. 2006b. "The iconiCITY of Ancient Rome." *Urban History* 33 (1): 20–38.

Fentress, E. 2000. "Introduction: Frank Brown, Cosa, and the Idea of a Roman City." In *Romanization and the City: Creation, Transformations, and Failures: Proceedings of a Conference Held at the American Academy in Rome to Celebrate the 50th Anniversary of the Excavations at Cosa, 14–16 May, 1998*, edited by E. Fentress, 11–24. Portsmouth, RI: Journal of Roman Archaeology.

Février, P.A. 1969. "Enceinte et colonie. De Nîmes à Vérone, Toulouse et Tipasa." *Rivista di Studi Liguri* 35: 277–286.

Fields, N., and B. Delf. 2010. *Ancient Greek Fortifications 500–300 BC.* Oxford: Osprey.

Fiorelli, G. 1860. *Pompeianarum antiquitatum historia.* Vol. 1. Naples.

———. 1861. *Giornale degli scavi di Pompei fasc. 1–4, 8–10, 13–15.* Naples: Tipografia Italiana.

———. 1862. *Pompeianarum antiquitatum historia.* Vol. 2. Naples.

———. 1864. *Pompeianarum antiquitatum historia.* Vol. 3. Naples.

———. 1873. *Gli scavi di Pompei dal 1861 al 1872.* Naples: Tipografia Italiana.

———. 1875. *Descrizione di Pompeii.* Naples: Tipografia Italiana.

Fischetti, L., and L. Conforti. 1907. *Pompei Past and Present*. Milan: Beccarini.

Fitzgerald Marriott, H.P. 1895. *Facts about Pompei: Its Masons' Marks, Town Walls, Houses, and Portraits*. London: Hazell, Watson, & Viney.

Fontaine, P. 2013. "Les enceintes préromaines de l'Italie centrale. Traditions régionales et influences extérieures (VIIIe s.–IIe S.)." In *Mura di legno, mura di terra, mura di pietra: Fortificazioni nel Mediterraneo antico*, edited by G. Bartoloni and L.M. Michetti, 267–295. Rome: Quasar.

Fontaine, P., and Helas, S. (ed.). 2016. *Le fortificazioni arcaiche del Latium vetus e dell'Etruria meridionale (IX–VI Sec. A.c.): Stratigrafia, cronologia e urbanizzazione: Atti delle giornate di studio, Roma, Academia Belgica, 19–20 settembre 2013*. Bruxelles: Brepols.

Forlati Tamaro, B. 1965. "Il restauro della Porta Leoni." *Notizie degli scavi di antichitá* 19s: 12–34.

Fowler, W. 1966. "Lustratio." In *Anthropology and the Classics; Six Lectures Delivered Before the University of Oxford*, 169–191. New York: Barnes & Noble.

Frederiksen, R., H. Lauter, and S. Müth. 2016. "Source Criticism: Fortifications in Written and the Visual Arts." In *Ancient Fortifications: A Compendium of Theory and Practice*, edited by S. Müth, P. Schneider, M. Schnelle, and P. De Staebler, 173–196. Oxford: Oxbow.

Frigerio, F. 1935. *La cerchia di Novum Comum. Antiche porte di città italiche e romane*. Como: Tipografia Editrice Cesare Nani.

Fröhlich, T. 1995. "La Porta di Ercolano a Pompei e la cronologia dell'opus vittatum mixtum." In *Archäologie und Seismologie. La regione vesuviana dal 62 Al 79 d.C. problemi archeologici e sismologici. Colloquium, Boscoreale 26.–27. novembre 1993*, edited by T. Fröhlich and L. Jacobelli, 153–159. Munich: Biering & Brinkmann.

Fulford, M., and A. Wallace-Hadrill. 1999. "Towards a History of Pre-Roman Pompeii. Excavations beneath the House of Amarantus (I.9.11–12), 1995–98." *Papers of the British School at Rome* 67: 37–144.

Fumagalli, P. 1828. *Pompeia: trattato pittorico, storico e geometrico: opera disegnata negli anni 1824 al 1827*. Florence.

Futrell, A. 1997. *Blood in the Arena: The Spectacle of Roman Power*. Austin: University of Texas Press.

Gabba, E. 1972. "Urbanizzazione e rinnovamenti urbanistici nell'Italia centro-meridionale del I sec. a.C." *Studi Classici e Orientali* 21; 73–112.

Gaius. 2009. *Digest of Justinian*. Translated by A. Watson. Philadelphia, PA: University of Pennsylvania Press.

Gall, J. le. 1975. "Les romaines et l'orientation solaire." *Mélanges de l'École Français de Rome* 87: 287–320.

Galsterer, H. 2006. "Coloni, galli ed autoctoni." In *Rimini e l'Adriatico nell'età delle guerre puniche: atti del convegno internazionale di studi Rimini, musei comunali, 25–27 marzo 2004*, edited by F. Lenzi, 11–18. Bologna: Ante Quem.

García y García, L. 1993. "Divisione fiorelliana e piano regolatore di Pompei." *Opuscula pompeiana* 3: 55–70.

———. 2006. *Danni di guerra a Pompei. Una dolorosa vicenda quasi dimenticata. Con numerose notizie sul museo pompeiano distrutto nel 1943*. Rome: L'Erma di Bretschneider.

Garlan, Y. 1968. "Fortifications et histoire Grècque." In *Problèmes de la guerre en Grèce ancienne*, edited by J.P. Vernant, 245–260. Paris: Mouton.

Garrison, J.D. 1992. *Pietas from Vergil to Dryden*. University Park: Pennsylvania State University Press.

Garrucci, P.R. 1851. "Intorno ad un iscrizione osca recentemente scavata in Pompei." *Estratto della Regale Accademia Ercolanese di Archeologia* 7: 21–38.

———. 1852. "Intorno ad una lapide viaria osca di Pompei, nuove osservazioni." *Bullettino Archeologico Napoletano* 11: 81–84.

Garzia, D. 2008. "Pompei. Regio VI, Insula 2. Aggere. Relazione di scavo settembre 2007." *Fastionline* 122: 1–3.

Gasparini, V., and J. Uroz Sàez. 2012. "Las murallas de Pompeya. Resultados del sondeo efectuado en Porta Nocera (2010) y su contextualizacion." *Vesuviana* 4: 9–68.

Gat, A. 2002. "Why City States Existed? Riddles and Clues of Urbanisation and Fortifications." In *A Comparative Study of Six City-State Cultures; an Investigation Conducted by the Copenhagen Polis Centre*, edited by M.H. Hansen, 125–139. Copenhagen: Det Kongelige Danske Videnskabernes Selskab.

Geertman, H. 2001. "Lo studio della città antica. Vecchi e nuovi approcci." In *Pompei: Scienza e società: 250 anniversario degli scavi di Pompei, convegno internazionale, Napoli, 25–27 novembre 1998*, edited by P.G. Guzzo, 131–135. Milan: Electa.

———. 2007. "The Urban Development of the Pre-Roman City." In *The World of Pompeii*, edited by J.J. Dobbins and P. Foss, 82–97. New York: Routledge.

Geis, Marion. 2006. "Die Stadttore von Thasos: mythologische Reliefs zwischen Archaik und Klassik." Ph.D. thesis, Universität Hamburg, Germany.

Gell, W. 1832. *Pompeiana: The Topography, Edifices and Ornaments of Pompeii*. London: Henry Bohn.

Gell, W., and J.P. Gandy. 1833. *Pompeiana: The Topography, Edifices, and Ornaments of Pompeii*. London: Henry Bohn.

Gellius, A. 1927. *The Attic Nights of Aulus Gellius*. Translated by J.C. Rolfe. New York: G.P. Putnam's Sons.

Giglio, M. 2016. "Considerazioni sull'impianto urbanistico di Pompeii." *Vesuviana* 8: 11–48.

Gisborne, M. 2005. "A Curia of Kings: Sulla and Royal Imagery." In *Imaginary Kings: Royal Images in the Ancient Near East, Greece and Rome*, edited by O. Hekster and R. Fowler, 105–123. Stuttgart: Steiner.

Giuliani, C. 1996. "Lacus Curtius." In *Lexicon topographicum urbis Romae*, edited by E.M. Steinby, 3:166–167. Rome: Quasar.

Goalen, M. 1995. "The Idea of the City and the Excavations at Pompeii." In *Urban Society in Roman Italy*, edited by T.J. Cornell and K. Lomas, 181–202. London: Routledge.

Goldsworthy, A.K. 2003. *In the Name of Rome: The Men Who Won the Roman Empire*. New Haven, CT: Yale University Press.

Graefe, R. 1979. *Vela erunt: Die Zeltdächer der römischen Theater und ähnlicher Anlagen*. Mainz am Rhein: Von Zabern.

Greco, E., and F. Longo. 2008. *Magna Grecia*. Rome: GLF editori Laterza.

Greco, E., F. Longo, D. Marchiandi, R. Di Cesare, and M.C. Monaco. 2014. "I quartieri occidentali." In *Topografia di Atene: sviluppo urbano e monumenti dalle origini al III secolo d.C. Tomo 4*, edited by E. Greco, 1271–1281. Athens; Paestum: Pandemos.

Greco, G. 2014. "Elea, che i romani chiamarono Velia." *Territori della cultura* 18: 16–37.

Grimaldi, M. 2014. *Pompei: la casa di Marco Fabio Rufo*. Naples: Valtrend.

Gros, P. 1992. "Moenia: Aspects defénsifs et aspects représentatifs des fortifications." In *Fortificationes Antiquae. Including the Papers of a Conference Held at Ottawa University, October 1988*, edited by S. van de Maele and J.M. Fossey, 211–224. Amsterdam: Gieben.

———. 1996. *L'architecture romaine: du début du IIIe siècle Av. J.-C. à la fin du haut-empire*. Paris: Picard.

Grünewald, T. 2004. *Bandits in the Roman Empire: Myth and Reality*. New York: Routledge.

Grünhagen, W. 1976. "Bemerkungen zum Minerva-Relief in der Stadtmauer von Tarragona." *Madrider Mitteilungen* 17: 209–225.

Gusman, P. 1900. *Pompei, the City, its Life & Art*. London: Heinemann.

Gustafsson, G. 2000. *Evocatio Deorum: Historical and Mythical Interpretations of Ritualised Conquests in the Expansion of Ancient Rome*. Uppsala: Uppsala University.

Guzzo, P.G. 2000. "Alla ricerca della Pompei sannitica." In *Studi sull'Italia dei Sanniti*, edited by R. Cappelli, 107–117. Milan: Electa.

———. 2007. *Pompei. Storia e paesaggi della città antica*. Milan: Electa.

Guzzo, P.G., and Guidobaldi, M.P. (ed.) 2008. *Nuove ricerche archeologiche nell'area vesuviana (scavi 2003–2006). Atti del convegno internazionale, Roma 1–3 febbraio 2007*. Rome: L'Erma di Bretschneider.

Habinek, T.N. 1998. *The Politics of Latin Literature Writing, Identity, and Empire in Ancient Rome*. Princeton, NJ: Princeton University Press.

Hackworth Petersen, L. 2006. *The Freedman in Roman Art and Art History*. Cambridge: Cambridge University Press.

Hansen, M.H. 2000a. "The Concepts of City-State and City-State Culture." In *A Comparative Study of Thirty City-State Cultures: An Investigation Conducted by the Copenhagen Polis Centre*, edited by M.H. Hansen, 11–34. Copenhagen: Kongelige Danske Videnskabernes Selskab.

———. 2000b. "The Hellenic Polis." In *A Comparative Study of Thirty City-State Cultures: An Investigation Conducted by the Copenhagen Polis Centre*, edited by M.H. Hansen, 141–188. Copenhagen: Kongelige Danske Videnskabernes Selskab.

Hartnett, J. 2008. "Si Quis Hic Sederit: Streetside Benches and Urban Society in Pompeii." *American Journal of Archaeology* 112 (1): 91–119.

Hauschild, T. 1979. "Die römische Stadtmauer von Tarragona." *Madrider Mitteilungen* 20: 204–250.

———. 1985. "Ausgrabungen in der römischen Stadtmauer von Tarragona. Torre de Minerva (1979) und die Torre de Cabiscol (1983)." *Madrider Mitteilungen* 26: 75–90.

———. 2006. "Die römische Tore des 2. Jhs. V. Chr. In der Stadtmauer von Tarragona." In *Stadttore: Bautyp und Kunstform: Akten der Tagung in Toledo vom 25. bis 27. September 2003*, edited by T.G. Schattner, 153–171. Mainz am Rhein: Von Zabern.

Haverfield, F. 1913. *Ancient Town-Planning*. Oxford: Clarendon Press.

Hellmann, M. 2010. *L'Architecture grecque. Habitat, urbanisme et fortifications*. Paris: Picard.

Henzen, G. 1852. "Iscrizione osca scoperta a Pompei." *Bullettino dell'Istituto di Corrispondenza Archeologica* 6: 87–91.

Hoffmann, A. 1979. "L'Architettura." In *Pompei 79. Raccolta di studi per il decimonono centenario dell'eruzione Vesuviana*, edited by F. Zevi, 111–115. Naples: Banco di Napoli.

———. 1990. "Elemente bürgerlicher Repräsentation. Eine Späthellenistische Hausfassade in Pompeji." In *Akten des 13. Internationalen Kongresses für Klassische Archäologie, Berlin*, 490–495. Mainz am Rhein: Von Zabern.

Holappa, M., and E.M. Viitanen. 2011. "Topographic Conditions in the Urban Plan of Pompeii: The Urban Landscape." In *The Making of Pompeii*, edited by S.J.R. Ellis, 169–190. Portsmouth, RI: Journal of Roman Archaeology.

Holliday, P.J. 2005. "The Rhetoric of 'Romanitas': The 'Tomb of the Statilii' Frescoes Reconsidered." *Memoirs of the American Academy in Rome* 50: 89–129.

Holloway, R.R. 2000. *Archaeology of Ancient Sicily*. New York: Routledge.

Hori, Y. 2010. "Pompeian Town Walls and Opus Quadratum." In *Pompeii: Report of the Excavation at Porta Capua, 1993–2005*, edited by H. Etani, 277–306. Kyoto: Paleological Association of Japan.

Hori, Y., O. Ajioka, and A. Hanghai. 2007. "Laser Scanning in Pompeian City Wall a Comparative Study of Accuracy of the Drawings from the 1930s to 40s." *International Archives of Photogrammetry, Remote Sensing and Spatial Information Sciences* 36 (5): 1–5.

Huet, V. 2007. "Le laraire de L. Caecilius Jucundus: Un relief hors norme?" In *Contributi di archeologia vesuviana. Raccolta critica della documentazione III, I culti di Pompeii*, edited by L. Bernabei, 142–150. Naples: L'Erma di Bretschneider.

Iorio, V. 2008. "La presenza della cinta muraria nei mosaici di Pompei e del suo ager ed in quelli di Ostia. Un confronto." In *Atti del XIII colloquio dell'Associazione italiana per lo studio e la conservazione del mosaico. Canosa di Puglia, 21–24 Febbraio 2007*, edited by C. Angelelli and F. Rinaldi, 289–298. Tivoli: Scripta Manent.

Ippel, A. 1925. *Pompeji*. Leipzig: Seeman.

Jackson, M.D., and F. Marra. 2006. "Roman Stone Masonry: Volcanic Foundations of the Ancient City." *American Journal of Archaeology* 110 (3): 403–436.

Jackson, M.D., F. Marra, R.L. Hay, C. Cawood, and E.M. Winkler. 2005. "The Judicious Selection and Preservation of Tuff and Travertine Building Stone in Ancient Rome." *Archaeometry* 47 (3): 485–510.

Jacobelli, L. 1994. "Alcune osservazioni sull'area di Punta della Campanella." In *Scritti di varia umanità in memoria di Benito Iezzi*, edited by M. Capasso and E. Puglia, 65–77. Sorrento: Franco di Mauro.

———. 2001. "Pompei fuori le Mura. Note sulla gestione e l'organizzazione dello spazio pubblico e privato." In *Pompei tra Sorrento e Sarno. Atti del terzo e quarto ciclo di conferenze di geologia, storia e archeologia. Pompei, gennaio 1999–maggio 2000*, edited by F. Senatore, 29–61. Rome: Bardi.

———. 2006. "Su un nuovo cippo L.P.P. trovato nell'area delle terme suburbane di Pompei." *Rivista di Studi Pompeiani* 17: 67–68.

Jacobelli, L., and P. Pensabene. 1995. "La decorazione architettonica del tempio di Venere a Pompei. Contributo allo studio e alla ricostruzione del santuario." *Rivista di studi pompeiani* 7: 45–76.

Jashemski, W.F. 1979. *The Gardens of Pompeii*. New Rochelle: Caratzas Brothers.

Jashemski, W.F., and F.G. Meyer. 2002. *The Natural History of Pompeii*. Cambridge: Cambridge University Press.

Johannowsky, W. 1982. "Nuovi rinvenimenti a Nuceria Alfaterna." In *La regione sotterrata dal Vesuvio. Studi e prospettive. Atti del convegno internazionale, 11–15 novembre 1979*, 835–862. Naples: Università di Napoli.

———. 1994. "Considerazioni sull'architettura militare del II sec. a.C. nei centri della Lega Nucerina." In *Nuceria Alfaterna e il suo territorio: dalla fondazione ai Longobardi*, edited by A. Pecoraro and G. Pugliese Carratelli, 123–135. Nocera: Aletheia.

———. 2009. "Osservazioni sulle fortificazioni della Campania e del Sannio con doppio cammino di ronda." *Vesuviana* 1: 13–15.

Jones, R., and D. Robinson. 2005a. "The Economic Development of the Commercial Triangle (VI.I.14–18, 20–21)." In *Nuove ricerche archeologiche a Pompei ed Ercolano. Atti del convegno internazionale, Roma 28–30 novembre 2002*, edited by P.G. Guzzo and M.P. Guidobaldi, 270–277. Naples: Electa.

———. 2005b. "The Structural Development of the House of the Vestals." In *Nuove ricerche archeologiche a Pompei ed Ercolano. Atti del convegno internazionale, Roma*

28–30 novembre 2002, edited by P.G. Guzzo and M.P. Guidobaldi, 257–269. Naples: Electa.

———. 2007. "Intensification, Heterogeneity and Power in the Development of Insula VI, 1." In *The World of Pompeii*, edited by J.J. Dobbins and P. Foss, 389–406. New York: Routledge.

Jouffroy, H. 1986. *La construction publique en Italie et dans l'Afrique romaine*. Strasburg: AECR.

Kähler, H. 1935. "Die Römische Stadttore von Verona." *Jahrbuch des Deutschen Archäologischen Instituts* 50: 138–197.

———. 1942. "Römischen Torburgen der frühen Kaiserzeit." *Jahrbuch des Deutschen Archäologischen Instituts* 57: 1–108.

Kampen, N. 1991. "Reliefs of the Basilica Aemilia: a Redating." *Klio* 73: 448–458.

Karlsson, L. 1992. *Towers and Masonry in the Walls of the Hegemony of Syracuse, 405–211 B.C.: A Study in Late Classical and Early Hellenistic Fortification Techniques*. Stockholm: Åström.

Kastenmeier, P., G. Di Maio, G. Balassone, M. Boni, M. Joachimski, and N. Mondillo. 2010. "The Source of Stone Building Materials from the Pompeii Archaeological Area and Its Surroundings." *Periodico di Mineralogia* Special Issue: 39–59.

Keenan-Jones, D. 2010. "The Aqua Augusta: Regional Water Supply in Roman and Late Antique Campania." Ph.D. thesis, Macquarie University, Australia.

———. 2015. "The Somma Vesuvius Ground Movements." *American Journal of Archaeology* 119 (2): 191–215.

Kern, H. 1981. *Labirinti: forme e interpretazioni 5000 anni di presenza di un archetipo: manuale e filo conduttore*. Milan: Feltrinelli.

Kern, P.B. 2000. *Ancient Siege Warfare*. London: Souvenir.

Kockel, V. 1983. *Die Grabbauten vor dem Herkulaner Tor in Pompeji*. Mainz am Rhein: Von Zabern.

Krauss, F., and R. Herbig. 1939. *Der korinthisch-dorische Tempel am Forum von Paestum*. Berlin: de Gruyter.

Krischen, F. 1941. *Die Stadtmauern von Pompeii und griechische Festungsbaukunst in Unteritalien und Sizilien*. Berlin: de Gruyter.

La Rocca, E. 1984. "Fabio o Fannio. l'Affresco medio-repubblicano dell'Esquilino come riflesso dell'arte rappresentativa e come espressione di mobilità sociale." *Dialoghi di Archeologia* 2: 31–53.

———. 1992. "Claudio a Ravenna." *La Parola del Passato* 47: 265–314.

———. 2000. "l'Affresco con veduta di città dal Colle Oppio." In *Romanization and the City: Creations, Transformations, and Failures: Proceedings of a Conference Held at the American Academy in Rome to Celebrate the 50th Anniversary of the Excavations at Cosa, 14–16 May, 1998*, edited by E. Fentress, 57–71. Portsmouth, RI: Journal of Roman Archaeology.

Laidlaw, A. 1985. *The First Style in Pompeii: Painting and Architecture*. Rome: Giorgio Bretschneider.

———. 2007. "Mining the Early Published Sources: Problems and Pitfalls." In *The World of Pompeii*, edited by J.J. Dobbins and P. Foss, 620–636. New York: Routledge.

Lancaster, L.C. 2005. *Concrete Vaulted Construction in Imperial Rome: Innovations in Context*. Cambridge: Cambridge University Press.

Laurence, R. 2007. *Roman Pompeii: Space and Society*. London; New York: Routledge.

Laurence, R., and D. J. Newsome (ed.). 2011. *Rome, Ostia, Pompeii: Movement and Space*. Oxford; New York: Oxford University Press.

Lavagne, H. 1987. "Un embléme de romanitas: le motif des tours et ramparts en mosa-ïque." In *Le monde des images en Gaule et dans les provinces voisines: Actes du colloque, Ecole Normale Supérieure, Sévres, 16 et 17 mai 1987*, 135–147. Paris: Éditions Errance.

Lawrence, A.W. 1979. *Greek Aims in Fortification*. Oxford: Clarendon Press.

Lawrence, A.W., and R.A. Tomlinson. 1996. *Greek Architecture*. New Haven, CT: Yale University Press.

Le Riche, J. 1827. *Vues des monuments antiques de Naples*. Paris.

Lejeune, M. 1964. "Vénus romaine et Vénus osque." In *Hommages à Jean Bayet*, edited by M. Renard, 383–400. Brusselles: Latomus.

Lepone, A. 2004. "Venus Fisica Pompeiana." *Siris* 5: 159–169.

Lepore, E. 1989. *Origini e strutture della Campania antica. Saggi di storia etno-sociale*. Bologna: Il Mulino.

Ling, R. 1990. "A Stranger in Town: Finding the Way in an Ancient City." *Greece and Rome* 37: 204–214.

———. 2005. *Pompeii. History, Life and Afterlife*. Stroud: History Press.

———. 2007. "Development of Pompeii's Public Landscape in the Roman Period." In *The World of Pompeii*, edited by J.J. Dobbins and P. Foss, 119–128. New York: Routledge.

Ling, R., P.R. Arthur, and L. Ling. 2005. *The Insula of the Menander at Pompeii: The Decorations*. Oxford: Oxford University Press.

Livy. 1912. *History of Rome*. Translated by Rev. Canon Roberts. New York: Dutton & Co.

Lo Cascio, E. 1991. "La società pompeiana dalla città sannitica alla colonia romana." In *Pompei*, edited by F. Zevi, 1:113–128. Naples: Banco di Napoli.

———. 1996. "Pompei dalla città sannitica alla colonia sillana: le vicende istituzionali." In *Les élites municipales de l'Italie péninsulaire des Gracques à Néron: Actes de la table ronde de Clermont-Ferrand (28–30 novembre 1991)*, edited by M. Cébeillac-Gervasoni, 111–123. Naples: Centre Jean Bérard.

Longobardi, G. 2002. *Pompei sostenibile*. Rome: L'Erma di Bretschneider.

Lorenzoni, S., E. Zanettin, and A.C. Casella. 2001. "La più antica cinta muraria di Pompei. Studio petro-archeometrico." *Rassegna di Archeologia Classica e Postclassica* 18 B: 35–49.

Los, A. 1995. "Quand et pourquoi a-t-on envoyé les prétoriens à Pompéi?" In *Nunc de suebis dicendum est: Studia archaeologica et historica Giorgio Kolendo ab amicis et discipulis dicata: studia dedykowane Profesorowi Jerzemu Kolendo W 60-Lecie Urodzin I 40-Lecie Pracy Naukowej*, edited by J. Kolendo, 165–170. Warsaw: Instytut Archeologii Uniwersytetu Warzawskiego.

Lugli, G. 1966. *Studi minori di topografia antica*. Rome: De Luca.

———. 1968. *La tecnica edilizia romana con particolare riguardo a Roma e Lazio*. New York: London: Johnson Reprint.

Lynch, K. 1960. *The Image of the City*. Cambridge, MA: MIT Press.

MacDonald, W.L. 1986. "Empire Imagery in Augustan Architecture." In *The Age of Augustus: Interdisciplinary Conference Held at Brown University, April 30–May 2, 1982*, edited by R. Winkes, 137–148. Providence, RI: Center for Old World Archaeology and Art.

———. 1988. *The Architecture of the Roman Empire*. Vol. 2. New Haven, CT: Yale University Press.

Mackenzie, W.M., and A. Pisa. 1910. *Pompeii*. London: A. & C. Black.

Magaldi, E. 1939. "Echi di Roma a Pompei III." *Rivista di Studi Pompeiani* 3: 21–60.

Maier, F.G. 1961. *Griechische Mauerbauinschriften*. Vol. 2. Heidelberg: Quelle und Meyer.

Maiuri, A. 1929. "Studi e ricerche sulle fortificazioni di Pompei." *Monumenti Antichi dell'Accademia dei Lincei* 33: 113–289.

———. 1939. "Pompei. Scavo della Grande Palestra nel quartiere dell'Anfiteatro." *Notizie degli Scavi di Antichità* (a.1935–1939), 165–238.

———. 1942. *l'Ultima fase edilizia di Pompei*. Rome: Istituto di Studi Romani.

———. 1943. "Pompei isolamento della cinta murale fra Porta Vesuvio e Porta Ercolano." *Notizie degli Scavi di Antichità*, 275–314.

———. 1948. *Bicentenario degli scavi di Pompei, l'inaugurazione dell'Antiquarium*. Naples: Macchiaroli.

———. 1958. "Portali con capitelli cubici a Pompei." *Rendiconti della Accademia di Archeologia, Lettere e Belle Arti, Napoli* 33: 203–218.

———. 1959. *Pompei ed Ercolano: Fra case e abitanti*. Milan: Martello.

———. 1960. "Pompei. Sterro dei cumuli e isolamento della cinta murale. Contributo all'urbanistica della città dissepolta." *Bollettino d'Arte* 45: 166–179.

———. 1973. *Alla ricerca di pompei preromana*. Naples: Società Editrice Napoletana.

Malnati, L., and G. Sassatelli. 2008. "La città e i suoi limiti in Etruria padana." In *La città murata in Etruria: atti del XXV convegno di studi etruschi ed italici, Chianciano Terme, Sarteano, Chiusi, 30 marzo–3 aprile 2005: in memoria di Massimo Pallottino*, edited by G. Camporeale, 429–470. Pisa: Fabrizio Serra.

Mansuelli, G. 1957. "La posizione storica dell'arco di Augusto in Rimini." In *Atti del V convegno nazionale di storia dell'architettura: Perugia, 23 settembre, 1948*, 161–171. Florence: Niccioli.

———. 1960. *Il monumento augusteo del 27 a.C: nuove ricerche sull'arco di Rimini*. Bologna.

Marett, R.R., A. Evans, A. Lang, G. Murray, F.B. Jevons, J.L. Myres, and W.W. Fowler. 1908. *Anthropology and the Classics; Six Lectures Delivered Before the University of Oxford*. New York: Barnes & Noble.

Mau, A. 1879. *Pompejanische Beiträge*. Berlin: Reimer.

———. 1882. *Geschichte der decorativen Wandmalerei in Pompeji*. Berlin: Reimer.

———. 1890. "Gli scavi fuori Porta Stabiana." *Mitteilungen des Deutschen Archäologischen Instituts, Römischen Abteilung* 5: 276–284.

———. 1902. *Pompeii, Its Life and Art*. Translated by F.W. Kelsey. New York; London: Macmillan.

Mazois, F. 1824a. *Les ruines de Pompéi*. Vol. 1. Paris: Didot.

———. 1824b. *Les ruines de Pompéi*. Vol. 2. Paris: Didot.

McK. Camp II, J. 2000. "Walls and the Polis." In *Polis and Politics: Studies in Ancient Greek History: Presented to Mogens Herman Hansen on His Sixtieth Birthday, August 20, 2000*, edited by M.H. Hansen, P. Flensted-Jensen, T.H. Nielsen, and L. Rubinstein, 40–57. Copenhagen: Museum Tusculanum Press.

McNicoll, A. 2007. *Hellenistic Fortifications from the Aegean to the Euphrates*. Edited by N. P Milner. Oxford: Clarendon Press.

Menichetti, M. 2000a. "Political Forms in the Archaic Period." In *The Etruscans*, edited by M. Torelli, 205–226. London: Thames and Hudson.

———. 2000b. "The Rituals of Archaic Power." In *The Etruscans*, edited by M. Torelli, 588–591. London: Thames and Hudson.

Miller, M. 1995. *Befestigungsanlagen in Italien vom 8. bis 3. Jahrhundert vor Christus*. Hamburg: Dr. Kovac.

Minervini, G. 1851. "Interpretazione di una epigrafe osca scavata ultimamente a Pompei." *Memorie della Regale Academia Ercolanese di Archeologia* 7: 1–19.

————. 1853. "Notizia dei più recenti scavi di Pompei." *Bullettino Archeolgico Napoletano* 24: 185–187.

Miniero, P., A. D'Ambrosio, A. Sodo, G. Bonifacio, V. Di Giovanni, G. Gasperetti, and R. Cantilena. 1997. "Il santuario campano in località Privati presso Castellammare di Stabia. Osservazioni preliminari." *Rivista di Studi Pompeiani* 8: 11–56.

Mogetta, M. 2016. "The Early Development of Concrete in the Domestic Architecture of Pre-Roman Pompeii." *Journal of Roman Archaeology* 29: 43–72.

Molesworth, B.G. 1904. *Pompei as It Was & as It Is; the Destruction of Pompei, Life in Italy in the First Century, Italian Villas of the Period of Pompei, and the Poetry, Painting and Sculpture of the Time*. London: Skeffington & Son.

Mols, S.T.A.M. 2005. "Il primo stile 'retró': dai propilei di Mnesicle a Pompei." In *Omni pede stare. Saggi architettonici e circumvesuviani in memoriam Jos de Waele*, edited by S.T.A.M. Mols and E.M. Moormann, 243–246. Naples: Electa.

Monnier, M. 1865. *Pompéi et les pompéiens*. Paris: Typographie Lahure.

————. 1870. *The Wonders of Pompeii*. New York: Scribner & Co.

Moormann, E. 2001. "Scene storiche come decorazioni di tombe romane." In *La peinture funéraire antique: IVe siècle Av. J.-C.-IVe Siècle Ap. J.-C.: Actes du VIIe colloque de l'Association internationale pour la peinture murale antique (AIPMA), 6–10 octobre 1998, Saint-Romain-En-Gal, Vienne*, edited by A. Barbet, 99–107. Paris: Errance.

————. 2011. *Divine Interiors Mural Paintings in Greek and Roman Sanctuaries*. Amsterdam: Amsterdam University Press.

Morel, J.P. 1982. "Marina di Ieranto, Punta della Campanella. Observations archéologiques dans la presqu'île de Sorrente." In *Απαρχαι. Nuove ricerche e studi sulla Magna Grecia e la Sicilia antica in onore di Paolo Enrico Arias*, edited by L. Beschi, M.L. Gualandi, and L. Massei 2:147–153. Pisa: Giardini.

Moretti, G. 2014. "La Porta Venere di Spello." *Bolletino di Foligno* 37: 229–237.

Morricone, M.L. 1973. "Mosaico." In *Enciclopedia dell'Arte Antica Classica e Orientale. Supplemento 1 1970*. Rome: Instituto della enciclopedia italiana.

Müller-Römer, F. 2008. "A New Consideration of the Construction Methods of the Ancient Egyptian Pyramids." *Journal of the American Research Center in Egypt* 44: 113–140.

Müth, S. 2016. "Urbanistic Functions and Aspects." In *Ancient Fortifications: A Compendium of Theory and Practice*, edited by S. Müth, P. Schneider, M. Schnelle, and P. De Staebler, 159–172. Oxford: Oxbow.

Müth, S., E. Laufer, and C. Brasse. 2016. "Symbolische Funktionen." In *Ancient Fortifications: A Compendium of Theory and Practice*, edited by S. Müth, P. Schneider, M. Schnelle, and P. De Staebler, 126–158. Oxford: Oxbow.

Müth, S., and U. Ruppe. 2016. "Regional begrenzte Phänomene." In *Ancient Fortifications: A Compendium of Theory and Practice*, edited by S. Müth, P. Schneider, M. Schnelle, and P. De Staebler, 231–248. Oxford: Oxbow.

Nappo, S.C. 1997. "Urban Transformation at Pompeii in the Late 3rd and Early 2nd C. B.C." In *Domestic Space in the Roman World: Pompeii and Beyond*, edited by R. Laurence and A. Wallace-Hadrill, 22:91–120. Portsmouth, RI: Journal of Roman Archaeology.

Nappo, S.C. 2007. "Houses of Regions I and II." In *The World of Pompeii*, edited by J.J. Dobbins and P. Foss, 347–372. New York: Routledge.

Naso, A. 1996. *Architetture dipinte: decorazioni parietali non figurate nelle tombe a camera dell'Etruria meridionale: VII-V sec. a.C.* Rome: L'Erma di Bretschneider.

Neil, S. 2016. "Materializing the Etruscans: The Expression and Negotiation of Identity during the Orientalizing, Archaic, and Classical Periods." In *A Companion to the Etruscans*, edited by S. Bell and A.A. Carpino, 15–27. Chichester: Wiley Blackwell.

Niccolini, F., and F. Niccolini. 1854. *Le case ed i monumenti di Pompei*. Vol. 1. Naples.

———. 1862. *Le case ed i monumenti di Pompei*. Vol. 2. Naples.

———. 1890. *Le case ed i monumenti di Pompei*. Vol. 3. Naples.

———. 1896. *Le case ed i monumenti di Pompei*. Vol. 4. Naples.

Nilsson, M. 2008. "Evidence of Palma Campania Settlement at Pompeii." In *Nuove ricerche archeologiche nell'area vesuviana (scavi 2003–2006). Atti del convegno internazionale, Roma 1–3 febbraio 2007*, edited by P.G. Guzzo and M.P. Guidobaldi, 81–86. Rome: L'Erma di Bretschneider.

Nissen, H. 1877. *Pompeianische studien vur Städtekunde des Altertums*. Leipzig: Breitkopf und Härtel.

Noack, F., and K. Lehmann-Hartleben. 1936. *Baugeschichtliche untersuchungen am Stadtrand von Pompeji*. Berlin: de Gruyter.

Oakley, S.P. 1995. *The Hill-Forts of the Samnites*. Rome: British School at Rome.

Ohlig, C.J.P. 2001. *De Aquis Pompeiorum. Das Castellum Aquae in Pompeji. Herkunft, Zuleitung und Verteilung des Wassers*. Nijmegen: Gieben.

———. 2004. "Städtebauliche Veränderungen im bereich des Pomeriums und der Porta Vesuvio unter dem Einfluss des Baues der Fernwasserversorgung in Pompei." *Babesch* 79: 75–109.

Onians, J. 1988. *Bearers of Meaning: The Classical Orders in Antiquity, the Middle Ages, and the Renaissance*. Princeton, NJ: Princeton University Press.

Ortalli, J. 1995. "Nuove fonti archeologiche per Ariminum. Monumenti, opere pubbliche e assetto urbanistico tra la fondazione coloniale e il principato augusteo." In *Pro poplo arimenese. Atti del convegno internazionale Rimini antica una respublica fra terra e mare (Rimini, ottobre 1993)*, edited by A. Calbi and G. Susini, 469–529. Faenza: Fratelli Lega.

———. 1999. "Colonia Augusta Ariminesis. Il volto della Città al tempo di Augusto." In *L'arco d'Augusto tra storia antica e nuove scoperte*, edited by G. Susini, 15–26. Rimini: Raffaelli.

Osanna, M. 2016. "Gesto rituale e spazio sacro nella Pompei di età sannitica." In *Sacrum facere. III seminario di archeologia del sacro. Lo spazio del sacro: Ambiente e gesti del rito*, edited by F. Fontana and E. Murgia, 179–201. Trieste: Edizioni Università di Trieste.

Overbeck, J. 1854. *Pompeji in seinen Gebäuden, Alterthümern und Kunstwerken für Kunst- und Alterthumsfreunde dargestellt*. Leipzig: Engelmann.

Overbeck, J., and A. Mau. 1884. *Pompeji in seinen Gebäuden, Alterthümern und Kunst- werken*. Leipzig: Engelmann.

Pagano, M., and R. Prisciandaro. 2006. *I primi anni degli scavi di Ercolano, Pompei e Stabiae: raccolta e studio di documenti e disegni inediti*. Rome: L'Erma di Bretschneider.

Palmer, R.E.A. 1980. "Customs on Market Goods Imported into the City of Rome." *Memoirs of the American Academy in Rome* 36: 217–233.

———. 2009. *The Archaic Community of the Romans*. Cambridge: Cambridge University Press.

Palombi, D., J. Tabolli, and G. Viani. 2013. "Sulla cronologia delle mura di Cora." In *Mura di legno, mura di terra, mura di pietra: Fortificazioni nel Mediterraneo antico*, edited by G. Bartoloni and L.M. Michetti, 525–556. Rome: Quasar.

Pappalardo, U. 2005. "Nuove ricerche nella Villa Imperiale a Pompei." In *Nuove ricerche archeologiche a Pompei ed Ercolano. Atti del convegno internazionale, Roma 28–30 novembre 2002*, edited by P.G. Guzzo and M.P. Guidobaldi, 331–338. Naples: Electa.

Pappalardo, U., R. Ciardello, and M. Grimaldi. 2008. "L'Insula Occidentalis e la Villa Imperiale." In *Nuove ricerche archeologiche nell'area vesuviana (scavi 2003–2006). Atti del convegno internazionale, Roma 1–3 febbraio 2007*, edited by P.G. Guzzo and M.P. Guidobaldi, 193–307. Rome: L'Erma di Bretschneider.

Paribeni, R. 1902a. "Relazione degli scavi eseguiti nel marzo 1902." *Notizie degli Scavi di Antichità*, 210–213.

———. 1902b. "Relazione degli scavi eseguiti nel mese d'ottobre 1902." *Notizie degli Scavi di Antichità*, 564–568.

———. 1903. "Relazione degli scavi eseguiti durante il mese di novembre." *Notizie degli Scavi di Antichità*, 25–33.

Pasquinucci, M., and S. Menichelli. 2000. "Le mura etrusche di Volterra." *Atlante Tematico Topografia Antica* 9: 39–53.

Pedroni, L. 2011. "The History of Pompeii's Urban Development in the Area Just North of the Altstadt." In *The Making of Pompeii*, edited by S.J.R. Ellis, 159–168. Portsmouth, RI: Journal of Roman Archaeology.

Pensa, M. 1998. "Immagini di città e porti: aspetti e problemi." In *XLIII corso di cultura sull'arte ravennate e bizantina: seminario internazionale di studi sul tema ricerche di archeologia e topografia: Ravenna, 22–26 Marzo 1997: in memoria del Prof. Nereo Alfieri*, edited by R. Farioli Campanati, 689–710. Ravenna: Edizioni del Girasole.

Pernice, E. 1938. *Pavimente und figürliche Mosaiken*. Berlin: de Gruyter.

Pesando, F. 2000. "Edifici pubblici 'antichi' nella Pompei augustea: il caso della palestra sannitica." *Mitteilungen des Deutschen Archäologischen Instituts, Römische Abteilung* 107: 155–175.

———. 2006a. *Gli ozi di Ercole: residenze di lusso a Pompei ed Ercolano*. Rome: L'Erma di Bretschneider.

———. 2006b. "Il secolo d'oro di Pompei. Aspetti dell'architettura pubblica e privata nel II secolo a.C." In *Sicilia ellenistica. Consuetudo italica. Alle origini dell'architettura ellenistica in Occidente. Spoleto, complesso monumentale di S. Nicolò, 5–7 novembre 2004*, edited by M. Osanna and M. Torelli, 227–241. Pisa: Edizioni dell'Ateneo.

———. 2006c. "Le residenze dell'aristocrazia sillana a Pompei: alcune considerazioni." *Ostraka* 15 (1): 75–96.

———. 2008a. "Case di età medio-sannitica nella Regio VI: tipologia edilizia e apparati decorativi." In *Nuove ricerche archeologiche nell'area vesuviana (scavi 2003–2006), atti del convegno internazionale, Roma 1–3 febbraio 2007*, edited by P.G. Guzzo and M.P. Guidobaldi, 159–173. Rome: L'Erma di Bretschneider.

———. 2008b. "Pompei nel III secolo a.C. le trasformazioni urbanistiche e monumentali." In *Iberia e Italia. Modelos romanos de integración territorial*, edited by J. Uroz, J.M. Noguera, and F. Coarelli, 221–246. Murcia: Tabularium.

———. 2009. "Prima della catastrofe. Vespasiano e le città Vesuviane." In *Divus Vespasianus. Il bimillenario dei Flavi*, edited by F. Coarelli, 378–385. Milan: Electa.

———. 2010a. "Appunti sull'evoluzione urbanistica di Pompei fra l'età arcaica e il III secolo a.C. ricerche e risultati nel settore nord-occidentale della città." In *Sorrento e la Penisola Sorrentina tra italici, etruschi e greci nel contesto della Campania antica. Atti della giornata di studio in omaggio a Paola Zancani Montuoro (1901–1987), Sorrento, 19 maggio 2007*, edited by F. Senatore and M. Russo, 223–245. Rome: Oebalus.

———. 2010b. "Quadratariorum notae pompeianae. Sigle di cantiere e marche di cava nelle domus vesuviane." *Vesuviana* 2: 47–75.

———. 2011. "Ruinae et parietinae pompeianae. Distruzioni e abbandoni a Pompei all'epoca dell'eruzione." *Vesuviana* 3: 9–30.

———. 2013. "Pompei in età sannitica. Tipologia, uso e cronologia delle tecniche edilizie." In *Tecniche costruttive del tardo ellenismo nel Lazio e in Campania (Atti del convegno, Segni 2 dicembre 2011)*, edited by F.M. Cifarelli, 117–126. Rome: Espera.

Pesando, F., and M.P. Guidobaldi. 2006. *Pompei, Oplontis, Ercolano, Stabiae*. Rome: Laterza.

Peterse, K. 2007. "Select Residences in Regions V and IX: Early Anonymous Domestic Architecture." In *The World of Pompeii*, edited by J.J. Dobbins and P. Foss, 373–388. New York: Routledge.

Peterse, K., and J. de Waele. 2005. "The Standardized Design of the Casa degli Scienziati (VI.14.43) in Pompeii." In *Omni pede Stare. Saggi architettonici e circumvesuviani in memoriam Jos de Waele*, edited by S.T.A.M. Mols and E. Moormann, 197–220. Naples: Electa.

Picard, C., and S. Risom. 1962. *Les murailles. Les portes sculptées à image divine*. Paris: de Boccard.

Pina Polo, F. 2003. "Minerva, custos urbis de Roma y de Tarraco." *Archivo Español de Arqueología* 76: 111–119.

Pinder, I. 2011. "Constructing and Deconstructing Roman City Walls: The Role of Urban Enceintes as Physical and Symbolic Borders." In *Places in between the Archaeology of Social, Cultural and Geographical Borders and Borderlands*, edited by D. Mullin, 67–79. Oxford: Oakville Oxbow Books.

Pirson, F. 2005. "Spuren antiker Lebenwirklichkeit." In *Lebenswelten. Bilder und Räume in der römischen Stadt der Kaiserzeit*, 129–145. Rome: Deutsches Archäologisches Institut Rom.

Platner, S.B., and T. Ashby. 1929. *A Topographical Dictionary of Ancient Rome*. Cambridge: Cambridge University Press.

Pliny. 1893. *The Natural History of Pliny*. Translated by J. Bostock and H.T. Riley. New York: Bell and Sons.

Plutarch. 1998. *Plutarch's Lives: With an English Translation*. Translated by B. Perrin. Cambridge, MA: Harvard University Press.

Pocetti, P. 1988. "Riflessi di strutture di fortificazioni nell'epigrafia Italica tra il II e I secolo a.C." *Athenaeum* 66: 303–328.

Poehler, E. 2006. "The Circulation of Traffic in Pompeii's 'Regio VI'." *Journal of Roman Archaeology* 19: 53–74.

———. 2012. "The Drainage System at Pompeii: Mechanisms Operation and Design." *Journal of Roman Archaeology* 25: 95–120.

———.2017. *The Traffic Systems of Pompeii*. Oxford, New York: Oxford University Press.

Poggesi, G., and P. Pallecchi. 2012. "La cinta muraria di Cosa." In *Quarto seminario internazionale di studi sulle mura poligonali, Palazzo Conti Gentili, 7–10 Ottobre 2009*, edited by L. Attenni and D. Baldassarre, 161–168. Ariccia: Aracne.

Pollitt, J.J. 2009. *Art in the Hellenistic Age*. Cambridge: Cambridge University Press.

Polybius. 1889. *The Histories of Polybius*. Translated by F. Hultsch and E.S. Shuckburgh. London; New York: Macmillan.

"Pompei, ancora un crollo–Corriere della Sera." 2011. News. *Www.corriere.it*. October 11. www.corriere.it/cronache/11_ottobre_22/pompei-crollo_9626fc92-fc8e-11e0-92e3-d0ce15270601.shtml?refresh_ce-cp.

Prayon, F. 2001. "Tomb Architecture." In *The Etruscans*, edited by M. Torelli, 335–343. London: Thames and Hudson.

Proietti, G. 1986. *Cerveteri*. Rome: Quasar.

Purcell, N. 1995. "Forum Romanum." In *Lexicon topographicum urbis Romae*, edited by E.M. Steinby, 2:325–336. Rome: Quasar.

Quaranta, B. 1851. *Intorno ad un'osca iscrizione incisa nel cippo disotterrato a Pompei nell'agosto de Mdcccl.* Naples: Stamperia Reale.

Quilici, L. 1966. "Telesia." In *Studi di urbanistica antica*, 85–106. Rome: De Luca.

Quilici, L., and S. Quilici Gigli. 1995. "Ricerca topografica a Ferentinum." In *Opere di assetto territoriale ed urbano*, edited by L. Quilici and S. Quilici Gigli, 159–244. Rome: L'Erma di Bretschneider.

———. 2001. "Sulle mura di Norba." In *Fortificazioni antiche in Italia: età repubblicana*, edited by L. Quilici and S. Quilici Gigli, 181–244. Rome: L'Erma di Bretschneider.

Ramanius, R. 2012. "Euergetism and City-Walls in the Italian City of Telesia." *Opuscula Pompeiana* 5: 113–122.

Rebecchi, F. 1978. "Antefatti tipologici delle porte a galleria. Su alcuni rilievi funerari di età tardo- repubblicana con raffigurazione di porte urbiche." *Bullettino della Commissione Archeologica Comunale di Roma* 86: 153–166.

———. 1987. "Les enceintes augustéenes en Italie." In *Les enceintes augustéennes dans l'occident Romain*, edited by M.G. Colin, 130–150. Nîmes: Musée Archéologique.

Reinach, S. 1922. *Répertoire de peintures grecques et romaines*. Paris: Leroux.

Reinicke, R. 1896. "Die Befestigungsturme von Pompeii." *Zeitschrift für Bildende Kunst mit dem beiblatt Kunst-Chronik und Kunstliteratur* 7: 81–85.

Rhodes, R.F. 1998. *Architecture and Meaning on the Athenian Acropolis*. Cambridge: Cambridge University Press.

Ribezzo, F. 1917. "La nuova eituns di Pompei." *Rivista Indo-greco-italica di Filologia, Lingua, Antichità* 1: 55–63.

Richards, J. 1994. *Facadism*. London: Routledge.

Richardson, L. 1988. *Pompeii. An Architectural History*. Baltimore: Johns Hopkins University Press.

Richmond, I.A. 1932. "Augustan Gates at Torino and Spello." *Papers of the British School at Rome* 12: 52–62.

———. 1933. "Commemorative Arches and City Gates in the Augustan Age." *The Journal of Roman Studies* 23: 149–174.

Robinson, M. 2008. "La stratigrafia nello studio dell'archeologia preistorica e protostorica a Pompei." In *Nuove ricerche archeologiche nell'area vesuviana (scavi 2003–2006). Atti del convegno internazionale, Roma 1–3 febbraio 2007*, edited by P.G. Guzzo and M.P. Guidobaldi, 125–138. Rome: L'Erma di Bretschneider.

Romanelli, D. 1817. *Viaggio a Pompei: a Pesto e di ritorno ad Ercolano ed a Pozzuoli.* Naples: Angelo Traci.

Rosada, G. 1990. "Mura porte e archi nella Decima Regio. Significati e corealzioni areali." In *La città nell'Italia settentrionale in età romana: morfologie, strutture e funzionamento dei centri urbani delle Regiones X E XI: Atti del convegno organizzato dal Dipartimento di Scienze*, 364–409. Trieste: Università di Trieste; Rome: Ecole française de Rome.

———. 1992. "Mura e porte: tra architettura e simbolo." In *Civiltà dei Romani: il rito e la vita privata*, edited by S. Settis, 124–139. Milan: Electa.

———. 2000. "Le porte urbiche a corridoio. La monumentalizzazione di una forma arcaica." *Histria Antiqua* 6: 181–194.

Rossi, A. 1982. *The Architecture of the City*. Cambridge, MA: MIT Press.

Rossini, L. 1831. *Le antichità di Pompei: delineate sulle scoperte fatte sino a tutto l'anno MDCCCXXX*. Rome.

Rowlands, M.J. 1972. "Defence: A Factor in the Organisation of Settlements." In *Man, Settlement and Urbanism: Proceedings of a Meeting of the Research Seminar on*

Archaeology and Related Subjects, edited by G.W. Dimbleby, R. Tringham, and P.J. Ucko, 447–461. London: Duckworth.

Ruiz de Arbulo, J. 2006. "Scipionum Opus and Something More: An Iberian Reading of the Provincial Capital (2nd–1st C. B.C.)." In *Early Roman Towns in Hispania Tarraconensis*, edited by L. Abad Casal, S. Keay, and S. Ramallo Asensio, 33–43. Portsmouth, RI: Journal of Roman Archaeology.

Russo, F., and F. Russo. 2005. *89 a.C.: assedio a Pompei: la dinamica e le tecnologie belliche della conquista sillana di Pompei*. Pompei: Flavius.

Russo, M. 1990. *Punta della Campanella: epigrafe rupestre osca e reperti vari dall'Athenaion*. Rome: Accademia Nazionale dei Lincei.

———. 1992. "Materiali arcaici e tardo-arcaici dalla stipe dell'Athenaion di Punta Campanella." *Annali. Sezione di Archeologia e Storia Antica* 14: 201–219.

———. 1998. "Il territorio tra Stabia e Punta della Campanella nell'antichità. La via Minervia, gli insediamenti, gli approdi." In *Pompei, il Sarno e la Penisola Sorrentina. Atti del primo ciclo di conferenze di geologia, storia e archeologia, Pompei, aprile–giugno 1997*, edited by F. Senatore, 23–98. Pompei: Rufus.

Ryberg, I.S. 1955. *Rites of the State Religion in Roman Art*. Rome: American Academy in Rome.

Rykwert, J. 1988. *The Idea of a Town: The Anthropology of Urban Form in Rome, Italy and the Ancient World*. Cambridge, MA: MIT Press.

Säflund, G. 1932. *Le mura di Roma repubblicana: saggio di archeologia romana*. Lund: Gleerup.

Sakai, S. 1992. "VE28 Reconsidered." *Opuscula Pompeiana* 2: 1–13.

———. 2000/2001. "La storia sotto il suolo del 79 d.C. considerazioni sui dati provenienti dalle attività archeologiche svolte sulle fortificazioni di Pompei." *Opuscula pompeiana* 10: 87–100.

———. 2004. "Il problema dell'esistenza della cosiddetta Porta Capua a Pompei." *Opuscula pompeiana* 12: 27–63.

Sakai, S., and V. Iorio. 2005. "Nuove ricerche del Japan Institute of Paleological Studies sulla fortificazione di Pompei." In *Nuove ricerche archeologiche a Pompei ed Ercolano. Atti del convegno internazionale, Roma 28–30 novembre 2002*, edited by P.G. Guzzo and M.P. Guidobaldi, 318–330. Naples: Electa.

Salmon, E.T. 1970. *Roman Colonization under the Republic*. Ithaca, NY: Cornell University Press.

Sampaolo, V. 1990. "I 7,7: Casa del Sacerdos Amandus." In *Pompei: Pitture e mosaici*, edited by G. Pugliese Carratelli and I. Baldassarre, 1:586–618. Rome: Istituto della Enciclopedia Italiana.

———. 1999. "IX 8, 3.7: Casa del Centenario." In *Pompei: Pitture e mosaici*, edited by G. Pugliese Carratelli and I. Baldassarre, 9:903–1104. Rome: Istituto della Enciclopedia Italiana.

Sánchez Real, J. 1985. "La exploración de la muralla de Tarragona en 1951." *Madrider Mitteilungen* 26: 91–121.

Sanzi di Mino, M.R. 1983. "Fregio pittorico dal colombario Esquilino." In *L'Archeologia in Roma capitale tra sterro e scavo*, edited by L. Quilici and G. Pisani Sartorio, 163–164. Venice: Marsilio Editori.

Sauron, G. 1998. "Le propriétaire de la villa dite de P. Fannius Synistor à Boscoreale vers 50 Av. J.C. P. Aninius C.f. ancien duovir de la colonie de Pompéi, ou son fils?" *Atti della Pontificia Accademia Romana di Archeologia. Rendiconti* 71: 1–28.

Savino, E. 1998. "Note su Pompei colonia sillana: popolazione, strutture agrarie, ordinamento istituzionale." *Athenaeum* 86 (2): 439–461.

Scagliarini Corlàita, D. 1979. "La situazione urbanistica degli archi onorari nella prima età imperiale." In *Studi sull'arco onorario romano*, edited by G.A. Mansuelli, 29–72. Rome: L'Erma di Bretschneider.

Scatozza, L.A. 1997. "Le terrecotte architettoniche del tempio dorico di Pompei. L'eredità arcaica." In *Deliciae fictiles, 2. Proceedings of the Second International Conference on Archaic Architectural Terracottas from Italy, Held at the Netherlands Institute in Rome, 12–13 June 1996*, edited by P.S. Lulof and E.M. Moormann, 189–197. Amsterdam: Thesis Publishers.

Scatozza Höricht, L. A. 2001. "Il sistema di rivestimento sannitico e altre serie isolate." In *Il tempio dorico del Foro Triangolare di Pompei*, edited by J.A.K.E. De Waele, 223–310. Rome: L'Erma di Bretschneider.

———. 2005. "L'Atena del Foro Triangolare e la fase sannitica di Pompei." In *Aeimnestos. miscellanea di studi per Mauro Cristofani*, edited by B. Adembri, 660–667. Florence: Centro Di.

Schläger, H. 1957. *Das Westtor von Paestum (Porta Marina)*. Munich.

Segenni, S. 2001. "Il prosciugmanto del Lago Fucino e le scoperte archeologiche." In *Il tesoro del lago: l'archeologia del Fucino e la collezione Torlonia*, edited by A. Campanelli and S. Agostini, 25–28. Pescara: Carsa.

Seiler, F., H. Beste, C. Piriano, and D. Esposito. 2005. "La Regio VI Insula 16 e la zona della Porta Vesuvio." In *Nuove ricerche archeologiche a Pompei ed Ercolano. Atti del convegno internazionale, Roma 28–30 novembre 2002*, edited by P.G. Guzzo and M.P. Guidobaldi, 216–234. Naples: Electa.

Senatore, 1999. "Necropoli e società nell'antica Pompei. Considerazioni su un sepolcreto di poveri." In *Pompei, il Vesuvio e la Penisola Sorrentina. Atti del secondo ciclo di conferenze di geologia, storia e archeologia, Pompei, ottobre 1997–febbraio 1998*, edited by F. Senatore, 91–121. Pompei : Edizioni Rufus.

———. 2001. "La Lega Nucerina." In *Pompei tra Sorrento e Sarno. Atti del terzo e quarto ciclo di conferenze di geologia, storia e archeologia. Pompei, gennaio 1999–maggio 2000*, edited by F. Senatore, 185–265. Rome: Bardi.

Serra Villaró, J. 1949. "La muralla de Tarragona." In *Archivo Español de Arqueología* 22: 221–236.

Sertà, C.A. 2001/2002. "La "ordinatio" epigrafica sulle stele pompeiane di T. Suedius Clemens fuori Porta Nocera e fuori Porta Marina." *Rivista di Studi Pompeiani* 12/13: 228–239.

Seston, W. 1966. "Les murs, les portes, et les tours des enceintes urbaines et le problème des res sanctae en droit romain." In *Mélanges d'archeologie et d'histoire offerts à Andre Piganoil*, edited by R. Chevallier, 1489–1498. Paris: S.E.V.P.E.N.

Sewell, J. 2010. *The Formation of Roman Urbanism, 338–200 B.C.: Between Contemporary Foreign Influence and Roman Tradition*. Portsmouth, RI: Journal of Roman Archaeology.

Sgobbo, I. 1942. "Un complesso di edifici sannitici e i quartieri di Pompei per la prima volta riconosciuti." *Rendiconti Accademia di Archeologia Lettere e Belle Arti di Napoli* 6: 15–41.

Siculus, Diodorus. 1954. *Diodorus Siculus Library of History*. Translated by C.H. Oldfather. Cambridge, MA: Harvard University Press.

Smith, C., and E. Tassi Scandone. 2013. "Diritto augurale romano e concezione giuridico-religiosa delle mura." In *Mura di legno, mura di terra, mura di pietra: Fortificazioni nel Mediterraneo antico*, edited by G. Bartoloni and L.M. Michetti, 455–474. Rome: Quasar.

Sogliano, A. 1898a. "Pompei. Relazione degli scavi fatti nel febbraio 1898." *Atti dell'Accademia Nazionale dei Lincei. Rendiconti della Classe di Scienze Morali Storiche e Filologiche* 5 (6): 68–70.

———. 1898b. "Relazione degli scavi fatti nel marzo 1898." *Notizie degli Scavi di Antich-ità*, 125–127.

———. 1901. "Relazione degli scavi fatti nel mese di agosto 1901." *Notizie degli Scavi di Antichità*, 357–362.

———. 1904. "Gli scavi di Pompei dal 1873 al 1900." In *Atti del congresso internazionale di scienze storiche* 5: 295–349.

———. 1906. "Relazione degli scavi fatti dal dicembre 1902 a tutto marzo 1905." *Notizie degli Scavi di Antichità* 3: 97–107.

———. 1918. "Porte, torri e vie di Pompei nell'epoca sannitica." *Rendiconti dell'Accademia di Archeologia Lettere e Belle Arti di Napoli* 6: 153–180.

———. 1932. "Sulla Venus Fisica Pompeiana." *Accademia di Archeologia di Napoli* 12: 359–374.

———. 1937. *Pompei nel suo sviluppo storico. Pompei preromana*. Rome: Athenaeum.

Spano, G. 1910. "Gli scavi fuori Porta Vesuvio." *Notizie degli Scavi di Antichità* 7: 399–418.

———. 1911. "Pompei." *Notizie degli Scavi di Antichità* 8: 374–377.

———. 1937. "Porte e regioni pompeiane e vie campane." *Rendiconti dell'Accademia di Archeologia Lettere e Belle Arti di Napoli* 17: 269–321.

———. 1959. "Nuove osservazioni intorno ai bassirilievi pompeiani ricordanti il terremoto del 63 d.C." In *Studi in onore di Riccardo Filangieri*, Vol. 1: 7–19. Naples: L'Arte Tipografica.

Spaziani, P. 2012. "Il circuito murario di Ferentino nel quadro delle fortificazioni centro italiche tra IV e II secolo a.C." In *Quarto seminario internazionale di studi sulle mura poligonali, Palazzo Conti Gentili, 7–10 Ottobre 2009*, edited by L. Attenni and D. Baldassarre, 229–239. Ariccia: Aracne.

———. 2015. "Brevi considerazioni sulla cosidetta Porta Pentagonale. Un esempio di aperture ad ogiva nelle mura di Ferentino." In *Studi sulle mura poligonali. Alatri atti del quinto seminario 30–31 ottobre 2010*, edited by L. Attenni, 33–39. Naples: Valtrend.

Stamper, J.W. 2008. *The Architecture of Roman Temples: The Republic to the Middle Empire*. Cambridge: Cambridge University Press.

Steffelbauer, I., and C. Ruggeri. 2013. *Die antiken Schriftzeugnisse über den Kerameikos von Athen*. Vol. 2. Vienna: Holzhausen Verlag

Stevens, S. 2016. "Candentia Moenia. The Symbolism of Roman City Walls." In *Focus on Fortifications: New Research on Fortifications in the Ancient Mediterranean and the Near East*, edited by R. Frederiksen, S. Müth, P. Schneider, and M. Schnelle, 288–299. Oxford: Oxbow.

Strabo. 1950. *Geography*. Translated by H.L. Jones. London: Heinemann.

Sun Tzu. 2012. *The Art of War: A New Translation*. Translated by J. Trapp. London: Amber Books.

Thiermann, E. 2005. "Ethnic Identity in Archaic Pompeii." In *SOMA 2003. Symposium on Mediterranean Archaeology*, edited by C. Briault, J. Green, and A. Kaldelis, 157–160. Oxford: Archaeopress.

Thomas, E. 2007. *Monumentality and the Roman Empire: Architecture in the Antonine Age*. Oxford: Oxford University Press.

Thomas, M.L., I. Van der Graaff, and P. Wilkinson. 2013. "The Oplontis Project 2012–13: A Report of Excavations at Oplontis B." *Fastionline* 295: 1–9.

Thucydides. 1812. *The History of the Grecian War: In Eight Books*. Translated by Thomas Hobbes. London: Longman, Hurst, Rees, Orme and Brown.

Tocco Sciarelli, G. 1988. "I Santuari." In *Poseidonia-Paestum. Atti del ventisettesimo convegno di studi sulla Magna Grecia, Taranto–Paestum 9–15 ottobre 1987*, edited by A. Satzio and S. Ceccoli, 361–452. Taranto: Istituto per la storia e l' archeologia della Magna Grecia.

———. 2009. *La cinta fortificata e le aree sacre: Velia*. Milan: Electa.

Torelli, M. 2000. "The Etruscan City-State." In *A Comparative Study of Thirty City-State Cultures: An Investigation Conducted by the Copenhagen Polis Centre*, edited by M.H. Hansen, 189–208. Copenhagen: Kongelige Danske Videnskabernes Selskab.

———. 2008. "Urbs ipsa moenia sunt (Isid. XV 2, 1). Ideologia e poliorcetica nelle fortificazioni etrusche di IV-II sec. a.C." In *La città murata in Etruria: atti del XXV convegno di studi etruschi ed italici, Chianciano Terme, Sarteano, Chiusi, 30 marzo–3 aprile 2005: in memoria di Massimo Pallottino*, edited by G. Camporeale, 265–278. Pisa: Fabrizio Serra.

Torelli, M., and M. Cipriani. 1999. *Paestum romana*. Paestum: Ingeneria per la Cultura.

Trachtenberg, M. 2010. *Building-in-Time: From Giotto to Alberti and Modern Oblivion*. New Haven, CT: Yale University Press.

Tréziny, H. 1986. "Les techniques grecques de fortification et leur diffusion à la périphérie du monde Grec d'occident." In *La fortification dans l'histoire du monde Grec*, edited by P. Leriche and H. Tréziny, 185–200. Paris: Centre national de la recherche scientifique.

Trinquier, J. 2004. "Les loups sont entrés dans la ville: de la peur du loup a la hantise de la cité ensauvagée." In *Les espaces du sauvage dans le monde antique: approches et définitions*, 85–118. Franche-Comté: Presses universitaires Franc-Comtoises.

Tucci, P.L. 2005. "Where High Moneta Leads Her Steps Sublime: The 'Tabularium' and the 'Temple of Juno Moneta'." *Journal of Roman Archaeology* 18: 6–33.

Turner, V.W. 1977. *The Ritual Process: Structure and Anti-Structure*. Ithaca, NY: Cornell University Press.

Tybout, R.A. 2007. "Rooms with a View. Residences Built on Terraces along the Edge of Pompeii (Regions VI, VII and VIII)." In *The World of Pompeii*, edited by J.J. Dobbins and P. Foss, 407–420. New York: Routledge.

Valchera, A. 2009. "Ferentino." In *Le mura megalitiche: il Lazio meridionale tra storia e mito*, edited by A. Nicosia and M.C. Bettini, 159–168. Rome: Gangemi.

Van Buren, A.W. 1925. "Further Studies in Pompeian Archaeology." *Memoirs of the American Academy in Rome* 5: 103–113.

———. 1932. "Further Pompeian Studies." *Memoirs of the American Academy in Rome* 10: 7–54.

Van Daele, B. 2004. *Het Romeinse leger*. Leuven: Davidsfonds.

Van Deman, E. 1922. "The Sullan Forum." *Journal of Roman Studies* 12: 1–31.

Van der Graaff, I. 2015. "The Recovered Tympanum of Cubiculum 11 at Villa A ('of Poppaea') at Oplontis (Torre Annunziata, Italy): A New Document for the Study of City Walls." In *Antike Malerei zwischen Lokalstil und Zeitstil*, edited by N. Zimmermann, 559–564. Vienna: Österreichischen Akademie der Wissenschaften.

———. Forthcoming. "The Excavations at the Porta Stabia" in *The Pompeii Archaeological Research Project: Porta Stabia. Volume I: Structure, Stratigraphy, and Space*, edited by S. Ellis, A. Emmerson and K. Dicus.

Van der Graaff, I., and S.J.R. Ellis. 2017. "Minerva, Urban Defenses, and the Continuity of Cult at Pompeii." *Journal of Roman Archaeology* 30: 283–300.

Van der Poel, H. 1981. *Corpus Topographicum Pompeianum, 5. Cartography*. Austin: University of Texas Press.

Van Gennep, A. 1961. *The Rites of Passage*. London: Routledge & Kegan Pau.

Van Tilburg, C. 2007. *Traffic and Congestion in the Roman Empire*. London: Routledge.

————. 2008. "Gates, Suburby and Traffic in the Roman Empire." *Babesch* 83: 133–147.

Varone, A. 1988. "Attivitá dell'ufficio scavi: 1987–1988." *Rivista di Studi Pompeiani* 2: 143–154.

Varone, A. 2008. "Per la storia recente, antica e antichissima del sito di Pompei." In *Nuove ricerche archeologiche nell'area vesuviana (scavi 2003–2006). Atti del convegno internazionale, Roma 1–3 febbraio 2007*, edited by P.G. Guzzo and M.P. Guidobaldi, 349–362. Rome: L'Erma di Bretschneider.

Varone, A., and G. Stefani. 2009. *Titulorum pictorum pompeianorum qui in CIL vol. IV collecti sunt: imagines*. Rome: L'Erma di Bretschneider.

Vetter, E. 1953. *Handbuch der italischen Dialekte*. Heidelberg: Winter.

Vitale, F. 2000. *Astronomia ed esoterismo nell'antica Pompei e ricerche archeoastronomiche a Paestum, Cuma, Velia, Metaponto, Crotone, Locri e Vibo Valentia*. Padua: Cooperativa Libraria Editrice dell'Universita di Padova.

Vitruvius. 1960. *Vitruvius: The Ten Books on Architecture*. Translated by M.H. Morgan. Cambridge, MA: Harvard University Press.

Von Blanckenhagen, H. 1968. "Daedalus and Icarus on Pompeian Walls." *Mitteilungen des Deutschen Archäologischen Instituts, Römischen Abteilung* 75: 106–145.

Von Gerkan, A. 1940. *Der Stadtplan von Pompeji*. Berlin: Archäologisches Institut des Deutschen Reiches.

Von Hesberg, H. 1992. *Römische Grabbauten*. Darmstadt: Wissenschaftliche Buchgesellschaf.

————. 1998. "Minerva Custos Urbis. Zum Bildschmuck der Porta Romana in Ostia." In *Imperium Romanum. Studien zu Geschichte und Rezeption. Festschrift für Karl Christ Zum 75. Geburtstag*, edited by P. Kneissl and V. Losemann, 370–378. Stuttgart: Steiner.

————. 2005. "Il recinto nelle necropoli di Roma in età repubblicana: origine e diffusione." In *Terminavit sepulcrum i recinti funerari nelle necropoli di Altino*, edited by G. Cresci Marrone and M. Torelli, 59–77. Rome: Quasar.

Von Mercklin, E. 1962. *Antike Figuralkapitelle*. Berlin: de Gruyter.

Von Rohden, H. 1880. *Die terrakotten von Pompeji*. Stuttgart: Spemann.

Wallace-Hadrill, A. 2005. "Excavation and Standing Structures in Pompei Insula I.9." In *Nuove ricerche archeologiche a Pompei ed Ercolano. Atti del convegno internazionale, Roma 28–30 novembre 2002*, edited by P.G. Guzzo and M.P. Guidobaldi, 101–108. Naples: Electa.

————. 2007. "The Development of the Campanian House." In *The World of Pompeii*, edited by P. Foss and J.J. Dobbins, 279–292. New York: Routledge.

————. 2008. *Rome's Cultural Revolution*. Cambridge: Cambridge University Press.

————. 2011. "Pompeian Identities: Between Oscan, Samnite, Greek, Roman, and Punic." In *Cultural Identity in the Ancient Mediterranean*, edited by E.S. Gruen, 415–427. Los Angeles: Getty Research Institute.

Wallat, K. 1993. "Opus Testaceum in Pompeji." *Mitteilungen des Deutschen Archäologischen Instituts, Römische Abteilung* 100: 353–382.

Warden, P.G. 2012. "The Importance of Being Elite: The Archaeology of Identity in Etruria." In *A Companion to the Archaeology of the Roman Republic*, edited by J.D. Evans, 354–368. Chichester: Wiley Blackwell.

Ward-Perkins, J.B. 1994. *Roman Imperial Architecture*. New Haven, CT: Yale University Press.

Ward-Perkins, J.B., and A. Claridge. 1978. *Pompeii A.D. 79: Essay and Catalogue*. Boston: Knopf.

Welch, K.E. 2009. *The Roman Amphitheatre: From Its Origins to the Colosseum*. Cambridge: Cambridge University Press.

Wilkins, H., L. Caracciolo, and F. Inghirami. 1819. *Suite de vues pittoresques des ruines de Pompeii et un précis historique de la ville, avec un plan des fouilles qui ont été faites jusqu'en février 1819 et une description des objets les plus intéressants*. Rome.

Winter, F.E. 1971. *Greek Fortifications*. Toronto: University of Toronto Press.

Wolf, M. 2004. "Tempel und Macht in Pompeji." In *Macht der Architektur, Architektur der Macht. Bauforschungskolloquium in Berlin vom 30. Oktober bis 2. November 2002*, edited by E.L. Schwander and K. Rheidt, 191–200. Mainz am Rhein: Von Zabern.

———. 2009. "Forschungen zur Tempelarchitektur Pompejis – der Venus-Tempel im Rahmen des pompejanischen Tempelbaus." *Mitteilungen des Deutschen Archäologischen Instituts, Römische Abteilung* 115: 221–355.

Zanker, P. 1988. *The Power of Images in the Age of Augustus*. Ann Arbor: University of Michigan Press.

———. 1998. *Pompeii. Public and Private Life*. Cambridge, MA: Harvard University Press.

———. 2000. "The City as Symbol: Rome and the Creation of an Urban Image." In *Romanization and the City: Creations, Transformations, and Failures*, edited by Elizabeth Fentress, 25–41. Portsmouth, RI: Journal of Roman Archaeology.

Zevi, F. 1982. "Urbanistica di Pompei." In *La regione sotterrata dal Vesuvio. Studi e prospettive. Atti del convegno internazionale, 11–15 novembre 1979*, 353–365. Naples: Università di Napoli.

———. 1991. "La città sannitica l'edilizia privata e la Casa del Fauno." In *Pompei*, edited by F. Zevi, 1:47–71. Naples: Banco di Napoli.

———. 1995. "Personaggi della Pompei sillana." *Papers of the British School at Rome* 63: 1–24.

———. 1996. "Pompei dalla città sannitica alla colonia sillana: per un'interpretazione dei dati archeologici." In *Les élites municipales de l'Italie péninsulaire des Gracques à Néron: actes de la table ronde de Clermont-Ferrand (28–30 novembre 1991*, edited by M. Cébeillac-Gervasoni. Naples: Centre Jean Bérard.

———. 2003. "Pompei, prima e dopo l'eruzione." In *Studi in onore di Umberto Scerrato per il suo settantacinquesimo compleanno*, edited by M.V. Fontana and B. Genito, 851–866. Naples: Università degli Studi di Napoli l'Orientale.

Index